Nuclear Magnetic
Resonance in Solids

NATO ADVANCED STUDY INSTITUTES SERIES

A series of edited volumes comprising multifaceted studies of contemporary scientific issues by some of the best scientific minds in the world, assembled in cooperation with NATO Scientific Affairs Division.

Series B: Physics

RECENT VOLUMES IN THIS SERIES

The series is published by an international board of publishers in conjunction with NATO Scientific Affairs Division

A	Life Sciences	Plenum Publishing Corporation
B	Physics	New York and London
C	Mathematical and Physical Sciences	D. Reidel Publishing Company Dordrecht and Boston
D	Behavioral and Social Sciences	Sijthoff International Publishing Company Leiden
E	Applied Sciences	Noordhoff International Publishing Leiden

Nuclear Magnetic Resonance in Solids

Edited by

Lieven Van Gerven

Laboratorium voor Vaste Stof-Fysika en Magnetisme
Katholieke Universiteit Leuven
Leuven, Belgium

SPRINGER SCIENCE+BUSINESS MEDIA, LLC

Library of Congress Cataloging in Publication Data

Nato Advanced Study Institute on Nuclear Magnetic Resonance in Solids, Leuven, Belgium, 1974.
Nuclear magnetic resonance in solids.

(NATO advanced study institutes series: Series B, physics; v. 22)
"Lectures presented at the NATO Advanced Study Institute on Nuclear Magnetic Resonance in Solids held in Leuven, Belgium, August 26-September 6, 1974."
"Published in cooperation with NATO Scientific Affairs Division."
Includes index.
1. Nuclear magnetic resonance—Addresses, essays, lectures. 2. Solids—Addresses, essays, lectures. 3. Matter—Properties—Addresses, essays, lectures. I. Van Gerven, Lieven. II. North Atlantic Treaty Organization. Division of Scientific Affairs. III. Title. IV. Series.
QC762.N37 1974   538'.3   77-1230
ISBN 978-1-4684-2810-0   ISBN 978-1-4684-2808-7   (eBook)
DOI 10.1007/978-1-4684-2808-7

Lectures presented at the NATO Advanced Study Institute on Nuclear Magnetic Resonance in Solids held in Leuven, Belgium, August 26—September 6, 1974

© Springer Science+Business Media New York 1977
Originally published by Plenum Press, New York in 1977
Softcover reprint of the hardcover 1st edition 1977

Preface

Plus est en vous ...
(Device of the Lords of Gruuthuse,
Brugge, 15th century)

This book is based on, and contains the lectures of the
NATO Advanced Study Institute *NUCLEAR MAGNETIC RESONANCE IN SOLIDS*,
which was held August 26 to September 6, 1974, at the University
of Leuven, Belgium.

The planning of the Leuven Institute developed over several
years. The idea came into being around 1971 in the *A.E.I.O.U.*, the
*Association Européenne pour une Interaction entre les Organismes
Universitaires*. The first practical steps were undertaken exactly
two years before it took place, on the evening of the 25th of
August 1972 in a restaurant on the Eerikinkatu in Turku, Finland,
during the 17th Congress AMPERE. Meeting – and eating, of course –
that evening, in that most excellent restaurant, were Raymond
Andrew, Karl Hausser, John Waugh and myself. From the beginning on
it was decided to take as general theme of the Institute, the recent
developments in high resolution and high sensitivity nuclear
magnetic resonance in solids.

NMR has been, from the beginning on – 1946, when it was detected for the first time by Felix Bloch's team at the West coast of
the U.S.A. and by Edward Purcell's team at the East coast – a
powerful tool for studying internal static and dynamic properties
of matter, in particular the finest details of structures and
motions. When investigating solids, however, there is a severe
restriction. Dipolar broadening, which is overwhelmingly important
in solids – unlike in liquids and gases, where it is cancelled by
rapid motions – has prevented for a long time high resolution and
high sensitivity radiospectroscopy in solids. A lot of efforts have
been made in recent years to solve the problem, or to get around it,
by sophisticated techniques e.g., introducing some (real or fictive)
random "motion" or interaction in the spin system. The results are
very promising, and quite a number of contributions in the School
and in the present book are devoted to these new developments. On
the other hand the directors of the Institute wanted to provide
the study of recent developments in NMR with a firm basis, and
finally they also wanted to present new applications of modern NMR
to all kind of solids, in chemistry and biology as well as in
physics. The results of these efforts, of the really interdisci-

plinary School which was the NATO Advanced Study Institute in
Leuven, are to be found in the present volume.

The presentation of this book gives me the opportunity to
remind of the organizations and the people, who contributed to the
organization of the Institute.

First of all our thanks are due to the *Scientific Affairs
Division of the North Atlantic Treaty Organization* and to Dr. T.
Kester, Director of the Division. The NATO Summer School program
has been criticized from different sides, very often too by students
organizations. However, as I remarked in my welcome speech at the
Leuven School, one could, if one wants, blame NATO for many things,
but one should not blame it for supporting and doing science. To a
representative of one Eastern European country at the Congress
AMPERE in Turku, who was not very happy about the possibility of
having the NATO Summer School in Leuven linked to the next Congress
AMPERE (the one in Nottingham), my reply was that I would welcome
very much a joint effort of NATO *and* the Warsaw Pact in organizing
the Leuven Summer School. Of course, this has not been realized!
Nevertheless, we were happy to welcome in Leuven many participants
and students also from Socialist Countries. Their presence made it
clear that, at least in science, there is no iron curtain, nor any
other curtain whatsoever.

Many others contributed to the success of the Institute, and in
this way to the present book. We are grateful for the sponsorship
of the *European Physical Society*, for the financial support of the
Ministery of Education and Flemish Cultural Affairs, and for the
grants provided by the *National Science Foundation* of the United
States of America. Moreover several local and national - even multi-
national, why not! - industrial companies supported the Institute.
We thank the N.V.'s *Philips, Brouwerijen Artois, Raychem, Bruker-
Spectrospin, Varian-Benelux, Metallurgie Hoboken-Overpelt, Agfa-
Gevaert, Bell Telephone Mfg. Co* and *Kredietbank*.

The Scientific Directors of the Institute, Professor E.R.
Andrew and myself, have been very fortunate in having an effective
Advisory Committee which helped us with the planning and the
scientific organization of the School. Dr. P. Grobet and Dr. P.
Van Hecke served as Secretaries; members of the International Ad-
visory Committee were: R. Blinc, K. Hausser, J.Jeener, G.J. Martin,
A. Redfield, C.P. Slichter, N. Trappeniers and J. Waugh. In addition
Miss M.A. Jennes, Miss H. Van Gerven, Mr. A. Werner, Dr. G.
Adriaenssens, Drs. A. Stesmans and Drs. G. Janssens have been very
helpful in organizing and running the School. We are most grateful
to all of them.

Many thanks also - last but not least - to the professors
of the Advanced Study Institute: H. Alloul (Orsay), R. Blinc
(Ljubljana), S. Clough (Nottingham), M. Guéron (Paris), U. Haeberlen
(Heidelberg), J. Jeener (Bruxelles), D. Kearns (Riverside),

W. Müller-Warmuth (Münster), R. Orbach (Los Angeles), Ch. Slichter (Urbana), J. Tegenfeldt (Cambridge), P. Van Hecke (Leuven), K. Wüthrich (Zürich). Professor H.B.G. Casimir, President of the European Physical Society, gave an Opening Address on *Resonance*. They have put our School at a high level ... without losing contact with the students. They made this book.

I also would like to thank those who helped me in preparing the manuscripts, Guy Adriaenssens, Luc Parijs and in particular Marie-Anne Jennes. She contributed to the Institute from the early correspondence to the final stages. Her continued dedication, both to the Institute and to this book guaranteed success.

To say a final word about the book, as editor I have taken the liberty not to print the lectures in chronological order, as they were given at the Institute, but to collect them in four chapters: *Principles, Systems, Methods* and .... *Life*. Read and judge!

I hope this book will help you to get more information out of your NMR-spectra and to challenge them: *Plus est en vous* ...

Lieven Van Gerven

This seal, designed by Rijkhard Van Gerven, symbolizes the NATO Advanced Study Institute on Nuclear Magnetic Resonance in Solids, its place and subject. It represents the contours of the Gothic Town Hall of Leuven(finished in 1463), together with a semi-exponential decay pattern of transversal magnetization in nuclear magnetic resonance(recorded in Leuven in 1963).

Contents

Chapter II: SYSTEMS:
PHONONS, NON-METALS, METALS

MAGNETIC RESONANCE AND RELAXATION: A PROBE OF THE
PHONON SPECTRUM

MAGNETIC RESONANCE AND STRUCTURAL PHASE TRANSITIONS

NMR STUDIES OF MOLECULAR SOLIDS, POLYMERS AND GLASSES

NMR IN METALS AND ALLOYS

During the course of the preparation of my lectures I
received the sad news of the death of a young, German scientist
for whom I hold the highest admiration and regard, Dr. Horst
Seidel. It was my great good fortune to have him as a guest in
my laboratory for a year. My students and I had the benefit of
his deep and contagious love of physics, his superb physical
insight, his clever approach to the design of experiments, his
profound ability to strip away theoretical complexity to reveal
the essential physical concepts behind natural phenomena, and
his humane and sensitive dealings with others in the laboratory.
For his Ph. D. thesis at Stuttgart with Professor Wolf, he made
a major breakthrough in the technique of electron-nuclear double
resonance and applied it to the understanding of the structure of
color centers. He is one of the major figures in the younger
generation of European scientists who revolutionized the style
of European experimental research. Building on the ancient
tradition of superb experimentation, he added the new dimension
of a thorough-going theoretical analysis of his own results.
He was a scientist who knew and contributed to the truly
international character of science.

This occasion, gathering scientists from many countries,
focusing attention on those of you at the start of your careers,
emphasizing the interplay of theory and experiment, and exploring
the uses of magnetic resonance in physics, chemistry, and
biology, strikes a chord exactly in tune with his spirit. I
therefore dedicate my lectures to the memory of our beloved
colleague, Horst Seidel.

C.P. SLICHTER

GENERAL ASPECTS OF NUCLEAR MAGNETIC RESONANCE IN SOLIDS

C.P. Slichter

University of Illinois,

Urbana-Champaign (Illinois, U.S.A.)

INTRODUCTION

It is a great honour to be invited to lecture at NATO Summer School on NMR in solids. I have done so before – many of today's prominent resonators were in the class. I look on this summer school as the start of both a personal and professional acquaintance.

We do resonance in solids to study problems in physics, chemistry, or biology. The subject is as broad as the problems of interest to scientists in those fields. The task of your lecturers will be to draw on their experience as physicists, chemists, and biologists to bring out resonance concepts and methods which have been fruitful. I will try to launch the main themes to be presented by the other speakers.

THE CONTRAST WITH HIGH RESOLUTION NMR

Most chemists and biologists think of high resolution when they think of NMR. They deal with liquid samples in order that they have resonance lines which are narrow as a result of motional narrowing. The narrowness enables one to resolve the chemical shifts associated with different nuclear sites, and the indirect spin-spin couplings from which we learn what groups of nuclei are near one another in a molecule.

But one also gives up some things.

(1) The problem of interest may inherently be a solid state problem, e.g., superconductivity or solid state rate

processes (like determining the mechanisms whereby atoms can diffuse in solids).

(2) The liquid motion averages out some things which contain information because their effect depends on the orientation of H_0 relative to crystal axes. Examples are:
    (a) dipolar interactions (which reveal the orientation of H_2O molecules in solids – the famous Pake doublet [1])
    (b) chemical shift anisotropy – basically the statement that local magnetic polarization is anisotropic
    (c) electric quadrupole coupling
    (d) anistropic coupling to electron spins (dipolar and pseudo dipolar).

(3) In solids the dipolar spin-spin coupling is often much stronger than the spin-lattice. This enables one to do trick experiments such as some of the double resonance schemes for getting very high sensitivity.

THE FLAVOR OF THE PROBLEMS STUDIED

We usually associate classification with the early stages of a scientific discipline – the period prior to establishment of a theoretical framework.

But names also evoke a whole host of associations, sometimes quite vivid. The words Jerusalem, Bethlehem and Mecca carry vivid connotations for most people. As scientists, names such as Leiden or the Cavendish Lab have special meaning and they stimulate our thoughts.

We might in this vein catalogue things to study – what aspects of solids people study by resonance. This listing is to stir the juices of the brain – to make you think "aha – that suggests a study".

Here's a physicist's list:
- static processes, rate processes
- perfect solids, imperfect solids (chemical impurities, mis-placed atoms, dislocations)
- insulators, semiconductors, conductors, superconductors
- diamagnets, paramagnets, ferromagnets, ferrimagnets, antiferro-magnets
- single crystals, powders, amorphous solids
- simple solids, multicomponent solids, alloys, molecular solids, polymers, biological molecules
- phase transitions (first order, second order).

Note these cut across the field in different ways. The above list being a physicist's list may be more like Sodom, Gomorrah, Waterloo, or Dunkirk to many of you.

I suggest you start contemplating during these two weeks
what would be some interesting problems to tackle with resonance
- try to end up with some ideas.

Make up your own set of names. We ought to have some chemists'
lists, and some biologists' and discuss them.

THE SPECIAL SOLID STATE ASPECTS OF THE COUPLING OF A NUCLEUS
TO ITS SURROUNDINGS - LINE POSITION AND SHAPE

A principal reason NMR is useful in solids is that the
position and width (in fact the complete shape) of the resonance
reveals substantial information. Relaxation times likewise tell
us a great deal, but owing to the limitations of time I have
decided to treat only one aspect of relaxation times, the
concept of what is called $T_{1\rho}$, a subject I will treat in the next
lecture.

It is well to start by listing the interactions [2, 3]
which determine these features, and to get down the central
resulting solid state features. The nucleus interacts by either
its magnetic dipole moment (with other nuclear magnetic dipole
moments, with magnetic fields arising from the orbital motion
of electrons, with electron spin magnetic moments, and of
course with the externally applied laboratory field which may
determine the proper quantization scheme to use), or by its
electric quadrupole moment (with the electric field gradients
arising from electrons and charges of other nuclei).

We treat these in order.

Couplings via the Nuclear Magnetic Dipole Moment

(1) Coupling to other nuclear magnetic dipole moments

The coupling to other nuclear magnetic dipole moments can be
represented in general by the dipolar Hamiltonian H_d [2]

$$H_d = \vec{\mu}_1 \cdot \vec{\mu}_2 / r_{12}^3 - 3(\vec{\mu}_1 \cdot \vec{r}_{12})(\vec{\mu}_2 \cdot \vec{r}_{12})/r_{12}^5 \qquad (1)$$

for a pair of nuclei. As is well known, this interaction is
frequently decomposed into terms A, B, C, D, E, and F following
Bloembergen, Purcell, and Pound [2, 4]. While such a decomposition
can always be done, it is usually done when there is an applied
magnetic field H_0 present, oriented in the z-direction. Only the
so-called secular terms A and B contribute to the line width of
a solid except in the case that motion is so rapid that H contains
significant frequencies as high as the Larmor frequency H_0 in the

static field H_o [2, 3]. If the interacting nuclei are unlike, only the term A is secular. For a pair of nuclei the A and B terms are

$$H_{dA} = (\gamma_1\gamma_2\hbar^2 I_{1z} I_{2z}/r_{12}^3)(1 - 3\cos^2\theta_{12}) \tag{2}$$

$$H_{dB} = (-\tfrac{1}{4} \gamma_1\gamma_2\hbar^2/r_{12}^3)(I_1^+ I_2^- + I_1^- I_2^+)(1 - 3\cos^2\theta_{12}) \tag{3}$$

where θ_{12} is the angle between $\vec{r}_{12}$ and the applied static field $\vec{H}_o$.

To see the nature of the effects of these terms, consider two unlike nuclei (e.g., an H and an F atom) coupled together, with an external field $\vec{k}H_o$ in the z-direction, giving a Hamiltonian

$$H = H_z + H_{dA} + H_{dB} \tag{4}$$

where H_z is the Zeeman energy

$$H_z = -(\gamma_1 I_{1z} + \gamma_2 I_{2z})\hbar H_o \tag{5}$$

We shall assume H_o to be much larger than the dipolar coupling, so that to a first approximation I_{1z} and I_{2z} commute with H. Thus as good quantum numbers we take the quantum numbers m_1 and m_2 ($m_1 = I_1, I_1 - 1, \ldots, -I_1, m_2 = I_2, I_2 - 1, \ldots, -I_2$) with corresponding eigenfunctions $|m_1 m_2\rangle$, and zero order eigenvalues of energy.

$$E = -(\gamma_1 m_1 + \gamma_2 m_2)\hbar H_o \tag{6}$$

As we have remarked the term H_{dB} is now non-secular since it connects states $|m_1 m_2\rangle$ with $|m_1 + 1, m_2 - 1\rangle$ or $|m_1 - 1, m_2 + 1\rangle$ which differ in energy. We therefore drop it.

Using first order perturbation theory we then get the eigenvalues of energy to be

$$E_{m_1 m_2} = \langle m_1 m_2 | H | m_1 m_2 \rangle$$

$$= \langle m_1 m_2 | H_z + H_{dA} | m_1 m_2 \rangle \tag{7}$$

$$= -(\gamma_1 m_1 + \gamma_2 m_2)\hbar H_o + (\gamma_1\gamma_2\hbar^2/r_{12}^3)(1 - 3\cos^2\theta_{12})m_1 m_2$$

(Problem: Verify the steps of (7)).

What NMR spectrum does this produce? We consider an alternating field of angular frequency ω applied in the x-direction (transverse to H_o) of amplitude $2H_1$ (in the usual notation) giving a time-dependent perturbation $H_{pert}(t)$:

$$H_{pert}(t) = - \hbar(\gamma_1 I_{1x} + \gamma_2 I_{2x})2H_1 \cos\omega t \qquad (8)$$

This perturbation will have non-vanishing matrix elements between $|m_1 m_2\rangle$ and either $|m_1 \pm 1, m_2\rangle$ or $|m_1, m_2 \pm 1\rangle$. (Problem: Verify this selection rule). That is, we flip either one or the other nucleus, but not both simultaneously. Which nucleus we flip will depend on ω – it must be chosen to correspond to the Bohr frequency condition associated with the energy of the transition.

Thus, if m_1 is being flipped from m_1 to $m_1 - 1$, we have

$$\hbar\omega = E_{m_1-1, m_2} - E_{m_1 m_2}$$

$$= \gamma_1 \hbar H_o + (\hbar^2 \gamma_1 \gamma_2 / r_{12}^3)(1 - 3\cos^2\theta_{12})m_2 \qquad (9)$$

Equation (9) shows us that there is a resonance line corresponding to each allowed value of m_2, hence a total of $2I_2 + 1$ lines. Since the matrix element of $H_{pert}(t)$ is independent of m_2 (Problem: Verify this statement) the intensity of each transition will depend only on the number of nuclei of species 2 in the sample which occupy the state m_2. For a sample in thermal equilibrium, this number is given by the Boltzmann factor. However, for nuclei the size of $k_B\theta$ (where k_B is the Boltzmann constant and θ the temperature) puts us in the high temperature limit where for nuclei to a first approximation the different m_2 states are equally populated.

The lines are equally spaced in frequency, with spacing

$$\delta\omega = (\gamma_1\gamma_2\hbar/r_{12}^3)(1 - 3\cos^2\theta_{12}) \qquad (10)$$

Thus the absorption spectrum would look like figure 1.

Note that the fact that there are $2I_2 + 1$ distinct transitions enables us to determine I_2 from counting lines. This fact helps enable us to identify the species to which nucleus 1 is coupled when several possible species of differing spins exist.

(Problem: Consider a sample containing hydrogen and chlorine with each proton coupled to a single chlorine nuclei. Assume Cl^{35} and Cl^{37} occur in their natural abundance, 75 % and 25 % respectively with γ_{35} 20 % larger than γ_{37}, and that all vectors

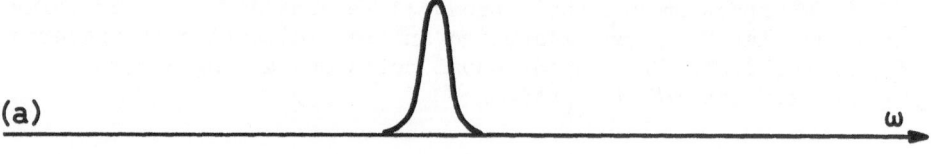

(a)   ω

(b)   ω

$\longleftarrow \delta\omega \longrightarrow$

Fig. 1.(a) Absorption spectrum of nucleus of type 1 uncoupled
           to other nuclei.

       (b) Absorption spectrum of nucleus of type 1 coupled by
           magnetic dipolar coupling to a nucleus of type 2.
           For this figure we have taken I_2 = 3/2 giving rise to
           the 4 equally spaced lines.

$\vec{r}_{12}$ are parallel. What facts would this enable you to state about
the number, position, and intensities of proton absorption lines?)

    Note also that if one varies the orientation of $\vec{H}_O$ with respect
to $\vec{r}_{12}$ (i.e. changes θ_{12}) $\delta\omega$ changes, reaching its maximum value
when $\vec{r}_{12}$ and $\vec{H}_O$ are parallel (θ = 0 or π). Thus if we know the
orientation of $\vec{H}_O$ with respect to the crystal axes, we can
determine the orientation of $\vec{r}_{12}$ with respect to the crystal axes.
Such use of NMR was discovered by Pake [1] who showed how to
use NMR to determine the H-H orientation of water molecules in
gypsum. Since this case has two identical nuclei, H_{dB} is secular
and must be included, as is shown by Pake in his article. (Problem:
See if you can compute the NMR spectrum for the H-H case without
reading Pake's paper - then check your answer with his paper. It
is interesting to note that his paper basically describes his
Ph.D. thesis).

(2) Coupling to electron orbital motion

    The electron orbital motion gives rise to electric currents
which produce magnetic fields. The interaction can be shown [2]
to give rise to a term H_{nL} to the nuclear Hamiltonian, representing
the nuclear - electron orbital coupling where

$$H_{nL} = -(1/c)\int (\vec{r} \times j(\vec{r})/r^3) d^3\vec{r} \cdot \gamma\hbar\vec{I} \qquad (11)$$

where $\vec{r}$ is the vector from the nucleus to points in the current distribution $\vec{J}(\vec{r})$; where $\vec{J}$ is the expectation value of the quantum-mechanical current density operator. The same current density operator comes into the calculation of the magnetic moment $\vec{M}$ associated with electron motion

$$\vec{M} = (1/2c)\int(\vec{r} \times \vec{J})d^3\vec{r} \qquad (12)$$

Equation (11) describes a variety of physical situations differing principally in (a) whether the bulk system possesses a permanent orbital moment on the lattice or a sublattice in the absence of H_o or rather needs an applied magnetic field, H_o, to induce a current flow, and (b) the effects of the structure and atomic make-up of the solid on the polarizability. The overwhelming majority of cases do not involve permanent orbital moments. Then $\vec{J}(\vec{r})$ becomes a linear function of the static field, and equation (11) can be written as

$$H_{nL} = - \sum_{\substack{\alpha,\alpha'= \\ x,y,z}} \gamma\hbar H_\alpha \sigma_{\alpha\alpha'} I_{\alpha'} \qquad (13)$$

where $\sigma_{\alpha\alpha'}$'s are components of a second rank symmetric $(\sigma_{\alpha\alpha'} = \sigma_{\alpha'\alpha})$ tensor, the so-called "chemical shift" tensor.

When the chemical shift tensor is expressed in terms of its principal axes and corresponding diagonal component σ_α we get

$$H_{nL} = - \sum_{\alpha=x,y,z} \gamma\hbar H_\alpha \sigma_\alpha I_\alpha \qquad (14)$$

$$= - \gamma\hbar\vec{H} \cdot \overset{\leftrightarrow}{\sigma} \cdot \vec{I}$$

where we have introduced the dyadic notation, $\overset{\leftrightarrow}{\sigma} \equiv \sum_{\alpha,\alpha'} \vec{i}_\alpha \sigma_{\alpha\alpha'} \vec{i}_{\alpha'}$.

In liquids we typically have rapid tumbling and can observe only the result of an average of principal axes randomly oriented with respect to H_o. The result is that the liquid chemical shift, σ_ℓ, is related to the instantaneous values along principal axes:

$$\sigma_\ell = (1/3)(\sigma_x + \sigma_y + \sigma_z) \qquad \text{x,y,z principal axes}$$

$$\qquad\qquad\qquad\qquad\qquad\qquad\qquad\qquad\qquad (15)$$

$$= (1/3)(\sigma_{xx} + \sigma_{yy} + \sigma_{zz}) \qquad \begin{array}{l}\text{x,y,z not necessarily} \\ \text{principal axes}\end{array}$$

The fact that chemical shifts must be anisotropic can be seen by reading section 4.5 of reference 2.

It is often the case in a solid that the symmetry of a nuclear site enables one to determine the directions of the principal axes. For example, for Na in NaCl the cube directions [100] are clearly principal axes. However, since they are equivalent, $\sigma_x = \sigma_y = \sigma_z$ and there can be no chemical shift anisotropy. For indium metal, which has hexagonal symmetry, the hexagonal axis is a principal axis, and any axis perpendicular to it is one as well, i.e. if we label the hexagonal axis z, $\sigma_x = \sigma_y \neq \sigma_z$. We have then "axial" symmetry for σ.

In principle one can measure the components of the full chemical shift tensor in a solid, something one cannot do in a liquid (except with a liquid crystal). This involves determining not only the principal values, but also the orientation of the principal axes in the molecule or solid. But the fact of solidification unfortunately also causes the "motional narrowing" to disappear. The resulting broad lines ordinarily mask the shift anisotropies. In principle one can make the chemical shifts bigger by increasing H_o, for example with a superconducting magnet (a shift anisotropy of $\sigma_x - \sigma_y = 10^{-4}$ typical of fluorine atoms gives a 6 gauss effect at 60 kG). In his lectures, Professor Van Hecke will describe another method, exceedingly ingenious in concept, which in effect enables one to overcome the masking of σ by dipolar broadening.

(3) Coupling to electron spin

Nuclear and electron spins couple together through their magnetic dipole moments with both the usual coupling of equation (1)(where we can let subscript 1 stand for a nucleus and subscript 2 label an electron) and also with the so-called Fermi contact term H_{nF} [2, 3] (where n stands for nucleus and F for Fermi)

$$H_{nF} = (8\pi/3)\gamma_n\gamma_e\hbar^2\vec{I} \cdot \vec{S} \, \delta(\vec{r}) \tag{16}$$

where we now introduce subscripts n and e for nucleus and electron, denote the spin operators of nucleus and electron respectively as $\vec{I}$ and $\vec{S}$, and where $\vec{r}$ is the radius vector from the nucleus to the electron. We have adopted the usual convention that γ_e is positive.

It is important to consider that one nucleus always couples to more than one electron since electron wave functions extend in space. Hence more useful is the expression

$$H_{nF} = \frac{8\pi}{3}\gamma_n\hbar\vec{I} \cdot \sum_{i=1}^{N}\gamma_e\hbar\vec{S}_i\delta(\vec{r}_i) \tag{17}$$

where i labels the N electrons. Rewriting, we get

$$H_{nF} = -\frac{8\pi}{3}\gamma_n \hbar \vec{I} \cdot \sum_{i=1}^{N} [-\gamma_e \hbar \vec{S}_i \delta(\vec{r}_i)] \tag{18}$$

$$= -(8\pi/3)\gamma_n \hbar \vec{I} \cdot \vec{M}_S$$

where $\vec{M}_S(\vec{r})$ is defined as

$$\vec{M}_S(\vec{r}) \equiv -\gamma_e \hbar \sum_{i=1}^{N} \vec{S}_i \cdot \delta(\vec{r}_i - \vec{r}) \tag{19}$$

$\vec{M}_e$ is an operator. Recalling that $-\gamma_e \hbar \vec{S}_i$ is the magnetic moment operator of the electron spin and recalling that *charge* density $\rho(\vec{r})$ can be represented by a charge density operator $\rho(\vec{r})$

$$\rho(\vec{r}) = \sum_i q_i \delta(\vec{r}_i - \vec{r}) \tag{20}$$

We see by analogy that $\vec{M}_e$ is the operator for *density of spin magnetization*.

In many cases, we are concerned with a nuclear-electron coupling in which we assume the electron state is given and are seeking to find the nuclear energy levels. Then we typically wish to replace the electron spin magnetization density operator by an appropriate thermal *average* of *electron expectation value* [2], which we denote $\vec{M}(\vec{r})$

$$\langle \vec{M}(\vec{r}) \rangle_{av} \equiv \vec{M}(\vec{r}) \tag{21}$$

where the <> indicates expectation value and the subscript "av" denotes a thermal average. Such an average of an expectation value gives us what classically we would call the spin magnetization density at point $\vec{r}$.

In terms of $\vec{M}(\vec{r})$ we have then

$$H'_{nF} = -(8\pi/3)\gamma_n \hbar \vec{I} \cdot \vec{M}(0) \tag{22}$$

for a nucleus at $\vec{r} = 0$, where the prime on H_{nF} denotes that we have done the thermal average and expectation value so that $\vec{M}(0)$ no longer contains electron coordinates. H'_{nF} involves only the nuclear spin as an operator.

In general $\vec{M}(\vec{r})$ is a linear function of H_0 and is non-uniform in space. (It may be non-uniform but periodic as with an ideal solid, or non-uniform and non-periodic as with disordered alloys).

In particular, since electrons spend more time near the positive nuclei than far from nuclei, $|\vec{M}(r)|$ is large near the nuclei. Thus we have that in general

$$M_\alpha(\vec{r}) = \sum_{\alpha'} \lambda_{\alpha\alpha'}(\vec{r}) H_{\alpha'} \qquad (23)$$

where α,α' are x,y, and z and $H_{\alpha'}$ is the α' component of $\vec{H}_o$.

If we perform a spatial average of $M_\alpha(\vec{r})$, denoted by the symbol $\langle M_\alpha(\vec{r}) \rangle_{\text{spat av}}$ we have

$$\langle M_\alpha(\vec{r}) \rangle_{\text{spat av}} = \sum_{\alpha'} \chi_{\alpha\alpha'} H_{\alpha'} \qquad (24)$$

where $\chi_{\alpha\alpha'}$ are the components of the susceptibility tensor. Clearly

$$\langle \lambda_{\alpha\alpha'}(\vec{r}) \rangle_{\text{spat av}} = \chi_{\alpha\alpha'} \qquad (25)$$

Using these ideas, we get then

$$H'_{nF} = -(8\pi/3) \sum_{\alpha,\alpha'} \gamma_n \hbar I_\alpha \lambda_{\alpha\alpha'}(0) H_{\alpha'} \qquad (26)$$

$$= -(8\pi/3)\gamma_n \hbar \vec{I} \cdot \overleftrightarrow{\lambda}(0) \cdot \vec{H}$$

where we introduce the dyadic notation $\leftrightarrow$ for λ. Frequently we write

$$H'_{nF} = -(8\pi/3) \sum_{\alpha,\alpha'} \gamma_n \hbar I_\alpha \chi_{\alpha\alpha'} A_{\alpha\alpha'} H_{\alpha''} \qquad (27)$$

where

$$A_{\alpha\alpha'} \equiv \lambda_{\alpha\alpha'}(0)/\langle \lambda_{\alpha\alpha'}(\vec{r}) \rangle_{\text{spat av}} \qquad (28)$$

$A_{\alpha\alpha'}$ represents the fact that the spin magnetization is non-uniform in space, and is in fact much larger at $\vec{r} = 0$ (the nuclear site) than its average value. It can be looked upon as a spin amplification factor. $A_{\alpha\alpha} \cong 200$ for Cu.

For an isotropic substance (or one in which susceptibility is isotropic and the nucleus is at a site of cubic symmetry) the tensor aspects disappear since $\chi_{\alpha\alpha'} = \delta_{\alpha\alpha'}\chi_o$ where χ_o is the static susceptibility and $\lambda_{\alpha\alpha'}$ is diagonal, so that

$$H'_{nF} = -(8\pi/3)\gamma_n \hbar \vec{I} \cdot \vec{H}\chi_o A \tag{29}$$

We can view equations (22) or (29) as saying that a nucleus experiences a magnetic field, $\vec{H}_{el\ spin}$, arising from the electron spins of magnitude

$$\vec{H}_{el\ spin} = (8\pi/3)\vec{M}(0) \tag{a}$$

$$= (8\pi/3)\overleftrightarrow{\lambda}(\vec{r}) \cdot \vec{H}_o \tag{b}$$

$$\equiv \overleftrightarrow{K} \cdot \overleftrightarrow{H}_o$$

$$(30)$$

where $\overleftrightarrow{K}$ is the so-called Knight shift tensor [2, 3] which for uniform isotropic substances becomes

$$\vec{H}_{el\ spin} = (8\pi/3)\chi_o A \vec{H}_o = K \vec{H}_o \tag{31}$$

In simple metals, the additional field of (31) shifts the nuclear resonance from its frequency in an insulator of the same species, since for the insulator there is zero spin magnetization density. The shift is named the Knight shift after its discoverer, Walter Knight (another important contribution which was a Ph.D. thesis).

Alloul will be discussing these shifts in his lectures as will I in my third lecture. We have both been concerned with dilute alloys of magnetic atoms (Fe, Co, etc.) in non-magnetic hosts (Cu, Al) for which the magnetic atoms destroy the periodicity of $\vec{M}(\vec{r})$. These expressions are also important for biological systems containing magnetic atoms such as heme groups, or for free radicals in chemistry with their unpaired spins.

Note that the field shift directly measures $\vec{M}(\vec{r})$ at nuclear sites (equation 30a).

For a substance with an isotropic Knight shift, A can be shown to be [2, 3]

$$A = \langle |\psi(0)|^2 \rangle_{E_F} \tag{32}$$

where $|\psi(0)|^2$ is the square of wave functions of one electron of wave vector $\vec{k}$ evaluated $\vec{r}$ at $\vec{r} = 0$, and the symbol $\langle \rangle_{E_F}$ implies an average over the Fermi surface. The wave function is normalized to the same volume as that to which χ_o is referred (i.e., one atom, a unit volume, etc.).

Coupling via the Nuclear Electric Quadrupole Moment

(1) The form of the interaction

The coupling of the nuclear electric quadrupole moment, Q, can be written in terms of the second derivatives of the electro-static potential, V, at the nucleus.

Defining

$$V_{xy} = \partial^2 V/\partial x \partial y \qquad (33)$$

etc., we get a quadrupole contribution to the Hamiltonian, H_Q, of [2, 3]

$$H_Q = [eQ/6I(2I - 1)]\underset{x,y,z}{\underset{\alpha,\beta=}{\Sigma}} V_{\alpha\beta} [\tfrac{3}{2}(I_\alpha I_\beta + I_\beta I_\alpha) - \delta_{\alpha\beta} I^2] \qquad (34)$$

Using principal axes we can eliminate all but terms involving V_{xx}, V_{yy}, and V_{zz}. Using Laplace 's equation ($\nabla^2 V = 0$) we then get

$$H_Q = [eQ/4I(2I - 1)][V_{zz}(3I_z^2 - I^2) + (V_{xx} - V_{yy})(I_x^2 - I_y^2)] \qquad (35)$$

A simple as well as common case involves choosing principal axes such that $V_{xx} = V_{yy} \neq V_{zz}$, a case called axial symmetry, for which

$$H_Q = [eQ/4I(2I - 1)]V_{zz}(3I_z^2 - I^2) \qquad (36)$$

A second frequent occurence is cubic symmetry for which $V_{xx} = V_{yy} = V_{zz}$. But since $\nabla^2 V = 0$, $V_{xx} = V_{yy} = V_{zz} = 0$, and the quadrupole coupling vanishes.

Solutions of the Hamiltonian containing an H_Q are more complex than for the simple Zeeman term, H_z, alone. For example, in some cases $|H_z| \gg |H_Q|$ so that H_Q can be treated as a perturbation. In other cases H_Q is small or the components $V_{\alpha\beta}$ so large that the Zeeman term can be treated as a perturbation on the solution of H_Q alone. Such a case is found with covalent bonds involving p or higher angular momenta as in many Cl compounds, or with certain non-cubic metals (e.g., Indium).

It is easy to show:when H_Q can be treated in only first order perturbation theory it does not change the center of gravity of the energy levels:

$$\sum_m \langle m|H_z|m\rangle = \sum_m \langle m|H_z + H_Q|m\rangle \tag{37}$$

where $|m\rangle$ are the eigenstates of H_z alone. (Problem: Prove that (37) is true).

Since the diagonal matrix elements of H_Q are the same for states of same magnitude but opposite sign of m, the first order energy shifts of +m and -m are the same. For a spin 3/2 nucleus the first order quadrupole effect must therefore give rise to shifts of $\pm\Delta$, where Δ is some constant, ordered as shown in figure 2.

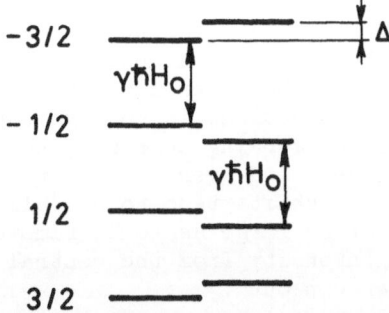

Fig. 2. The first order shifts of +3/2 and -3/2 are identical as are those of +1/2 and -1/2. Then to preserve the center of gravity +3/2 and +1/2 must displace in opposite directions.

By making use of the fact that a trace (sum of the diagonal elements of a matrix) can be evaluated in any representation, and substituting the *exact* eigenstates $|\xi\rangle$ for the zero order states $|m\rangle$, one can prove that the fact that H_Q does not shift the center of gravity is an *exact* theorem. (Problem: Prove, if one treats H_Q by perturbation theory, that *each order* of perturbation satisfies the center of gravity theorem).

(2) First order shifts, second order shifts, and the corresponding intensity wipeout

An ideal solid has a perfect lattice periodicity, but such solids are never found in nature. There are always misplaced atoms, dislocations, strains, chemical impurities, and surfaces which destroy the periodicity. These imperfections produce deviations in the electric field gradients, $\delta V_{\alpha\beta}(\vec{r})$, which vary throughout the lattice.

To first order they shift all energy levels, and thus all absorption lines. For I = 1/2 + an integer, they do *not* shift the +1/2 ↔ -1/2 (which we shall name the *central* transition for such spins, but not for integral spins)[5, 6]. Bloembergen and Rowland [7, 8] found that in most cases the broadening was not observed because either δV was too small at some nuclear sites to be seen, or so large at other sites that the intensity was shifted far out from the main line, removing these transitions from the intensity. Thus the effect of a low concentration of imperfections is to remove intensity from the absorption. If the concentration is high enough one is left with only the central transition. This phenomena is called "first order wipeout". This work was part of Rowland's Ph.D. thesis.

The second order perturbation shifts the m = +1/2 and m = -1/2 levels equal but opposite amounts. (Problem: Prove this statement). Thus it shifts the central transition. Therefore, when the concentration of imperfections is so high that a typical nucleus has a good chance of being near one, the $\delta V_{\alpha\beta}(\vec{r})$'s must be treated to second order, and even the central transition is broadened. Here again the shifts tend to be "all or nothing", i.e. too small to broaden, or so large as to be removed from the transition. One loses intensity from the central transition, and speaks of "second order wipeout".

In recent years, a number of investigators have realized that the shifted nuclei are likely to represent close shells to an imperfection (1st, 2nd, 3rd, etc. neighbours) and thus give rise to discrete lines, satellites of the main line, which though weak should be observable. I will discuss some of this work in my last lecture.

REFERENCES

1. G.E. PAKE, J.Chem.Phys. 16, 327 (1948)

2. C.P. SLICHTER, "Principles of Magnetic Resonance", Harper and Row Publishers, New York (1963)

3. A. ABRAGAM, "The Principles of Nuclear Magnetism", Oxford Press Publishers, Oxford (1961)

4. N. BLOEMBERGEN, E.M. PURCELL, and R.V. POUND, Phys.Rev. 73, 679 (1948)

5. R.V. POUND, Phys.Rev. 79, 685 (1950)

6. B. FELD and W.E. LAMB, Phys.Rev. 67, 15 (1945)

7. N. BLOEMBERGEN, "Defects in Crystalline Solids", Bristol Conference, 1954, Physical Society, London (1955)

8. N. BLOEMBERGEN and T.J. ROWLAND, Acta Met. 1, 731 (1953).

INTRODUCTION TO SPIN TEMPERATURES AND THEIR RELATION TO THE BLOCH EQUATIONS

C.P. Slichter

University of Illinois

Urbana-Champaign (Illinois, U.S.A.)

A PREDICTION FROM THE BLOCH EQUATIONS

One of the most useful simplifications of the general problem of N nuclei coupled to one another, to applied fields, and to the lattice is the famous Bloch equations

$$dM_z/dt = \gamma(\vec{M} \times \vec{H}(t))_z + (M_o - M_z)/T_1 \tag{1}$$

$$dM_{x,y}/dt = \gamma(\vec{M} \times \vec{H}(t))_{x,y} - M_{x,y}/T_2 \tag{2}$$

where "x,y" stands for *either* the x component or the y component, and where $\vec{H}(t)$ is the applied magnetic field consisting of a static component H_o in the z-direction and a time dependent term of arbitrary orientation.

Let us consider a simple resonance experiment with a rotating magnetic field of angular frequency ω, transverse to H_o, tuned exactly to resonance

$$\omega = \gamma H_o \tag{3}$$

It is convenient to transform to a reference frame rotating at ω with H_1 along the x axis. In this frame the Bloch equations become

$$dM_z/dt = -\gamma M_y H_1 + (M_o - M_z)/T_1 \tag{4}$$

$$dM_x/dt = -M_x/T_2 \tag{5}$$

$$dM_y/dt = \gamma M_z H_1 - M_y/T_2 \qquad (6)$$

Suppose we now orient $\vec{M}$ along H_1 so that at $t = 0$ $M_x = M$, $M_y = M_z = 0$. From equation (5) we see that M_x will decay to zero in a time T_2. The low H_1 steady-state solution of the Bloch equations shows that they describe a Lorentzian line with a frequency width

$$\Delta\omega = 1/T_2 \qquad (7)$$

For solids typically

$$\Delta\omega \stackrel{\sim}{=} \gamma H_{neighbour}$$

$$\stackrel{\sim}{=} \gamma\mu/a^3 = \gamma^2 \sqrt{I(I+1)} \sqrt{Z}/a^3 \qquad (8)$$

where $H_{neighbour}$ is the nuclear magnetic dipole field due to neighbours, and "a" is the distance to the Z nearest neighbour distance. For typical solids $\Delta\omega$ is of order of a few tens of kilocycles (e.g., $\Delta\omega \cong 2\pi \times 10$ kc for Al metal).

Further examination of equation (4) and equation (6) shows that they do not involve M_x; thus if M_y and M_z are initially zero they would remain zero were it not for the term involving T_1. If $T_1 \gg T_2$, therefore, and for times up to about T_1 after $\vec{M}$ has been oriented along H_1, we still have $M_y = M_z = 0$. (Problem: Verify this intuitive statement by solving equations (4) and (5)).

Therefore these equations predict that when $\vec{M}$ is aligned along H_1, it will decay to zero in a time T_2, typically of order 10^{-4} to 10^{-5} seconds.

Experimentally this prediction (decay in T_2 when $\vec{M}$ is along $\vec{H}$) is found to be correct for liquids, but it is *not* correct for solids. Rather, for solids it is found that as long as H_1 is turned on, the decay rate of M_x is much more like T_1 than a time T_2 which characterizes the linewidth. The failure of the Bloch equations to describe the situation in solids was discovered by Redfield [1](it was his first work in NMR, done as post doc with Bloembergen immediately after Redfield got his Ph.D. at Illinois) who found that the Bloch equations did not properly describe saturation of NMR in metals. In seeking an explanation of his data, Redfield showed that for a solid Bloch equations violate the second law of thermodynamics, but that one could predict what happens if one introduces the concept of a spin temperature viewed in the rotating frame to analysis of the problem of N coupled nuclei.

THE CONCEPT OF SPIN TEMPERATURE IN THE LABORATORY FRAME IN THE
ABSENCE OF ALTERNATING MAGNETIC FIELDS

The idea of a spin temperature was most fully developed in
connection with the work on paramagnetic relaxation of the Leiden
school [2]. It was Van Vleck [3] however who really recognized
its power. A typical system we might consider is a group of N
spins of spin I, gyromagnetic ratio γ, acted on by an external
field H_O, and coupled together by a magnetic dipolar interaction
represented by a dipolar Hamiltonian, H_d, and the Zeeman
Hamiltonian H_z. The solutions of the Schrödinger equations are
then wave functions ψ_n of energy E_n:

$$H\psi_n = (H_z + H_d)\psi_n = E_n\psi_n \tag{9}$$

Unfortunately equation (9) is exceedingly difficult to solve,
depending as it does on the coordinates of 10^{22} spins. Here,
so to speak, is the fundamental problem: the energies and wave
functions of the system have never been solved.

One can assume, however, that if the spin system is in thermal
equilibrium with a reservoir of temperature θ, that the various
states n of the total system would be occupied with fractional
probabilities p_n given by the Boltzmann factor

$$p_n = (1/Z)e^{-E_n/k\theta} \tag{10}$$

where Z is the partition function

$$Z = \sum_n e^{-E_n/k\theta} \tag{11}$$

Quantities such as the average energy $\bar{E}$ and magnetization $\bar{M}_z$
would then be given by

$$\bar{E} = \sum_n p_n E_n \tag{12}$$

$$\bar{M} = \sum_n \gamma\hbar(n|I_z|n)p_n$$

It was Van Vleck [3] who recognized that expressions such as
equation (12) could be evaluated without solving the Schrödinger
equation. For example consider the calculation of the partition
function.

$$Z = \sum_n e^{-E_n/k\theta} = \sum_n (n|e^{-H/k\theta}|n) = \text{Tr}[e^{-H/k\theta}] \tag{13}$$

where Tr stands for the sum of the diagonal matrix elements. However, it is a well-known theorem of quantum mechanics that the trace is independent of the particular representation used to evaluate it. One can therefore use a convenient representation, such as the one in which the wave functions are eigenfunctions of the z-component of spin of individual nuclei, I_{zk}. This procedure makes it unnecessary to solve equation (9) in order to compute the macroscopic properties.

For nuclei, and often for electrons, one can also often expand the exponential in a power series and keep only the leading terms. We refer to this as the high-temperature approximate. Using it we have

$$Z = \text{Tr}(1 - H/k\theta + H^2/2k^2\theta^2 + \ldots)$$

$$= (2I + 1)^N + (1/2k^2\theta^2)\text{Tr}H^2 + \ldots \tag{14}$$

where we have used the fact that $\text{Tr}H = 0$, as can be readily verified for both H_z and H_d.

Using these methods one finds

$$\bar{E} = -C(H_o^2 + H_L^2)/\theta \tag{15}$$

where

$$C = N\gamma^2\hbar^2 I(I + 1)/3k \tag{16}$$

is the Curie constant, and H_L is a quantity we call the local field, which is of the order of the field one nucleus produces at a neighbour (several gauss) and is defined by

$$CH_L^2 = \text{Tr}H_d^2/k(2I + 1)^N \tag{17}$$

Since the trace of equation (17) can be computed, H_L may be considered to be precisely known. One finds

$$H_L^2 = \gamma^2\hbar^2 I(I + 1) \sum_j (1/r_{jk})^6 \tag{18}$$

One can compute the magnetization $\vec{M}$ and finds it obeys Curie's law

$$\vec{M} = C\vec{H}_o/\theta \tag{19}$$

Finally, the entropy σ is found to be

$$\sigma = (\bar{E} + k\theta\ln Z)/\theta = N \ln(2I + 1) - C(H_o^2 + H_L^2)/2\theta^2 \qquad (20)$$

ADIABATIC AND SUDDEN CHANGES

The significance of these results is more fully realized if we consider the behaviour of the spin system when the static field H_o is changed. We consider two cases (a) adiabatic changes, and (b) sudden changes [4].

(a) Adiabatic Changes

To be adiabatic, a change in H_o must satisfy two conditions. The first is that there should be no heat flow into or out of the spin system. The condition will be satisfied if we make the changes on a time scale rapid compared to the spin-lattice relaxation time, T_1. Frequently one has T_1's of seconds, and by cooling may achieve T_1's of hours. Such long times are practically infinite.

The second condition we must satisfy is that after each small change in H_o we must allow a new temperature to be reached before making another small change. This condition is typically that we must change H_o slowly on a time scale defined by the precession period of nuclei in the local field of neighbours $(1/\gamma H_L)$.

This time scale is a few tenths of a millisecond. Between these time intervals there is a readily achievable range for which H_o can be changed adiabatically.

For an adiabatic change we have a constant entropy. Thus from equation (20) we get $(H_o^2 + H_L^2)/\theta^2$ remains constant. If, then, we start with an initial magnetization M_i in an initial static field $H_i \gg H_L$, we get a magnetization M at field H given by

$$M = M_i H/\sqrt{H^2 + H_L^2} \qquad (21)$$

This equation is plotted in figure 1. We note if we lower the field till it is zero, that we find $M = 0$. However, since the field changes are *reversible* (and in fact at all times keep the entropy constant) we can recover M_i by raising H from zero back up to its initial value H_i. Note that the recovery of M in such a process does *not* involve spin-lattice relaxation

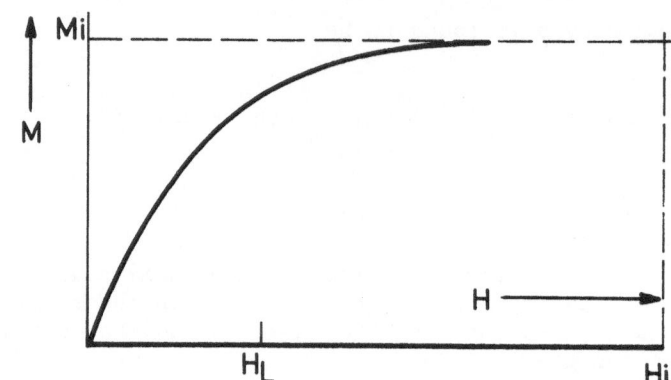

Fig. 1. M vs. H for adiabatic changes.

since, as we have postulated, everything is done on a time scale
short compared to T_1. This curve is often a surprise to those of
us who first learned magnetism by studying magnetic resonance,
since we learned that one needs T_1 to produce magnetization from
an unmagnetized sample. Actually, if one uses equations (19) and
(20) one sees that if one starts in zero field with spins which
are in thermal equilibrium with the lattice ($\theta = \theta_\ell$), the mere
act of adiabatically turning on the static field to $H_o \gg H_L$
will produce a magnetization of $M = CH_L/\theta_\ell = (M_oH_L/H_o)$ where M_o
is the magnetization one gets when the spin-lattice relaxation
produces thermal equilibrium between the spins and lattice
($\theta = \theta_\ell$) in the field H_o.

    One of the remarkable features of equation (21) or figure 1
is that when H = 0, M = 0. This is a very general consequence
of Curie's law. What is remarkable is that the degree of order
of the spin system is just the same when M = 0 as when $M = M_i$,
for, although it is clear that a system with net magnetization
is ordered, it is not clear how there can be order when M = 0.
The answer to this paradox is that even when H = 0, spins still
experience magnetic fields owing to the presence of their
neighbours. A typical spin will point either with or against
the local field. For a highly ordered system, there will be
a substantial excess pointing with the local field rather than
against it. Since the local fields at different nuclear sites
have random orientation (we here rule out highly ordered spin
arrangements such as in a ferromagnet) there is no resultant
macroscopic magnetization resulting from the alignment along the
microscopic local fields.

(b) Sudden Switching

    We have considered a process which is reversible. Now we

turn to one where things happen suddenly, resulting in irreversibility. Suppose we describe the system by a wave function ψ. Then we can say that a sudden change is one which takes place so rapidly that the wave function does not change. For example, suppose H_O changes at $t = 0$. Then, using 0- and 0+ to denote times just before $t = 0$, and just after $t = 0$, we have

$$\psi(t = 0-) = \psi(t = 0+) \tag{22}$$

We can compute the *energy* at $t = 0+$ by means of the relation

$$\bar{E}(0+) = \langle\psi(0+), H(0+)\psi(0+)\rangle \tag{23}$$

which becomes

$$\bar{E}(0+) = \langle\psi(0-), H(0+)\psi(0-)\rangle \tag{24}$$

If the system had been in thermal equilibrium at temperatures θ at $t = 0-$, we should perform a thermal average of equation (24), obtaining

$$\bar{E}(0+) = \langle\psi(0-), H(0+)\psi(0-) \text{ thermal average at } t = 0- \tag{25}$$

Let us now denote the eigenstates of $H(0-)$ by the notation $|\beta\rangle$ with eigenvalues E_β. Then in general

$$\psi(0-) = \Sigma C_\beta|\beta\rangle \tag{26}$$

where C_β are a set of coefficients describing the occupancy of the states $|\beta\rangle$. Since we assume a Boltzmann distribution, the thermal average values of C_β must give

$$(C_\beta{}^*C_\beta)_{\substack{\text{thermal average} \\ \text{at } t = 0-}} = e^{-E_\beta/k\theta}/Z \tag{27}$$

$$(C_\beta{}^*C_{\beta'})_{\substack{\text{thermal average} \\ \text{at } t = 0-}} = 0 \text{ if } \beta \neq \beta' \tag{28}$$

where equation (28) is the "random phase" approximation of statistical mechanics.

Using (26), (27) and (28) we then get

$$\bar{E}(0+) = \sum_{\beta,\beta'} (C_\beta{}^*C_{\beta'})_{\substack{\text{thermal ave} \\ \text{at } t = 0-}} \langle\beta|H(0+)|\beta'\rangle$$

$$= \Sigma_\beta e^{-\langle\beta|H(0-)|\beta\rangle/k\theta}\langle\beta|H(0+)|\beta\rangle/Z(0-) \tag{29}$$

which, using the invariance of the trace and the fact that $|\beta>$ is an eigenfunction of $H(0-)$, can be written as

$$\bar{E}(0+) = \mathrm{Tr}H(0+)e^{-H(0-)/k\theta_i}/Z(0-) \qquad (30)$$

At $t = 0+$, the spin system is not in general in thermal equilibrium. For example, if one had suddenly turned H_o from H_i to zero, the magnetization would be the same at $t = 0+$ as it was at $t = 0-$ but H would be zero. We see that this state of affairs ($H = 0$, $M \neq 0$) violates Curie's law and can not correspond to thermal equilibrium.

If, however, we wait for a long enough time (we still consider $T_1 = \infty$) we expect this complex system to reach an internal equilibrium, that is to be characterized by a temperature. The magnitude of the temperature is found by equating the energy of equation (15) to that at $t = 0+$, since from $t = 0+$ on the spin energy can not change (the energy is conserved when H is independent of time):

$$\bar{E}(0+) = - CH_L^2/\theta \qquad (31)$$

$$= \mathrm{Tr}\rho(t = 0-)H(t = 0+)$$

where

$$\rho(t = 0-) = \exp[-(H_d - \gamma\hbar H_i I_z)/k\theta]/Z \qquad (33)$$

and

$$H(t = 0+) = H_d \qquad (34)$$

If we denote the initial temperature as θ_i in the static field H_i, we find that the final temperature θ_f resulting from the sudden turn-off of H is

$$\theta_f = \theta_i \qquad (35)$$

If we had, however, turned H to zero *adiabatically* we would find, using equation (13) for $H_i \gg H_L$, that

$$\theta_f = \theta_i H_L/H_i \qquad (36)$$

Thus, the adiabatic turn-off cools the spins, whereas the sudden turn-off does not.

If now we turn on H slowly back to its initial value H_i, we would recover the full magnetization if we had done an adiabatic turn-off, but using equations (19), (20) and (36), we

find the sudden turn-off would enable us to achieve a final
magnetization

$$M_f = M_i H_L / H_i \qquad\qquad (38)$$

Thus, although the cycle of magnetic field has returned to its
initial value, we have had an irreversible loss in magnetization
accompanying the sudden turn-off. Note that if we bring H to zero
and turn on in *another* direction, we get a magnetization along
the new direction.

MAGNETIC RESONANCE AND SATURATION

The analysis of magnetic resonance by Bloembergen, Purcell,
and Pound [5] using standard perturbation theory is given rather
compactly in the differential equation for the population
difference, n, between the two energy levels of a system of spin
1/2 particles,

$$dn/dt = - 2W(\omega)n + (n_o - n)/T_1 \qquad\qquad (39)$$

$W(\omega)$ is the probability/second that a spin will be flipped by
the radio frequency field, H_1. Standard perturbation theory shows

$$W(\omega) = \pi\gamma^2 H_1^2 g(\omega)/2 \qquad\qquad (40)$$

where $g(\omega)$ is a function normalized to unit area having the same
dependence on frequency as the absorption line - that is, it
expresses the fact that the frequency of H_1 must be close to
resonance for H_1 to induce transitions. n_o is the thermal
equilibrium population difference, and T_1 the spin-lattice
relaxation time.

It is always possible, at least conceptually, to consider T_1
infinite, in which case equation (39) is especially simple to
solve.

$$n = Ae^{-2Wt} \qquad\qquad (41)$$

where A is a constant of integration.

It is well to recall the conditions for the validity of
equation (39). They are two:
(i) The perturbation matrix elements inducing transitions must
    be small compared to the width of the final state energy
    levels. This means $H_1 \ll H_L$.
(ii) The wave function must not change much.

We note,however, that equation (41) predicts that $n \to 0$ as $t \to \infty$. However, to satisfy condition (ii) we expect that we must consider times less than $1/W$. Of course, it is always easy to meet condition (i), but no matter how weak H_1, if we wait long enough we violate (ii).[Note we are requiring here also that $1/W \ll T_1$, otherwise the T_1 term would rescue condition (ii) even for times long compared to $1/W$].

We have the interesting problem, therefore, that we do not know how to integrate the equations of motion beyond a time for which n is almost its value at $t = 0$.

The solution of this problem was found by Alfred Redfield [1] and described in a paper which surely displays one of the profoundest insights in physics of any ever to be produced in the field of magnetic resonance. The essence of Redfield's approach is to note that a *resonant* time-dependent perturbation, no matter how weak, will eventually produce large effects. He therefore eliminates the time dependence by transforming to a reference system in which the Hamiltonian is essentially time independent. For such a system, the energy is conserved. More-over, the system is highly complex consisting of many interacting spins. One can thus predict that after a sufficiently long time the system will be found in a state of internal equilibrium. That is, it will be in one of its most probable states. Phrased alternatively, the energy states will be occupied according to a Boltzmann distribution at some temperature, θ.

We consider the Hamiltonian, H, given by

$$H = H_z(t) + H_d \qquad\qquad (42)$$

where $H_z(t)$ is the Zeeman interaction with the static field H_o and the rotating field of amplitude H_1 and angular frequency ω, rotating in the sense of the nuclear precession. The rotating field makes H_z time dependent. We are, of course, considering T_1 infinite.

The Schrödinger equation is then

$$-\hbar \partial \psi / i \partial t = H\psi \qquad\qquad (43)$$

To transform away the time dependence of H_1, one goes to a reference frame rotating with H_1. Formally this is accomplished by introducing the new wave function ψ' given by

$$\psi' = e^{-i\omega I_z t} \psi \qquad\qquad (44)$$

Substituting into equation (43), we get a new equation obeyed by ψ'

$$-\hbar \partial \psi' / i \partial t = H' \psi' \tag{45}$$

where

$$H' = - \gamma \hbar [(H_o - \omega/\gamma) I_z + H_1 I_x] + H_d^o \tag{46}$$

$$+ \text{ terms oscillating at } \pm \omega, 2\omega.$$

The term H_d^o is that part of the dipolar coupling which commutes with I_z. Physically, it is the part of H_d which is unchanged by rotation about the z-axis. (The two statements are of course equivalent since one can consider I_z as generating rotations).

The form of H_d^o is

$$H_d^o = (\gamma^2 \hbar^2 / 4) \sum_{j,k} (1 - 3 \cos^2 \theta_{jk})(3I_{zj}I_{zk} - I_j I_k)/r_{jk}^3 \tag{47}$$

where θ_{jk} and r_{jk} are coordinates of nucleus j with respect to nucleus k.

The term in the square bracket in equation (46) may be considered as the coupling of the spins to an effective static field $\vec{H}_e$

$$\vec{H}_e = \vec{k}(H_o - \omega/\gamma) + \vec{i}H_1 \tag{48}$$

In the absence of the time dependent terms, the energy levels of H' would be split by H_e and the dipolar couplings, so that we expect typical splittings to be

$$\Delta E \underset{\sim}{\sim} \gamma \hbar \sqrt{H_e^2 + H_L^2} \tag{49}$$

where the square root is a convenient way of including the two limiting cases of $H_e \gg H_L$ and $H_e \ll H_L$.

The time dependent terms will connect states of order $\gamma \hbar H_o$ apart in energy. Unless $H_o \underset{\sim}{\sim} H_L^2 + H_e^2$, a very low resonance field indeed, the time dependent terms are far from resonance and can be neglected since they are unable to produce transitions. We thus obtain a Hamiltonian which we call H, omitting primes for simplicity of notation,

$$H = - \gamma \vec{I} \cdot \vec{H}_e + H_d^o \tag{50}$$

$$= H_z + H_d^o$$

Now, in the absence of H_1, H_z and H_d^o commute, since the fact that $[I_z, H_d^O] = 0$ was the definition of H_d^O. Under this circumstance H_z and H_d^O would separately be constants of the motion. However, if $H_1 \neq 0, [H_z, H_d^O] \neq 0$, and the Zeeman and dipolar systems can then exchange energy. Since H is independent of time the total energy is conserved. Moreover, the system is very complex. Redfield therefore postulates that no matter what the state of the system at t = 0, a long time later it will be in a state of internal equilibrium described by a Boltzmann distribution. In other words, there will eventually be a temperature θ which can be assigned to the spins, giving thermal averages for the coefficients $C_\beta \, {}^{*}C_{\beta'}$

$$(C_\beta \, {}^{*}C_{\beta'})_{\substack{\text{thermal} \\ \text{average}}} = <\beta|\exp(-H/k\theta)|\beta>\delta_{\beta\beta'}/\text{Tr} \exp(-H/k\theta) \quad (51)$$

where H is the effective Hamiltonian of equation (51) and $|\beta>$ the eigenstate of H. Of course, we expect that after a long enough time Redfield's hypothesis would be fulfilled (unless there were some hidden selection rules which we have overlooked, such as the fact that H_z and H_d^O are perfectly isolated from one another if H_1 is zero). But the really important question becomes, how long does it take to reach equilibrium? The answer to this question clearly depends on the size of H_1, since H_1 is needed to prevent isolation of the dipolar and Zeeman systems. We shall return to the question later, but for the present consider that the time is short enough to make the establishment of a temperature practical.

REDFIELD THEORY NEGLECTING LATTICE COUPLING

The significance of equation (51) can be appreciated by calculating again the energy E, entropy σ, and magnetization $\bar{M}$. We find easily that

$$\bar{E} = -C(H_e^2 + H_L'^2)/\theta \quad (52)$$

$$\vec{M} = C\vec{H}_e/\theta \quad (53)$$

$$\sigma = N \ln(2I + 1) - C(H_e^2 + H_L'^2)/2\theta^2 \quad (54)$$

where C is the Curie constant, and where

$$CH_L'^2 = \text{Tr}(H_d^O)^2/k(2I + 1)^N \quad (55)$$

Evaluation of the trace of equation (55) gives, for a system with only one species,

$$H_L'^2 = <\Delta H^2>/3 \qquad\qquad (56)$$

where $<\Delta H^2>$ is the second moment of the resonance line. Following our earlier convention of omitting primes, we shall now use H_L for H_L', using the prime only when we wish to distinguish between the local field in the laboratory reference frame and the rotating reference frame.

It is important to notice that the Redfield assumption leads to Curie's law [equation (53)], and that the vector nature of the law shows that the nuclear magnetization is parallel to the effective field. Thus, if one is exactly at resonance, where $\vec{H}_e = \vec{i}H_1$, the magnetization is perpendicular to the static field.

Since the form of equation (52), (53), (54) is identical to that of the corresponding equations in the laboratory frame equations (9), (19) and (20) can be immediately applied to the rotating frame.

A. Adiabatic Demagnetization in the Rotating Frame

An adiabatic demagnetization in the rotating frame can be performed readily [6]. Suppose initially $H_1 = 0$, and that we have a sample magnetized to its thermal equilibrium value $\vec{k}M_0$. Let us turn H_0 far above the resonance value ω/γ, and turn on H_1. (We assume $H_0 - \omega/\gamma$ to be much bigger than H_L and H_1). We now have an effective field which is virtually parallel to $\vec{M}$. We now approach resonance sufficiently slowly to satisfy the criterion for a reversible change, although, since we assume $T_1 = \infty$, much faster than the time to reach thermal equilibrium with the lattice.

We then have that

$$M = M_0 H_e / \sqrt{H_e^2 + H_L^2} \qquad\qquad (57)$$

We can experimentally measure M by suddenly turning off H_1, leaving M to precess freely about H_0. The induced voltage immediately after turn off is proportional to M_x. One can measure M_z by noting that after turning off H_1, M_x decays to zero within a time of order $1/\gamma H_L$, but M_z does not change. One can thus wait till M_x has decayed, and then applied a $\pi/2$ pulse which rotates M_z into the x-y plane for inspection. The theoretical values of M_x and M_z are

$$M_x = MH_1/H_e \qquad\qquad (58)$$
$$= M_0 H_1 / \sqrt{H_e^2 + H_L^2}$$

$$M_z = Mh_o/H_e \qquad (59)$$

$$= M_o h_o / \sqrt{H_e^2 + H_L^2}$$

where h_o is the component of the effective field in the z-direction:

$$h_o \equiv (H_o - \omega/\gamma) \qquad (60)$$

Notice that if one does an adiabatic demagnetization exactly to resonance, the value of magnetization

$$M_x = M_o H_1 / \sqrt{H_1^2 + H_L^2}$$

will persist indefinitely (actually, when relaxation to the lattice is included, it decays, but on a time scale typically of order T_1). This result is in direct conflict with the well known result of the Bloch equations that M_x decays in T_2, where $T_2 \approx 1/\gamma H_L$. The fact that M does not shrink as long as H_e is kept constant, and that M precesses in step with H_1 is often described by the graphic term "spin locking".

The fact that the spins were locked to H_e was one of the most surprising results of the Redfield theory. It is, of course, nothing but the rotating frame equivalent of the statement that the magnetization in the usual laboratory frame adiabatic demagnetization has a one to one correspondence with H, as expressed in equation (21).

Note that if one pulses on H_1 when $(H_o - \omega/\gamma) \gg H_1$ and H_L, and then changes H_o so that one passes through resonance till one is far on the other side of the resonance, $(H_o - \omega/\gamma$ negative), one will have turned M to be pointing antiparallel to H_o. Moreover, although near to resonance one might have $M < M_o$, by the time one is far from resonance one would have $M \cong M_o$. This method provides a simple means of turning over the magnetization.

One can see from equation (58) that if one demagnetizes exactly to resonance, the magnetization will be the full M_o, provided $H_1 \gg H_L$, but if $H_1 \lesssim H_L$ one will not achieve the full magnetization. Were one to reduce H_1 slowly to zero after arriving at resonance, M would shrink to zero. In this manner all of the order represented by M_o prior to demagnetization would have been put into the dipolar system, that is, into alignment of spins along their local fields. One could, at a later time, slowly turn H_1 back on again, thereby recovering the magnetization.

B. Sudden Pulsing

A situation frequently encountered in experiment and interesting to contrast with adiabatic demagnetization is the effect of a sudden change in H_e [6]. We shall treat an especially simple case, that of suddenly turning on H_1. We assume we are off resonance by an amount h_o, and that the system is initially magnetized to M_o along H_o.

The sudden turn-on of H_1 is so fast that the system has the same wavefunction or density matrix just after turn-on that it had before. The dipolar energy, E_d, which depends on the relative orientation of spins is thus the same at $t = 0+$ as at $t = 0-$. Moreover, since the dipolar Hamiltonian H_d^o is the same in both laboratory and rotating frame,

$$\bar{E}_d = -CH_L^2/\theta_\ell = -M_o H_L^2/H_o \qquad (61)$$

where θ_ℓ is the lattice temperature, and the Zeeman energy is

$$\bar{E}_z = -\vec{M} \cdot \vec{H}_e \qquad (62)$$

$$= -M_o h_o$$

$$\bar{E} = -M_o(h_o + H_L^2/H_o) \overset{\sim}{=} -M_o h_o \qquad (63)$$

A long time later a spin temperature will be established together with a magnetization, M, parallel to H_e giving

$$\bar{E} = -C(H_e^2 + H_L^2)/\theta = -M(H_e^2 + H_L^2)/H_e \qquad (64)$$

But the total energy is conserved once H_1 has been turned on, so that, equating equation (63) and equation (64), we get

$$M = M_o h_o H_e/(H_e^2 + H_L^2); \quad M_x = M_o H_1 h_o/(H_e^2 + H_L^2); \quad M_z = M_o h_o^2/(H_e^2 + H_L^2) \qquad (65)$$

This equation shows that exactly at resonance, M would vanish. The null is very sharp, M_x varying linearly with h_o, so that observation of the null provides a precise method of observing exact resonance.

If $H_e^2 \gg H_L^2$, equation (65) has a simple geometrical meaning. After H_1 is pulsed on, $\vec{M}$ precesses about $\vec{H}_e$. The component of $\vec{M}$

parallel to $\vec{H}_e$ can not decay without energy exchange to the lattice, but the component perpendicular can decay since the local field gives a spread in precession frequencies. Thus, after several times $1/\gamma H_L$, $\vec{M}$ will be parallel to $\vec{H}_e$, and will have a magnitude given by the projection of the initial M_o on H_e.

THE APPROACH TO EQUILIBRIUM FOR WEAK H_1

    We saw in section V that standard perturbation theory predicted that following the turn-on of a weak H_1 the population difference n would go to zero for long times, although we recognized that we could not rigorously apply perturbation theory to times greater than $1/W$. The requirement of a weak H_1 was necessary in order that perturbation theory be valid for at least short times. Of course, since M_z is proportional to n, this implies M_z would go to zero. In section V however, we saw that M_z would, under these conditions, go to an equilibrium value

$$M_z\Big|_{equil} = M_o h_o^2/(h_o^2 + H_L^2) \tag{66}$$

where we have assumed $H_1^2 \ll h_o^2$ and $H_1^2 \ll H_L^2$ (although we note that the equilibrium expressions in section V were not limited to weak H_1). It is therefore clear, as we suspected, that for long times perturbation theory does not give correct predictions. For short time intervals, however, it must be correct. Recalling the proportionality between M_z and n, we have that

$$dM_z/dt = -2W(\omega)M_z \tag{67}$$

for timesshort compared to $1/W(\omega)$. How can we describe M_z for longer times?

    The solution to this problem was worked out by Provotorov [7] in an elegant paper which utilized powerful general techniques. Rather than outlining his analysis,we will give an alternate derivation of his result.

    We note that in the absence of H_1, the Zeeman interaction in the rotating frame is just

$$H_z = -\gamma \hbar h_o I_z \tag{68}$$

Let us assign a temperature θ_z to this Zeeman Hamiltonian, and θ_d to the dipolar H_d^o. Immediately before turning on H_1, the dipolar system is at the lattice temperature, θ_ℓ, since H_d^o is the same in the rotating or laboratory frames. The Zeeman temperature in the rotating frame, θ_z, is very cold compared to θ_ℓ since we have

$$M_o = CH_o/\theta_\ell \qquad \text{from the laboratory frame}$$

$$(69)$$

$$= Ch_o/\theta_z \qquad \text{from the rotating frame}$$

In as much as $H_o \gg h_o$, we find $\theta_z \ll \theta_\ell$. Turning on H_1 couples the two reservoirs together and they approach the final equilibrium value θ given by the analysis of section V . The coupling of H_1 produces transitions between the energy levels of the Zeeman and dipolar systems, which we assume are governed by simple rate equations for the population of the various states. (This assumption is quite common in all cross-relaxation calculations. Provotorov makes it implicitly in his work when he evaluates the relaxation times). Since there are many states, a large number of coupled rate equations result. As has been shown by Schumacher [8] when two systems are characterized by temperatures, the many equations reduce to two coupled linear rate equations, one for $(1/\theta_z)$, the other for $(1/\theta_d)$. But the conservation of energy gives a relationship between θ_z and θ_d:

$$-(Ch_o^2/\theta_z) - (CH_L^2/\theta_d) = \text{constant} \qquad (70)$$

Equation (70) is a first integral of the coupled equations, so that one of the resultant time constants is infinite, and a *single* exponential results. Since $M_z \propto 1/\theta_z$, this means M_z relaxes according to a *single* exponential towards its equilibrium value. Using the fact that $M_z = M_o$ initially, we get an equation for M_z as a function of time:

$$M_z - M_z \Big|_{\text{equil.}} = (M_o - M_z{}_{\text{equil}}) \, e^{-t/\tau} \qquad (71)$$

The only unknown in this equation is τ. We can, however, easily calculate it as follows. Taking the derivative of equation (70) evaluating it at $t = 0$, and comparing with equation (67) which must be valid initially where perturbation theory is correct, we get

$$\frac{1}{\tau} = 2W(\omega)M_o/(M_o - M_z)\Big|_{\text{equil}} \qquad (72)$$

Using equation (66) for $M_z\Big|_{\text{equil}}$ we get

$$1/\tau = 2W(\omega)(h_o^2 + H_L^2)/H_L^2 \qquad (73)$$

$$= \pi\gamma^2 H_1^2 [(h_o^2 + H_L^2)/H_L^2] g(\omega)$$

where we have used equation (40) for $W(\omega)$. This result is exactly that found by Provotorov, as indeed it must be, since we have made exactly the same approximations as he. The complete time development of the magnetization is therefore:

$$M_z = [M_o/(h_o^2 + H_L^2)]\{h_o^2 + H_L^2 \exp[-\gamma^2 H_1^2(h_o^2 + H_L^2)g(\omega)t/H_L^2]\} \quad (74)$$

This expression is remarkable since it involves the successful integration of the equations of motion well beyond the time $(1/W)$ for which perturbation theory is usually valid.

CONDITIONS FOR VALIDITY OF THE REDFIELD HYPOTHESIS

We have noted that the concept of spin temperature in the rotating frame has as a basic requirement neglect of certain time dependent terms in the Hamiltonian transformed to the rotating frame. This means, in essence, $H_o \gg \sqrt{H_e^2 + H_L^2}$. On the basis of the previous section we can now add a second requirement. We saw that in the absence of H_1, the equality of dipolar and Zeeman temperatures in the laboratory reference frame implied inequality in the rotating frame. Spin-lattice relaxation will always attempt to hold the two temperatures at the lattice temperature in laboratory frame. On the other hand, the presence of H_1 attempts to equalize them in rotating frame. We see that the spin temperature will be established in the rotating frame only if θ_z and θ_d approach each other in the rotating frame in a time much less than T_1. Thus we find that a spin temperature is established in the rotating frame provided

$$T_1/\tau \gg 1 \qquad \text{or}$$

$$\pi\gamma^2 H_1^2[(h_o^2 + H_L^2)/H_L^2] \, g(\omega)T_1 \gg 1 \qquad\qquad (75)$$

This is almost exactly the conventional condition for saturation. Note that the longer T_1, the smaller the H_1 which will satisfy equation (75). In particular, frequently equation (75) is satisfied for $H_1 \ll H_L$.

SPIN-LATTICE EFFECTS

So far we have considered the case of magnetic resonance on a time scale short compared to the spin-lattice relaxation time. In many instances one performs transient experiments which satisfy this condition. However, an equally important case arises when one does experiments on a time scale *long* compared to T_1 as when one employs steady-state apparatus. One can still satisfy equation (75), the criterion for validity of the Redfield theory, but the spin temperature θ in the rotating frame is now determined by the coupling to the lattice. As we shall see this statement by no means implies that $\theta = \theta_\ell$, but only that it is a *function* of θ_ℓ. Fortunately it is very simple to generalize our previous treatment to include the lattice. As a matter of fact, in Redfield's famous paper he considered just this case[1], but we have put off consideration of the lattice relaxation till this stage of the lectures to simplify the discussion.

In general, when the spin system exchanges energy with the lattice, the internal equilibrium of the spin system in the rotating frame is momentarily disturbed. The basic assumption we make is that the cross-relaxation between the dipolar and Zeeman systems is so rapid compared to T_1, that, following an exchange of energy of the spins with the lattice, a new spin temperature is rapidly established. Thus, the lattice always finds the spin-system described by a temperature in the rotating frame. We consider that there are three basic relaxation equations for the classical magnetizations M_x and M_z and for the expectation value of the dipolar energy $\langle H_d^O \rangle$, which we write down phenomenologically. They are

$$\partial M_z / \partial t = (M_o - M_z)/T_a \tag{76}$$

$$\partial M_x / \partial t = -M_x/T_b \tag{77}$$

$$\partial \langle H_d^O \rangle / \partial t = (\langle H_d^O \rangle_\ell - \langle H_d^O \rangle)/T_c \tag{78}$$

where T_a and T_b and T_c are relaxation times corresponding to exchange of energy with the lattice, and where $\langle H_d^O \rangle_\ell$ is the value of $\langle H_d^O \rangle$ when the spin temperature is equal to the lattice. We have used partial derivative signs in these equations to emphasize that they represent the changes in these quantities induced by lattice coupling only. Thus, although T_a is the usual T_1, T_b is *not* the usual $T_2 (\sim 1/\gamma H_L)$. T_b is generally of order T_1, a much longer time. A good analogy to these equations is to think of a Boltzmann equation in statistical mechanics for which there are a number of collision terms, each of which changes the distribution function f and for each one of which one could compute $\partial f/\partial t$. Equations (76), (77), and (78) guarantee that in the absence of H_1, the system (Zeeman plus dipolar energy) would reach thermal equilibrium with the lattice in the laboratory reference frame.

Using the fact that

$$\bar{E} = -M_z h_o - M_x H_1 + <H_d^o> \tag{79}$$

we can find the rate at which the lattice coupling changes $\bar{E}$ by taking the time derivative of equation (79).

$$d\bar{E}/dt = -(h_o \partial M_z/\partial t) - (H_1 \partial M_x/\partial t) + (\partial/\partial t)<H_d^o> \tag{80}$$

Employing equations (76), (77) and (78), assuming that M always lies along H_e, and using equations (52), (53) we find readily an equation for the magnitude of M

$$dM/dt = (M_{eq} - M)/T_{1\rho} \quad \text{or} \quad d(1/\theta)/dt = 1/T_{1\rho}\theta_{eq} - 1/\theta \tag{81}$$

where we have introduced a notation $T_{1\rho}$ and where

$$M_{eq} = M_o H_{eff}(h_o/T_a)/(h_o^2/T_a + H_1^2/T_b + H_L^2/T_c) \tag{82}$$

$$1/T_{1\rho} = (h_o^2/T_a + H_1^2/T_b + H_L^2/T_c)/(h_o^2 + H_1^2 + H_L^2) \tag{83}$$

[we have here neglected the term $<H_d^o>$ of equation (78)]

Note in particular that $M_{eq} = 0$ exactly at resonance, is positive when $H_O > 0$ (i.e. M is parallel to H_e) but is negative for $h_o < 0$ (that is, M is antiparallel to H_e). The last case corresponds to a negative spin temperature in the rotating frame. Equation (82) shows that the equilibrium θ is far from the lattice temperature, θ_ℓ, and may even be of the opposite sign. Since M_O is inversely proportional to θ_ℓ, it is still true that θ_ℓ determines θ, even though they are quite different. The negative temperature one sometimes finds is a simple manifestation of the fact that M_z always tends to be positive, whether h_o is positive or negative. In fact, one can say that the equilibrium is reached as follows: the strong internal coupling of the spin system (which guarantees a spin temperature) keeps $\vec{M}$ along H_e, since Curie's law is a vector law. The lattice is attempting (a) to make the z-component of M be M_o, but (b) the x-component be zero. (a) would make M bigger than M_o so that its projection on the z-axis is M_o, whereas (b) would make M be zero. The lattice is thus fighting itself since (a) and (b) are inconsistent. The equilibrium value of equation (82) results.

SPIN LOCKING, $T_{1\rho}$, AND SLOW MOTION

Equations (71) and (81) show that from an arbitrary initial condition a spin temperature in the rotating frame is established in a time τ without exchange of energy with the lattice. But this is only a quasiequilibrium value since over the time $T_{1\rho}$ the spin temperature changes as energy is exchanged with the lattice to drive $\vec{M}$ to M_{equil} of equation (82). During this process $\vec{M}$ lies along $\vec{H}_{eff}$. Thus if one starts with $\vec{M}$ along H_1, as we discussed before, it will not decay in a time characterized by the inverse of the line-width, but rather with the time $T_{1\rho}$ which requires energy exchange with the lattice. In principle, by going to low enough temperatures one should be able to make $T_{1\rho}$ as long as one pleases. If one does this, magnetization along $\vec{H}_{eff}$ in the rotating frame, following a time τ to establish a spin temperature, will remain without decay for as long as one has chosen to make $T_{1\rho}$. This time may be seconds in metals or even hours in insulators. This persistence is sometimes described by saying the spins are locked to $\vec{H}_{eff}$. It was first demonstrated by my student Holton [6] who, having done a Ph.D. thesis on ENDOR, had to wait two weeks to take the thesis examen and decided that it would be fun to demonstrate the spin-temperature ideas I have described earlier. In this interval he modified a pulse rig and used it to demonstrate the quantitative validity of equations (7) and(14). Spin locking is the basis of Hahn's famous double resonance technique[9, 10].

Even though we can lock the magnetization along $\vec{H}_{eff}$ for a time $T_{1\rho}$ when H_1 is on, if we *remove* H_1 *suddenly, M will decay to zero in a time of order of the inverse line width*. That is, the Bloch equations with the usual meaning of T_2 give a rough qualitative description of what happens.

In contemplating the Redfield theory, I always find it helpful to go back to the situations of the start of the lecture:the laboratory frame without alternating fields. There we see that, starting with a magnetized system,we can turn H_O to zero slowly and later turn it back up to its original value. When $H_O = 0$, $M = 0$, but the full M is recovered when H_O is turned back on,all without exchange of energy with the lattice. In zero field the order is manifested by the preferential alignment of nuclear moments along the direction of the local fields of their neighbours.

Consider a nucleus #1 with a neighbour nucleus #2. Suppose now that #2 makes a sudden jump, as with diffusion. The duration of the jump is perhaps 10^{-12} to 10^{-13} seconds, very fast compared to nuclear precession frequencies. Thus the local field at #1 arising from #2 changes suddenly in both magnitude and direction.

Nucleus #1 is thus somewhat randomized in orientation
relative to the local field. If the mean time a given nucleus sits
between jumps is τ_{jump}, we see that the alignment of nuclei in
the local fields of neighbours, that is the ordered state, can
only persist a time of τ_{jump}. Thus to carry out a full
demagnetization and remagnetization cycle with full recovery
of the initial magnetization, we must remain in zero field for
a time less than τ_{jump}. Of course, even were there *no* jumping, any
T_1 process would change the entropy of the spin system, so that
in any event we must complete the cycle in a time less than T_1.
We can conclude :

(i) the demagnetization-remagnetization cycle can be used to
monitor the zero field T_1

(ii) when there is jumping, we can detect it when $\tau_{jump} < T_1$.

The same considerations hold true in the rotating frame.
When there is jumping there is a contribution to T_c of

$$\left. 1/T_c \right)_{jump} = 2(1 - p)/\tau_{jump} \tag{84}$$

where p is a numerical quantity expressing the fact that a jump
does not totally randomize the local field at neighbour sites.
The demonstration of these effects was carried out by my student
David Ailion for his Ph.D. thesis [11]. A similar concept and
experimental demonstration was carried out independently by
Dr. Irving Lowe and his student D.C. Look [12].

Comparison of the usual criteria for motional narrowing
$(\tau_{jump} < 1/\Delta\omega_{rigid\ lattice})$ and for an effect on T_1
$(\tau_{jump} \sim 1/\omega_{Larmor})$ show that $T_{1\rho}$ will typically detect much slower
rates of jumping than the others, thereby opening a whole new
range of motion rates to investigation. The detection of motion
by observing $T_{1\rho}$ is often called the slow motion regime [11].

REFERENCES

1. A.G. REDFIELD, Phys.Rev. 98, 787 (1955)
              Phys.Rev. 128, 2251 (1962)

2. C.J. GORTER, "Paramagnetic Relaxation", Elsevier Publishing
   Co., Amsterdam (1947)

3. J.H. VAN VLECK, J.Chem.Phys. 5, 320 (1937)

4. L.C. HEBEL and C.P. SLICHTER, Phys.Rev. 113, 1504 (1959)

5. N. BLOEMBERGEN, E.M. PURCELL and R.V. POUND, Phys.Rev. 73,
   679 (1948)

6. C.P. SLICHTER and W.C. HOLTON, Phys.Rev. 122, 1701 (1961)

7. B.N. PROVOTOROV, Soviet Phys. JETP 14, 1126 (1962)

8. R.T. SCHUMACHER, Phys.Rev. 112, 837 (1958)

9. S.R. HARTMANN and E.L. HAHN, Phys.Rev. 128, 2042 (1962)

10. F.M. LURIE and C.P. SLICHTER, Phys.Rev. 133, A1108 (1964)

11. C.P. SLICHTER and D. AILION, Phys.Rev. 135, A1099 (1964)
    D. AILION and C.P. SLICHTER, Phys.Rev. 137, 2995 (1966)

12. D.C. LOOK and I.J. LOWE, J. Chem.Phys. 44, 2995 (1966)

SINGLE CRYSTALS, POWDERS, AND ANISOTROPY EFFECTS

C.P. Slichter

University of Illinois

Urbana-Champaign (Illinois, U.S.A.)

A SINGLE CRYSTAL EXAMPLE

The System to be Analyzed

I have remarked on the fact that the rapid motion of liquids characteristic of high resolution NMR causes us to lose certain types of information. Let us now examine what that information is, and illustrate its significance. To do so I will pick a concrete example - one which my student Dr. Thomas Stakelon has recently carried out for his Ph. D. thesis, completed just before the start of this summer school. Like Dr. Alloul, we have been interested in the general problem of magnetic atoms, such as those of the iron group of the periodic table, in non-magnetic metals such as Cu or Al. Dr. Stakelon has studied the NMR of Cu nuclei which are near neighbours to Fe, Co, or Ni atoms in single crystals of Cu. My student Thomas Aton is looking at the spectrum of single crystals of Cu containing Mn and Cr. In my lecture I will discuss the form of the interactions, and the number and intensity of spectral lines which should result. I will not discuss what this tells us about theories of magnetic atoms in non-magnetic hosts. My lecture will serve partly to introduce the work of Dr. Alloul.

For concreteness let us consider a cobalt atom in Cu. Cobalt goes into copper substitutionally. Since copper is a face centered cubic crystal, the cobalt atom sees cubic surroundings. It has 12 first neighbours, in the [110] and equivalent directions, 6 second neighbours in the [100] and equivalent directions. For the various tensors we will encounter it is convenient to define

coordinate axes at each site which, by symmetry, will turn out
to be principal axes of the relevant tensors. For some positions,
(the 3rd neighbour, for example) symmetry alone will not suffice
to determine the principal axes.

Figure 1   illustrates the axes at the Co, the first and
the second neighbour positions.

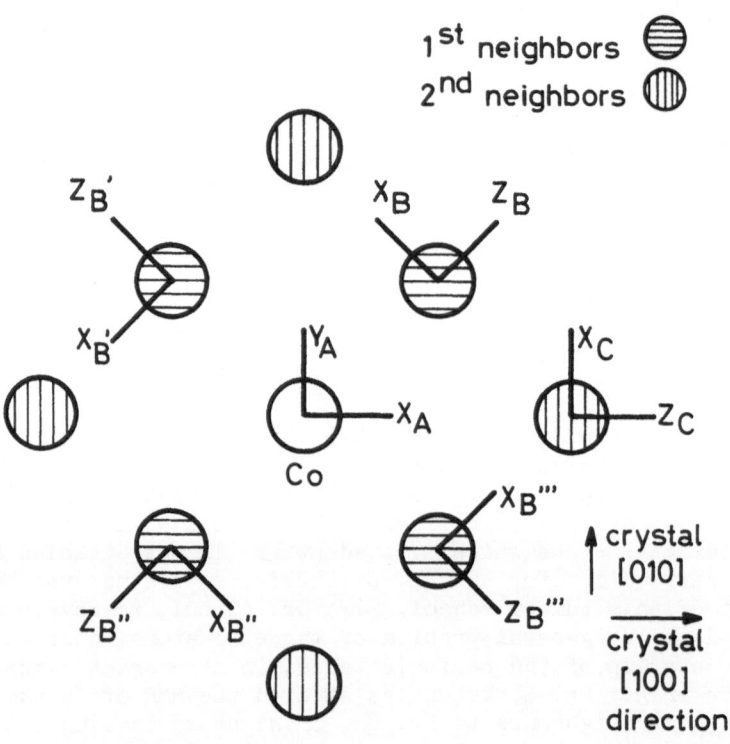

Fig. 1. The immediate vicinity of a Co atom present substitutional-
        ly in Cu. The cross-hatching shows atoms in a crystal
        (001) plane (see definition of the crystal directions in
        the figure). 4 of 12 first neighbours and 4 of the 6
        second neighbours are shown.

Note that each of the four first neighbours in the plane of the
figure has its own distinct axes $(x_B, y_B, z_B), (x_{B'}, y_{B'}, z_{B'})$ etc.

The Spin Hamiltonian

The total nuclear spin Hamiltonian of the system consists of a sum of the Hamiltonians of the individual nuclei plus the nuclear-nuclear couplings. It turns out that the latter terms principally result in a line broadening mechanism if one is considering ordinary NMR. Dr. James Boyce [1] for his Ph.D. thesis did a nuclear double resonance study of $\underline{Cu}Co$, utilizing spin echo double resonance to detect the weak $\overline{Cu}$ first neighbour resonance by observing its effect on the Co spin echo. That effect is dependent on the fact that the Co precession rate is changed when the Cu neighbour is flipped as a result of the Cu-Co secular dipolar coupling.

In the discussion below we will not discuss the nuclear dipolar coupling further.

We will make one assumption about the magnetic character of the center: in the absence of an applied field, $\vec{M}(\vec{r}) = 0$, and in the presence of an applied field $\vec{H}_o$, $\vec{M}(\vec{r})$ is a linear function of $\vec{H}_o$.

The Co spin Hamiltonian

Since the Co nucleus is at a site of cubic symmetry, clearly X_A, Y_A, and Z_A are principal axes for all tensors. We can conclude immediately that

(i) there is no quadrupole coupling since

$$V_{X_A X_A} = V_{Y_A Y_A} = V_{Z_A Z_A} = 0$$

(ii) the chemical shift is isotropic

$$\sigma_{X_A X_A} = \sigma_{Y_A Y_A} = \sigma_{Z_A Z_A} \equiv \sigma_{Co}$$

(iii) the Knight shift is isotropic

(iv) the dipolar coupling to the electron spin vanishes because the factor $(3 \cos^2\theta - 1)/r^3$ coupling an electron spin at coordinates (r, θ, ϕ) relative to a nucleus averages to zero over the angular coordinates. Thus we get

$$H_{Co} = -\gamma_{Co}\hbar\vec{H}_o \cdot \vec{I}_{Co} - \gamma_{Co}\sigma_{Co}\hbar\vec{H}_o \cdot \vec{I}_{Co} - \gamma_{Co}K_{Co}\hbar\vec{H}_o \cdot \vec{I}_{Co}$$

$$\text{chemical shift} \qquad \text{Knight shift} \qquad (1)$$

$$= - \gamma_{Co}\hbar\vec{H}_o(1 + \sigma_{Co} + K_{Co}) \cdot \vec{I}_{Co}$$

Narath [2] has made extensive studies of the nuclear resonance
of the magnetic atoms in non-magnetic hosts, and has shown
that often K is large and negative owing to core polarization
(note that the d-electrons themselves can not contribute to K),
and that σ is large and positive owing to the narrow width of
the d-levels of a magnetic atom.

The first neighbour spin Hamiltonian

The Cu neighbour we have labelled "B" in figure 1 has
principal axes X_B, Y_B, and Z_B for all tensors. In contrast to the
Co atom, at the Cu site the electron spin-nuclear spin dipolar
coupling produces a non-zero average. To a first approximation
we might consider this to arise from a net magnetic dipole
localized on the Co atom of magnitude $\chi_{Co}H_O$. However, the $1/r^3$
aspect of the dipole-dipole coupling brings out the effect of
relatively small spin density near to the Cu nucleus. These
effects can be lumped together with the Fermi contact term to
give a non-isotropic Knight shift. We have, therefore,

$$H_B = -\gamma_B \hbar \vec{H}_o \cdot \vec{I}_B - eQ_B[V_{Z_B Z_B}(3I_{Z_B}^2 - I_B^2) + (V_{X_B X_B} - V_{Y_B Y_B})$$

$$(I_{X_B}^2 - I_{Y_B}^2)]/4I_B(2I_B - 1) \qquad (2)$$

$$-\gamma_B \hbar (H_{X_B}\sigma_{X_A X_B}I_{X_B} + H_{Y_B}\sigma_{Y_B Y_B}I_{Y_B} + H_{Z_B}\sigma_{Z_B Z_B}I_{Z_B}) \left.\right\} \text{chemical shift}$$

$$-\gamma_B \hbar (H_{X_B}K_{X_A X_B}I_{X_B} + H_{Y_B}K_{Y_B Y_B}I_{Y_B} + H_{Z_B}K_{Z_B Z_B}I_{Z_B}) \left.\right\} \text{Knight shift}$$

$$= -\gamma_B \hbar \vec{H}_o \cdot (\overset{\leftrightarrow}{\Pi} + \overset{\leftrightarrow}{\sigma}_B + \overset{\leftrightarrow}{K}_B) \cdot \vec{I}_B + H_{Q_B}$$

where H_{Q_B} is the quadrupolar term, $\overset{\leftrightarrow}{\Pi}$ the unit dyadic, and where
the last line simply serves to express the result more compactly.

The second neighbour spin Hamiltonian

The Cu neighbour we have labelled "C", one of the six second
neighbours of the Co, has principal axes X_C, Y_C, Z_C. However, since
the crystal has four-fold symmetry about Z_C, the directions X_C and

Y_C are equivalent for the second rank tensors we are concerned with. That is, they all possess axial symmetry about Z_C. We therefore get the following spin Hamiltonian

$$H_C = -\gamma_C \hbar \vec{H}_o \cdot \vec{I}_C + eQ_C V_{Z_C Z_C} (3I_{Z_C}^2 - I_C^2)/4I_C(2I_C - 1) \qquad (3)$$

$$-\gamma_C \hbar [(H_{X_C} I_{X_C} + H_{Y_C} I_{Y_C})\sigma_\perp + H_{Z_C} I_{Z_C} \sigma_\|]$$

$$-\gamma_C \hbar [(H_{X_C} I_{X_C} + H_{Y_C} I_{Y_C})K_\perp + H_{Z_C} I_{Z_C} K_\|]$$

where we have introduced notations "$\perp$" and "$\|$" as subscripts to emphasize the axial symmetry of the tensors.

The Resultant Spectrum and What It Tells Us

The Co spectrum

Co has a spin of 7/2. The spin Hamiltonian of equation (1) corresponds to the usual Zeeman effect, but with H_o replaced by a field $H_o(1 + \sigma_{Co} + K_{Co})$. Thus the position is shifted without any splittings arising.

In practice intensity experiments show that frequently the alloy contains enough strains so that there is first order quadrupole wipeout. One observes in practice then only the $+1/2 \leftrightarrow -1/2$ transition.

The first neighbour spectrum

The first neighbour Hamiltonian, equation (2), consists of a magnetic term and a quadrupolar term. There are two isotopes of Cu (Cu^{63} and Cu^{65}) both with spin 3/2. The quadrupole term produces first order shifts of the $3/2 \leftrightarrow 1/2$ and $-1/2 \leftrightarrow -3/2$ transitions, but not of the central transition. For impurity atoms in Al, for which first order wipeout is much smaller than for Cu, the first order splittings are seen [3]. They are harder to see in Cu. We shall omit them from discussion for lack of time. Their dependence on the orientation of H_o with respect to crystal axes gives symmetry information much like that we discuss below for the magnetic interactions.

Thus we focus on the central transition ($+1/2 \leftrightarrow -1/2$) of the Cu atom. H_Q effects this transition in second order perturbation. Let us define angles θ_B, ϕ_B by the relations

$$H_{X_B} = H_o \sin\theta_B \cos\phi_B \qquad = H_o \cos(X_B, Z_o)$$

$$H_{Y_B} = H_o \sin\theta_B \sin\phi_B \qquad = H_o \cos(Y_B, Z_o) \qquad (4)$$

$$H_{Z_B} = H_o \cos\theta_B \qquad = H_o \cos(Z_B, Z_o)$$

where Z_o is the field direction.

It is easy to show that H_Q, taken to second order, shifts the $+1/2 \leftrightarrow -1/2$ transition by a frequency $+\delta\omega$ of

$$\hbar\delta\omega_{1/2,-1/2} = [f^2(\theta_B, \phi_B) - g^2(\theta_B, \phi_B)]/H_o \qquad (5)$$

where the $1/H_o$ comes from the energy denominator of the second order perturbation theory, and where, owing to the angular variation of f and g, for some orientations of H_o, $\delta\omega_{1/2,-1/2}$ is positive whereas for others it is negative. (Problem: show that equation (5) is correct).

The Knight shift and chemical shift contribute a shift in the effective magnetic field. It is convenient to quantize the nuclear spin along the direction of H_o. Calling this component I_{Z_o}, we express I_{X_B}, I_{Y_B}, and I_{Z_B} in terms of spin components I_{X_o}, I_{Y_o}, and I_{Z_o}, where X_o and Y_o are arbitrary directions perpendicular to Z_o. Only the I_{Z_o} component of I_{X_B}, I_{Y_B}, or I_{Z_B} will have diagonal elements in the I_{Z_o} quantization scheme.

We have that

$$I_{X_B} = I_{X_o} \cos(X_B, X_o) + I_{Y_o} \cos(X_B, Y_o) + I_{Z_o} \cos(X_B, Z_o)$$

$$I_{Y_B} = I_{X_o} \cos(Y_B, X_o) + I_{Y_o} \cos(Y_B, Y_o) + I_{Z_o} \cos(Y_B, Z_o)$$

$$I_{Z_B} = I_{X_o} \cos(Z_B, X_o) + I_{Y_o} \cos(Z_B, Y_o) + I_{Z_o} \cos(Z_B, Z_o) \qquad (6)$$

Only the terms in I_{Z_o} will contribute first order energy shifts in the I_{Z_o} representation, since I_{X_o} and I_{Y_o} have zero diagonal matrix elements in the I_{Z_o} representation. Using equation (4) and equation (6) we get that the magnetic term in H_B can be written

$$H_B\Big)_{magnetic} = -\gamma_B \hbar H_o I_{Z_o} [(1 + K_{X_B X_B} + \sigma_{X_B X_B})\sin^2\theta_B \cos^2\phi_B$$

$$+ (1 + K_{Y_B Y_B} + \sigma_{Y_B Y_B})\sin^2\theta_B \sin^2\phi_B \qquad (7)$$

$$+ (1 + K_{Z_B Z_B} + \sigma_{Z_B Z_B})\cos^2\theta_B] + \text{off diagonal terms}$$

This is equivalent to our having an effective applied field $H_{eff}(\theta_B, \phi_B)$ acting on a bare nucleus, with

$$
\begin{aligned}
H_{eff}/H_o = &(1 + K_{X_B X_B} + \sigma_{X_B X_B})\sin^2\theta_B \cos^2\phi_B \\
&+ (1 + K_{Y_B Y_B} + \sigma_{Y_B Y_B})\sin^2\theta_B \sin^2\phi_B \\
&+ (1 + K_{Z_B Z_B} + \sigma_{Z_B Z_B})\cos^2\theta_B
\end{aligned}
\tag{8}
$$

From the two terms equation (5) and equation (8) we see that for a fixed orientation of H_o the central transition will consist of a sum of 2 terms, one the magnetic term proportional to H_o, the other the second order quadrupole term inversely proportional to H_o. We can determine the individual contributions from observing the field dependence [4, 5].

Let us now consider what the NMR spectrum looks like. For simplicity let us consider H_o so large that we can neglect the second order quadrupole term. Suppose H_o were along the [001] crystal direction. We have then $\theta_B = \pi/2$, so that for nucleus B

$$
H_{eff} = H_o(1 + K_{Y_B Y_B} + \sigma_{Y_B Y_B})
\tag{9}
$$

The same effective field acts on all four first neighbours in the [001] plane (i.e., the plane of figure 1). There are 8 other first neighbours, however, for which $\sin^2\theta = \cos^2\theta = 1/2$, and $\cos^2\phi = 1$, $\sin^2\phi = 0$. For them

$$
H_{eff} = H_o[(1 + K_{X_B X_B} + \sigma_{X_B X_B})/2 + (1 + K_{Z_B Z_B} + \sigma_{Z_B Z_B})/2]
\tag{10}
$$

Therefore for this orientation of H_o the spectrum from the first shell will consist of two lines. One will have a relative intensity "8" at frequency $\omega = \gamma H_{eff}$ in which H_{eff} is given by equation (9). The other will have intensity "4" and occur at $\omega = \gamma H_{eff}$ given by equation (10).

If H_o were applied along the [110] direction we would get 3 lines of relative intensities 8 to 2 to 2.
(Problem: Compute the corresponding values of H_{eff}).
(Problem: Determine the number and relative intensity of lines arising from the first neighbour shell when H_o lies along the [111] direction).

The second neighbour spectrum

The expressions for the second neighbour are readily obtained

from the results for the first neighbour if we simply recognize
that the second neigbour has axial symmetry. We get, therefore,
a second order quadrupole shift of the central transition which is
independent of ϕ_C:

$$\hbar\delta\omega_{1/2,-1/2} = [f^2(\theta_C) - g^2(\theta_C)]H_o^{-1} \qquad (11)$$

and a magnetic term equivalent to an effective magnetic field,
H_{eff}, acting on a bare nucleus of

$$H_{eff} = H_o [1 + (K_\perp + \sigma_\perp)\sin^2\theta_C + (K_{||} + \sigma_{||})\cos^2\theta_C] \qquad (12)$$

Again we can distinguish the quadrupolar and magnetic
shifts experimentally by studying the field dependence of the
shifts.

Again we consider the strong magnetic field limit for
simplicity, since it enables us to neglect the electric quadrupole
term.

If H_o lies along the [001] direction, we have 2 second
neighbour nuclei with $\theta_C = \pi/2$. Thus we get

$$\begin{aligned}
&2 \text{ nuclei with } H_{eff} = (1 + K_{||} + \sigma_{||})H_o \\
&4 \text{ nuclei with } H_{eff} = (1 + K_\perp + \sigma_\perp)H_o
\end{aligned} \qquad (13)$$

If H_o lines along the [111] direction, all six nuclear axes make
equal angles with H_o and we get

$$6 \text{ nuclei with } H_{eff} = [1 + 2(K_\perp + \sigma_\perp)/3 + (K_{||} + \sigma_{||})/3]H_o \qquad (14)$$

(Problem: Compute the angular dependence and intensity of the 2nd
neighbour lines for H_o perpendicular to the [001] axis in terms
of the angle between H_o and the [100] crystal axis).

The examples we have seen illustrate that the number, position,
and intensity of the NMR lines, seen as the orientation of $\vec{H}_o$ is
varied with respect to the crystal axes, is a characteristic of
the particular shell. *Thus angular studies enable us to identify
which shell of neighbours gives rise to a given set of lines.*
There are ambiguities since, for example, the [100] and [200]
neighbours obviously have the same symmetry.

In discussing Stakelon's data we discovered a useful theorem.
The fact that the nuclei of a given shell have cubic symmetry
about the central atom (the Co) makes the magnetic anisotropy shift
of the center of gravity of the lines from that shell independent
of the orientation of H_o. This theorem helps assign spectral lines
to a particular shell when spectra of several shells intermingle.

Another important aid in identifying shells is to note that, when H_O lies along one of the principal axes, the resonance frequency has only second order variation with field orientation, whereas far from a principal axis the frequency has a first order variation. (Problem: Show that this theorem is true for the 2nd neigbours for variations in the orientation of H_O in the [001] plane close to [110] direction).

THE SPECTRA OF POWDER SAMPLES

Many solid samples are difficult or impossible to obtain in single crystal forms. Examples are powdered metals (where the powder is chosen so the alternating fields will not be screened out by eddy currents), polymers, glasses, and many biological systems. Since the position of the NMR absorption lines varies with orientation of H_O relative to the local crystal axes, the resultant spectrum is smeared over a range of frequencies. As we have seen a typical situation is that the resonance frequency, ω, of a given nucleus goes as

$$\omega = \omega_o + \omega_a \cos^2\theta \cos^2\phi + \omega_b \cos^2\theta \sin^2\phi + \omega_c \sin^2\theta \qquad (15)$$

where ω_o, ω_a, ω_b, and ω_c are three constants, and where (θ, ϕ) are the angles which describe the orientation of the static field with respect to the crystal axes. Equation (5) describes anisotropic Knight shifts and chemical shifts, line shifts due to nuclear dipole coupling between unlike nuclei, and first order quadrupole shifts. It does not describe second order quadrupole shifts of a central transition, but the general principles we outline still apply. If we define the axes such that $\omega_c > \omega_b > \omega_a$, equation (15) describes a total frequency spread of $\omega_c - \omega_a$ in the powder. When $\omega_c - \omega_a$ is much larger than the single crystal linebreadth, the powder line will be much broader than a single crystal line, and consequently much weaker in intensity. We may then wonder whether or not the intensity of the powder pattern will be large enough to be seen. Fortunately the situation is often much less discouraging than these remarks suggest, since the intensity of powder lines tends to peak strongly for certain orientations. We will illustrate this fact. The reader would find it useful to read a paper dealing with ESR of Fe^{+++} ions in glass [6] (in which further references can be found) to see an analogous problem in electron spin resonance.

For simplicity we consider the case of axial symmetry with $\omega_a = \omega_b < \omega_c$. (If $\omega_a = \omega_b > \omega_c$ the spectrum is the mirror image, about its average value, of the case we will analyze).

We consider, then, that

$$\omega = \omega_o + \omega_c \cos^2\theta + \omega_a \sin^2\theta \qquad (16)$$

$$= \omega_o + \omega_a + (\omega_c - \omega_a)\cos^2\theta$$

We consider the crystal axes in the sample at the position of the nuclear species of interest to have random distribution inangle with respect to H_o. Thus, if there are N nuclei, the number $dN(\theta, \phi)$ within a solid angle $d\Omega$ of any particular orientation θ, ϕ is

$$d^2N(\theta, \phi) = Nd\Omega/4\pi$$

$$\qquad (17)$$

$$= N \sin\theta d\theta d\phi/4\pi$$

For the case of axial symmetry it is convenient to integrate over $d\phi$ and use the expression

$$dN(\theta) = N\sin\theta d\theta/2$$

$$\qquad (18)$$

$$= -Nd(\cos\theta)/2$$

Now equation (16) describes a range of ω values which correspond to $\cos^2\theta$ ranging from 0 to 1. Thus

$$\omega_o + \omega_a \leq \omega \leq \omega_o + \omega_c \qquad (19)$$

When $\omega = \omega_o + \omega_a$, H_o is perpendicular to the symmetry axis. From equation (16) we get

$$\cos\theta = \pm \sqrt{(\omega - \omega_o - \omega_a)/(\omega_c - \omega_a)} \qquad (20)$$

where the two signs correspond to the fact that θ ranges from zero to π. Substitution into equation (18) gives

$$dN(\theta) = \frac{N}{4}d\omega/\sqrt{\omega - \omega_o - \omega_a} \sqrt{\omega_c - \omega_a} \qquad (21)$$

for the number within a range $d\omega$ near a given angle θ. However, for $\theta' = \pi-\theta$ the identical frequency occurs. Thus if we define $dN(\omega)$ as the number at frequency ω, thus counting the contributions from both θ and $\pi - \theta$, we get

$$dN(\omega) = \frac{N}{2}d\omega/\sqrt{\omega - \omega_o - \omega_a} \sqrt{\omega_c - \omega_a} \qquad (22)$$

$$\equiv g(\omega)d\omega$$

where $g(\omega)$ gives the number of nuclei whose resonance falls within a range $d\omega$ at a frequency ω, hence gives the line shape of the powder.

Note that $g(\omega)$ goes infinite as ω approaches $\omega_o + \omega_a$, but the infinity is integrable (i.e. the integrated intensity remains finite). At $\omega = \omega_o + \omega_c$, $g(\omega)$ is finite but its derivative is infinite since $g(\omega)$ drops discontinuously to zero for $\omega > \omega_o + \omega_c$. (Note equation (21) applies only for $\omega_o + \omega_a \leq \omega \leq \omega_o + \omega_c$. Outside this range $g(\omega)$ is zero).

The resulting absorption line is shown in figure 2.

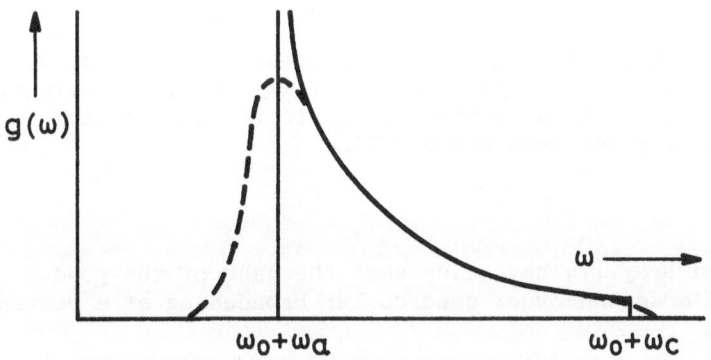

Fig. 2. Solid line: The line shape function $g(\omega)$ vs. ω for the case $\omega_c > \omega_a$, assuming axial symmetry in equation (16). Dashed line: The line shape function when nuclear dipolar broadening is included.

The actual resonance always has nuclear dipolar broadening, which removes the infinities in $g(\omega)$ or its derivatives. Using field modulation, the NMR spectra of such a line has a prominent peak at $\omega \cong \omega_o + \omega_a$, and a small peak at $\omega_o + \omega_c$. These peaks are distinguishable from the usual derivative peaks (which, being derivatives of symmetric lines are perfectly antisymmetric).

As we see in figure 3, the peak at $\omega_o + \omega_a$ is far from antisymmetric, being mostly of one sign.

Drain [3] has discussed the form of the singularities in $g(\omega)$ and its derivatives which one gets for the powder pattern of various first order quadrupolar interactions. Baugher, Kriz,

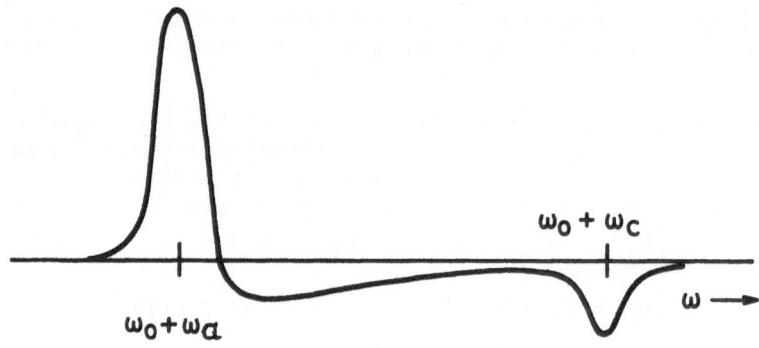

Fig. 3. Derivative signal, $dg(\omega)/d(\omega)$, for the dashed absorption
curve of figure 2. With poor signal to noise, only the
positive peak at $\omega \cong \omega_c + \omega_a$ is seen. Next easiest to see
is the negative peak near $\omega_0 + \omega_c$. The flat, slightly
negative signal between peaks is frequently hard to
distinguish from the baseline.

Taylor, and Bray [7] have discussed the case of the powder
pattern of a second-order quadrupolar broadening of a central
transition.

There is a simple explanation of the infinity at $\omega \cong \omega_0 + \omega_a$.
This value of ω corresponds to H_0 being nearly perpendicular
to the symmetry axis. If we think of a sphere with points on it
specifying the orientation of H_0, and with a polar axis
corresponding to the symmetry axis, all the points on the equator
correspond to $\omega = \omega_0 + \omega_a$. The polar axis (a single point)
corresponds to $\omega = \omega_0 + \omega c$. In fact, there is a band on both sides
of the equator for which $\omega \cong \omega_0 + \omega_a$, hence clearly a strong
statistical weighting for H_0 perpendicular to the symmetry axis.

This geometrical factor produces a strong peak at $\omega \cong \omega_0 + \omega_a$.
Such peaks are substantially more prominent than would result
from a random smear of intensity between $\omega_0 + \omega_a$ and $\omega_0 + \omega_c$.
Thus complex spectra of powders can often be seen. With modern
computor methods very complicated spectra can be simulated,
enabling one to get much of the single crystal wealth of
information from powder samples.

REFERENCES

1. J.B. BOYCE, Thesis, University of Illinois (unpublished)
   D.V. LANG, J.B. BOYCE, D.C. LO and C.P. SLICHTER, Phys.Rev.
   Lett. 29, 776 (1972)

2. A. NARATH, Crit.Rev.Solid State Sci. 3, 1 (1972)

3. L.E. DRAIN, J.Phys.C.(Proc.Phys.Soc.) 1, 1690 (1968)
              Proc.Phys.Soc. 88, 111 (1966)

4. D. LO, D.V. LANG, J.B. BOYCE and C.P. SLICHTER, Phys.Rev.B 8,
   973 (1973)

5. D.V. LANG, J.B. BOYCE, D.C. LO, and C.P. SLICHTER, Phys.Rev.B
   9, 3077 (1974)

6. T. CASTNER, W.C. HOLTON, G.S. NEWELL and C.P. SLICHTER, J.Chem.
   Phys. 32, 668 (1960)

7. J.F. BAUGHER, H.M. KRIZ, P.C. TAYLOR, and P.J. BRAY, Journal
   of Magnetic Resonance 3, 415 (1970)
   K. NARITA, J. UMEDA and H. KUSUMOTO, J.Chem.Phys. 44, 2719
   (1966)

NMR PARAMETERS FOR STUDYING STRUCTURE AND MOTION

W. Müller-Warmuth

Westfälische Wilhelms-Universität

Münster, Deutschland

INTRODUCTION

In a rigid lattice, information is obtained from the NMR frequency, from the intensity of a line and especially from the shape of the spectrum or the linewidth. Relaxation times become of importance if motions are included in the study. The resonance frequency may be influenced by the phenomena to be discussed in this lecture; however it depends principally upon the type of nulcei employed. The intensity is proportional to the concentration of the respective nuclear spins. Various factors contribute to the shape of the NMR spectrum and these will be the main topic of the present lecture.

The NMR Hamiltonian can be written

$$H = H_z + H_D + H_Q + H_\sigma \tag{1}$$

with H_z being the Zeeman interaction energy of the nuclear spins I in the external magnetic field H. Taking H along the z-direction,

$$H_z = -\gamma \hbar H I_z \tag{2}$$

It is generally assumed that the Zeeman term predominates sufficiently so that one may use perturbation theory in order to derive approximate solutions of the problem.

The magnetic dipole-dipole interaction between nuclear spins I_i and I_k separated by $\vec{r}_{ik}$ is represented by the Hamiltonian [1].

$$H_D = \sum_{i<k} [(\gamma_i \gamma_k \hbar^2)/r_{ik}^3] (\vec{I}^i . \vec{I}^k) - 3[(\vec{I}^i . \vec{r}_{ik})(\vec{I}^k . \vec{r}_{ik})]/r_{ik}^2 \quad (3)$$

$\gamma_i \gamma_k \hbar^2/r_{ik}^3$ is the order of magnitude of the interaction energy between two spins corresponding to a local magnetic field at the ith nucleus of $\gamma_k \hbar/r_{ik}^3$ (= 3.5 Gauss or 15 kHz for a proton with r = 2 Å).

Nuclei with spin I greater than 1/2 possess an electric quadrupole moment which may interact with the gradient of an electric field at the nucleus. In a coordinate system (ξ, η, ζ), in which the field gradient tensor is diagonal, the quadrupole Hamiltonian is [1]

$$H_Q = [e^2 qQ/4I(2I-1)][3 I_\zeta^2 - \vec{I}^2 + \eta (I_\xi^2 - I_\eta^2)] \quad (4)$$

with the principal values $V_{\xi\xi}$, $V_{\eta\eta}$, $V_{\zeta\zeta}$ of the field gradient tensor and

$$eq \equiv V_{\zeta\zeta}; \quad \eta \equiv (V_\xi - V_\eta)/V_\zeta$$

The order of magnitude of this interaction is characterized by the quadrupole coupling constant $e^2 qQ$. If q is produced by an isolated charge e at a distance r = 1 Å, then $q \approx V_{rr} = 2e/r^3 = 9.6 \cdot 10^{14}$ cgs leading with $Q \approx 10^{-25}$ cm^2 to a frequency splitting of $e^2 qQ/h \approx 7$ MHz.

A chemical shift arises when the external field induces a screening of the nuclei, which in a solid depends upon the direction of the field. The chemical shift Hamiltonian is [2]

$$H_\sigma = \gamma\hbar \vec{I} \sigma \vec{H}_o \quad (5)$$

with σ being the chemical shift tensor. The frequency shift is of the order of $\sigma_{zz} H_o$, where σ_{zz} is the component of σ along H_o. σ_{zz} is typically 10^{-3}-10^{-6} depending upon the size of the atom.

In addition to the interactions of eq. (1), which determine the NMR line shape, a "Knight shift" or a "paramagnetic shift" may occur. These shifts can formally be treated in exactly the same way as the chemical shift. They arise in metallic or (strongly) paramagnetic materials, respectively, from either the spin polarization of the conduction electrons or from paramagnetic centres polarized by the magnetic field. In these cases σ in eq. (5) has to be replaced by the appropriate tensor.

In order to discuss the different NMR parameters for studying structure, in this paper we shall treat the various interactions separately and shall refer only occasionally to combined effects.

DIPOLAR INTERACTIONS

The Hamiltonian of eq. (3) is conveniently decomposed in the following fashion

$$H_D = \sum_{i<k} (\gamma_i \gamma_k \hbar^2 / r_{ik}^3)(A_{ik} + B_{ik} + C_{ik} + D_{ik} + E_{ik} + F_{ik}) \quad (6)$$

where A_{ik} and B_{ik} (B only in the case of like spins) commute with the Zeeman operator whilst the terms C, D, E and F are off-diagonal. As far as the NMR of rigid solids at the resonance frequency $\omega_o = \gamma H_o$ is concerned, generally we may confine ourselves to the secular part H_D' of the magnetic interaction energy containing

$$A_{ik} = I_z^i I_z^k (1 - 3 \cos^2 \theta_{ik}) \quad \text{and}$$

$$\quad (7)$$

$$B_{ik} = -(1/4)(I_+^i I_-^k + I_-^i I_+^k)(1 - 3 \cos^2 \theta_{ik})$$

In eqs. (6) and (7) I_x and I_y are expressed in the terms of the raising and lowering operators I_+ and I_-, and the vector $\vec{r}_{ik}$ is expressed in spherical coordinates r, θ, ϕ . Then

$$H_D' = \sum_{i<k} F_{ik}[I_z^i I_z^k - \frac{1}{4}(I_+^i I_-^k + I_-^i I_+^k)] = \frac{1}{2} \sum_{i<k} F_{ik}(3 I_z^i I_z^k - \vec{I}^i \vec{I}^k)$$

$$F_{ik} = (\gamma_i \gamma_k \hbar^2 / r_{ik}^3)(1 - 3 \cos^2 \theta_{ik}) \quad (8)$$

Two-spin-systems

The simplest case for calculating the spectrum occurs, if only a small number of nuclei interact, as for instance pairs of proton spins in materials containing water of crystallization. First order perturbation theory may then immediately be applied

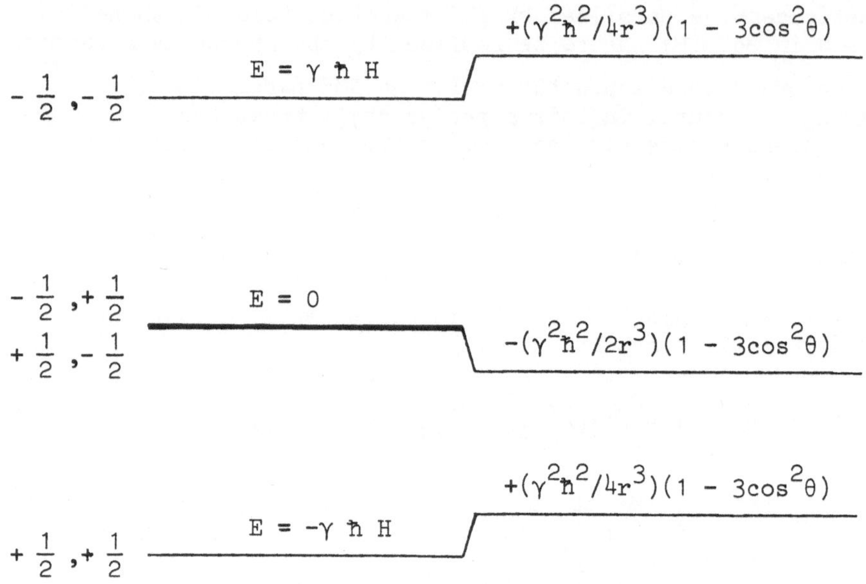

$$-\tfrac{1}{2}\,,-\tfrac{1}{2} \underline{\hspace{3cm}} E = \gamma\,\hbar\,H \underline{}\Big/\underline{\hspace{2cm}} +(\gamma^2\hbar^2/4r^3)(1 - 3\cos^2\theta)$$

$$
\begin{aligned}
-\tfrac{1}{2}\,,+\tfrac{1}{2} \\
+\tfrac{1}{2}\,,-\tfrac{1}{2}
\end{aligned}
\underline{\hspace{3cm}} E = 0 \underline{}\Big\backslash\underline{\hspace{1.5cm}} -(\gamma^2\hbar^2/2r^3)(1 - 3\cos^2\theta)
$$

$$+\tfrac{1}{2}\,,+\tfrac{1}{2} \underline{\hspace{3cm}} E = -\gamma\,\hbar\,H \underline{}\Big/\underline{\hspace{2cm}} +(\gamma^2\hbar^2/4r^3)(1 - 3\cos^2\theta)$$

Scheme 1. Zeeman energy (left) and dipolar coupling (right) of the
triplet states of a two-spin-system.

and the energy levels of scheme 1 are obtained.

The NMR spectrum for a given pair with the angle θ of its
separation relative to the z-direction shows a doublet

$$\omega = \omega_o \pm \gamma h\,(1 - 3\cos^2\theta)$$

$$h \equiv \frac{3}{4}\,\frac{\gamma\hbar}{r^3} \tag{9}$$

In a polycrystalline sample the individual crystal axes are
distributed randomly over all directions. The spectrum is smeared
out. The resonance condition (9) must be averaged over all
possible orientations. The shape function $g(\omega)$ is determined
by the probability that the spin pairs contribute to a resonance
in the interval ω to $\omega + d\omega$. This means [3]

$$g(\omega)\ d\omega = \frac{1}{\Omega} \int_{\omega}^{\omega+d\omega} d\Omega(\omega) \tag{10}$$

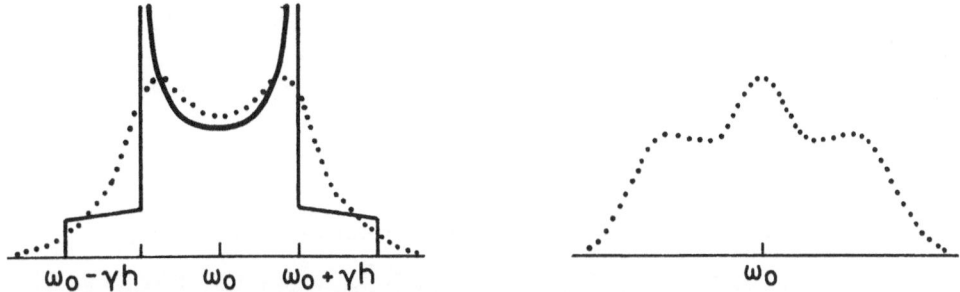

Fig. 1. Calculated line shape for a polycrystalline material
        containing two-spin-systems. The dotted lines for a
        two-spin (left) and a three-spin-system (right) are
        obtained after taking into account the interaction
        of other neighbours.

with $d\Omega = \sin\theta \, d\theta \, d\phi$ being an element of the solid angle $\Omega = 4\pi$.
For the case of eq. (9), where ω is only a function of $\cos\theta$ and
not of ϕ, $g(\omega)$ can be calculated using (10)

$$g(\omega) = -\frac{1}{2} \frac{d \cos\theta}{d\omega} = \begin{cases} \dfrac{1}{4\sqrt{3}\ \gamma h} \cdot \dfrac{1}{\sqrt{1 + \dfrac{\omega - \omega_o}{\gamma h}}} & \text{if } \gamma h < \omega - \omega_c < 2\gamma h \\[4ex] \dfrac{1}{4\sqrt{3}\ \gamma h} \cdot \dfrac{1}{\sqrt{1 - \dfrac{\omega - \omega_o}{\gamma h}}} & \text{if } 2\gamma h < \omega - \omega_o < \gamma h \end{cases}$$

$$(11)$$

The total spectrum is shown in figure 1. Actually, each of the
two lines is broadened by the interaction with other neighbours.
The three-spin and the four-spin cases have also been treated [1].

Line Shapes and Moments of a Resonance Curve

     In case the nuclei are not localized in small groups, unresolved

broad lines appear, which can sometimes be approximated by a
Gaussian curve

$$g(\omega) = \sqrt{\ln 2/\pi}(1/\Delta\omega) \; e^{-\ln 2 \; (\omega-\omega_o)^2/(\Delta\omega)^2} \tag{12}$$

or a Lorentzian (which results from the Bloch equations)

$$g(\omega) = (\Delta\omega/\pi) \; [(\Delta\omega)^2 + (\omega-\omega_o)^2]^{-1} \tag{13}$$

In eqs. (12) and (13) $\Delta\omega$ is the half width of the NMR line
at half intensity.

The nth moment M_n of a curve $g(\omega)$ with respect to ω_o is
defined as

$$M_n = \int_{-\infty}^{\infty} (\omega-\omega_o)^n \; g(\omega) \; d\omega \tag{14}$$

In case of a Gaussian line shape (12), integration of eq. (14)
yields for the second moment

$$M_2 = (\Delta\omega)^2/2 \ln 2 \tag{12a}$$

and $M_4/(M_2)^2 = 3$. No second or higher moment can be defined for a
Lorentzian shape, for the integral diverges. If the curve is only
considered in the interval $-\omega' < \omega - \omega_o < \omega'$ the second moment is
approximated by

$$M_2 \; \tilde{} \; 2\Delta\omega.\omega'/\pi \qquad \text{and} \quad M_4/(M_2)^2 \; \tilde{} \; \pi\omega'/6\Delta\omega \tag{13a}$$

The second (and eventually also the fourth) moment can be derived
from either the experimental absorption spectrum, eq. (12) and (13)
or from the Fourier transform of $g(\omega-\omega_o)$ (Fig. 2).

$$G(t) = G(0) \int_{-\infty}^{\infty} g(\omega-\omega_o) \; \cos(\omega - \omega_o)t \; d\omega \tag{15}$$

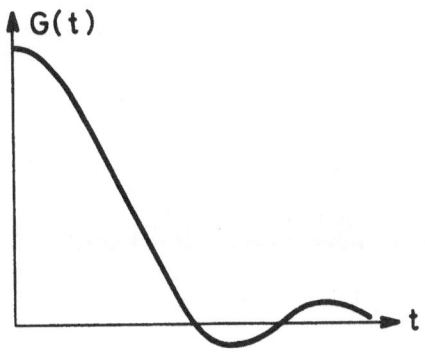 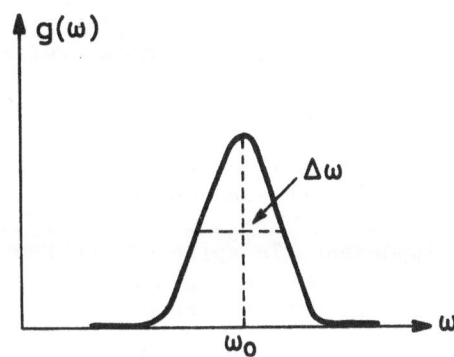

Fig. 2

The latter is the free precession decay after application of a 90° r.f. pulse. It is directly related to the theory by [1]

$$G(t) \sim \mathrm{Tr}\ \{e^{(i/\hbar)H_D t}\ I_x e^{-(i/\hbar)H_D t}\ I_x \} \qquad (16)$$

$$(I_x = \sum_i I_x^i)$$

Comparison of eqs. (14), (15) and (16) yields

$$M_2 = [-1/G(0)][d^2 G(t)/dt^2]_{t=0} = - \mathrm{Tr}\{[H_D', I_x]^2\}/\hbar^2 \mathrm{Tr}\{I_x^2\} \qquad (17)$$

Evaluation of the traces of eq.(17) with eq. (8) uses the commutator rules of the spin operators. Finally,

$$M_2\ (I\text{-}I) = \frac{3}{4}\ \gamma^4 \hbar^2\ I(I+1)\ \sum_k\ (1-3\cos^2\theta_{ik})^2\ /\ r_{ik}^6 \qquad (18)$$

is obtained for the interaction with like spins I, and

$$M_2(I\text{-}S) = \frac{1}{3}\gamma_I^2\gamma_S^2\ \hbar^2\ S(S+1)\ \sum_l\ (1-3\cos^2\theta_{il})^2\ /\ r_{il}^6 \qquad (19)$$

for the interaction with unlike spins S. For a polycrystalline material or a powder of crystallites of random orientations $(1 - 3\cos^2\theta_{ik})^2$ has to be replaced by its mean value 4/5, which is obtained by averaging over all directions.

Adiabatic Fast Passage and Dispersion Signal

In addition to studies where either absorption or pulse experiments are employed, application of strong radio-frequency fields $H_1 \gg H_{loc}$ to NMR in solids is becoming more and more important. In spite of the fact that these experiments are based on the spin temperature concept which will be discussed elsewhere, some parameters are of importance for studying dipolar solids.

Local dipolar magnetic fields and second moments can be obtained from an adiabatic fast passage experiment obeying the condition

$$1/\gamma H_{loc} \ll T \ll T_1, \ T_{1D}$$

The NMR signal is then proportional to the spin component in phase with the r.f. field ("dispersion") [4] (M_o = equilibrium nuclear magnetization, $\Delta = H - H_o$)

$$M_o \, H_1 \, / \sqrt{\Delta^2 + H_1^2 + H_L'^2}$$

and the (total) width at half intensity is

$$\delta H = \sqrt{12} \sqrt{H_1^2 + H_L'^2} \tag{20}$$

The "local field in the rotating frame" H_L' is defined by

$$H_L'^2 \ (\text{I-I}) = \text{Tr}\{H_D'^2\}/\gamma^2\hbar^2 \ \text{Tr}\{I_z^2\} = (1/3\gamma^2) \ M_2(\text{I-I}) \tag{21}$$

and is closely related to the second moment of an absorption line. In case of magnetic interaction with unlike spins, H_D' has to be replaced by an effective Hamiltonian containing not only terms for the I-S interaction, but also for the interaction between S spins themselves. $H_L'^2$ turns out to be

NMR PARAMETERS FOR STUDYING STRUCTURE AND MOTION

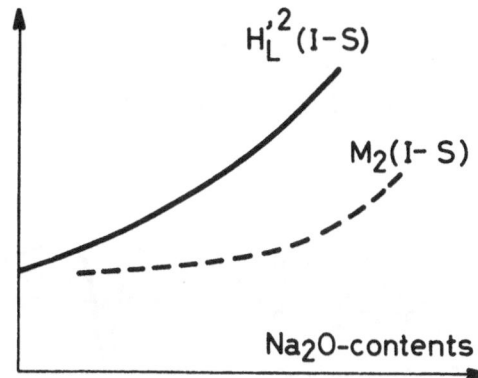

Fig. 3. Local field $H_L'^2$ (Si-Na) (obtained from an adiabatic fast
passage experiment) and M_2 (Si-Na) of the ^{29}Si NMR of a
silicate glass as a function of the soda contents. The
difference between the two curves corresponds to M_2 (Na-Na)
which otherwise cannot be obtained.

$$H_L'^2(I-S) = (1/\gamma_I^2) \, M_2(I-S) + \frac{1}{3}(N_S/N_I)[\gamma_S^2 S(S+1)/\gamma_I^2 I(I+1)]$$

$$(1/\gamma_S^2) \, M_2(S-S) \qquad (22)$$

An example of the application of eq. (22) is shown in figure 3.
Another possibility for obtaining information on local fields
is the so-called adiabatic fast modulation method [4, 6]. Here
again $H_1 \gg H_{loc}$, and a field modulation $H_m \sin\omega_m t$ with
$\omega_m \gg 1/T_1$ is applied. The dipolar dispersion signal is then
proportional to [7]

$$S = \{M_0 H_1 \Delta \sqrt{\Delta^2 + H_1^2 + H_L'^2} \,/[\Delta^2 + (T_1/T_{1x})H_1^2 + (T_1/T_{1D})H_L'^2]\} \, (1/\pi)$$

$$\int_0^{2\pi} \sin x \, dx \,/\sqrt{(\Delta + H_m \sin x)^2 + H_1^2 + H_L'^2} \qquad (23)$$

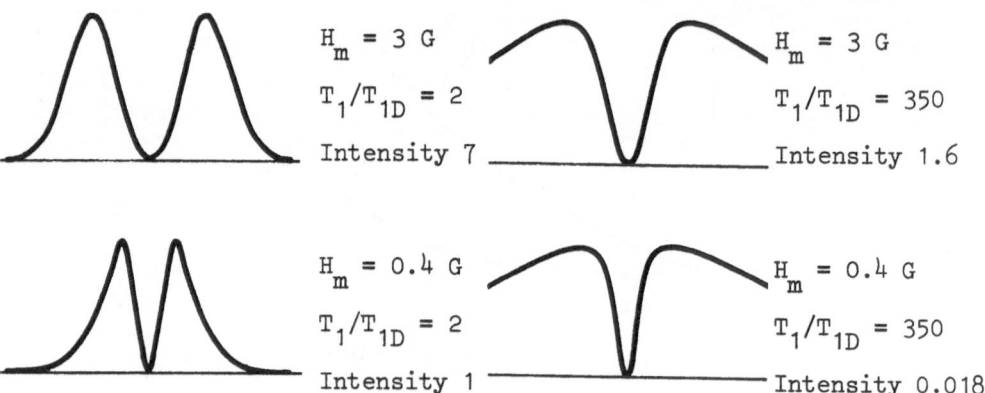

Fig. 4. Theoretical line shapes, eq. (23), for a fixed $H_L' = 1$ G and
different values of H_m and T_1/T_{1D}. In the limit of small
H_m values the distance between the maxima is proportional
to $H_L' \sqrt{T_1/T_{1D}}$.

T_1 is the normal or Zeeman spin-relaxation time, T_{1x} the
relaxation time in the rotating frame, and T_{1D} the dipolar
relaxation time. In contrast to other experimental methods
the signal is strongly dependent upon T_1/T_{1D}. Figure 4 shows
some theoretical line shapes [7].

Application of similar techniques in order to determine T_1
and T_{1D} of solids by means of continuous-wave experiments is
discussed in the literature [8].

QUADRUPOLE EFFECTS [1, 9]

In a high magnetic field, H_Q of eq. (4) can be treated as a
perturbation of the Zeeman Hamiltonian. Then in the first order,
the m → m-1 NMR transition appears at

$$\omega = \omega_o - \frac{1}{4}\omega_Q \left[(3\cos^2\theta - 1) - \eta\sin^2\theta\,\cos2\phi\right](2m - 1) \qquad (24)$$

with
$$\omega_Q \equiv 3\,e^2q\,Q\,/\,2\,I\,(2I - 1)\hbar$$

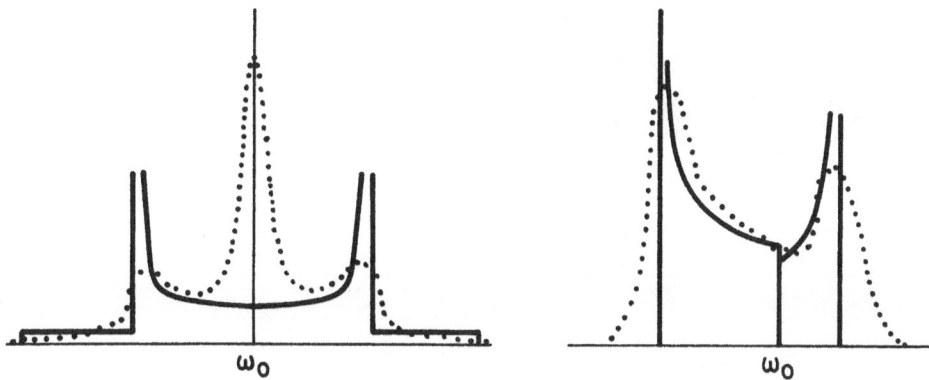

Fig. 5. Line shape for I = 3/2 in polycrystalline or powdered samples with axially symmetric field gradients. Left hand side: first-order quadrupole perturbation, right: second order effect. With dipolar broadening superimposed the dotted curves result.

For a given orientation – θ and ϕ are the angles of the principal coordinate system of the field gradient tensor relative to $H = H_z$ – in single crystals 2 I discrete resonances are observed. If I is half integral, second order calculation yields in the case where $\eta = 0$ for the central transition $1/2 \leftrightarrow - 1/2$

$$\omega = \omega_o - (\omega_Q^2/16\omega_o)[I(I+1) - \tfrac{3}{4}] \, (9 \cos^2\theta - 1)(1 - \cos^2\theta) \quad (25)$$

The frequency shift is a function of the angle between the magnetic field and the crystal axis.

Powder patterns are obtained from eqs. (24) and (25) for $\eta = 0$ using eq. (10) and the procedure outlined for the dipolar two-spin system. Figure 5 shows the resulting absorption line shapes.

The separation between the satellites at the right and left hand sides of the central resonance is exactly ω_Q, see eq. (24). The peaks of the central line splitting (Fig. 5, right) are separated by

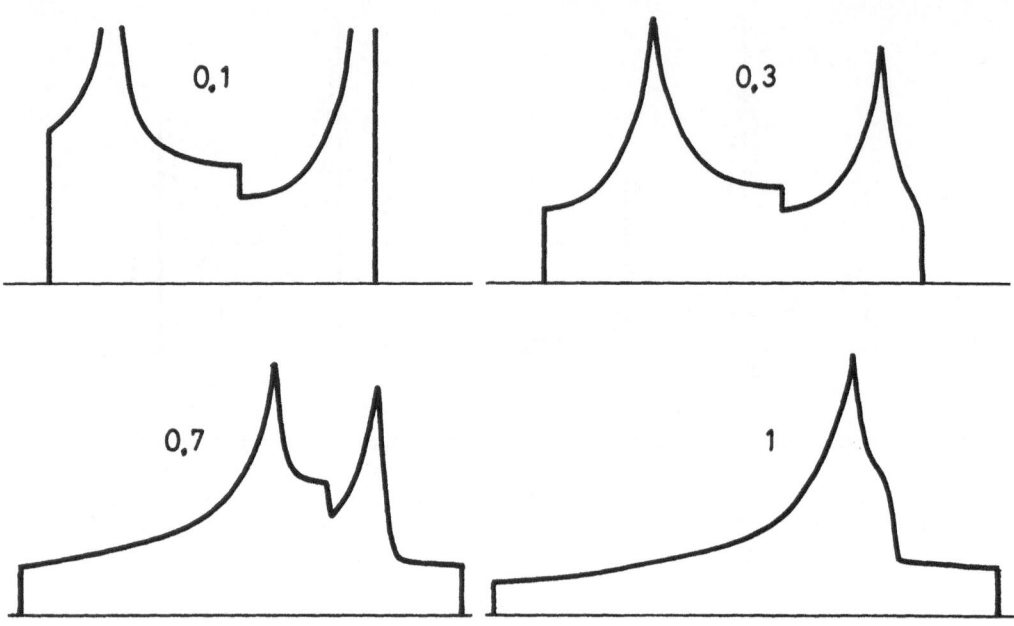

Fig. 6. Powder pattern of the central resonance for various values
of η as obtained by computer calculation [10].

$$\frac{25}{9} \frac{\omega_Q^2}{16\omega_o} \left[I(I+1) - \frac{3}{4} \right]$$

Figure 6 shows computed line shapes [10] for second order
quadrupole effects of powdered samples with η ≠ 0.

In cubic crystals the field gradient vanishes when the
crystal is perfect. If the electric field gradient at the nuclei
in question is sufficiently small, satellite splitting as in
figure 5 is observed. As the gradient increases there can also
be a shift and a splitting of the central line. The test for the
second-order effect is the inverse proportionality to the applied
frequency or field. In actual situations, there may occur a
distribution of splittings or shifts because the field gradient
differs slightly from nucleus to nucleus. Then the satellites
become unobservably broad. The central line, however, in first
order is only dipolarly broadened. In second order, when the
distribution of gradients is important, the central resonance may
become structureless and strongly broadened asymmetrically as in
figure 7, first curve.

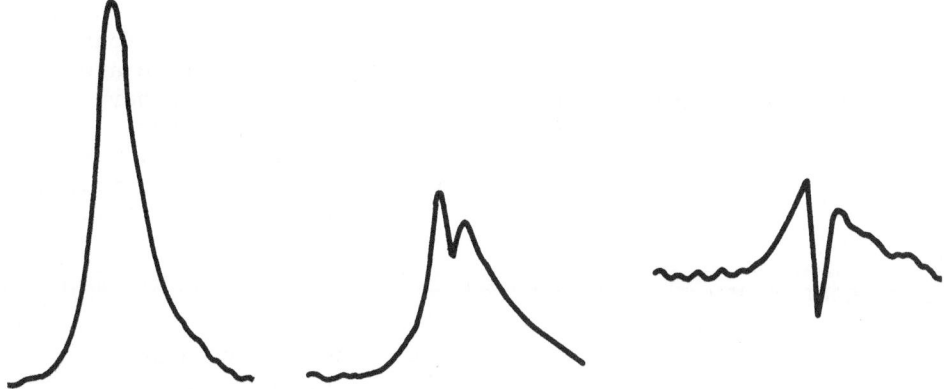

Fig. 7. Dispersion spectra of ^{23}Na in mixed alkali silicate glasses
[7]. From left to right the composition of the glasses
is changed so as to increase the T_1/T_{1D} ratio drastically.
The first curve, equivalent to the absorption line is
determined by second-order quadrupole effects with a
distribution of field gradients. The third curve is
reminiscent of figure 4; dipolar effects predominate.
Similar effects have been observed in the presence of
motion, as the temperature is varied.

Quadrupole effects in NMR of solids can also be studied
using the adiabatic fast modulation method which – in favourable
cases – provides better sensitivity. There, the line shape function
$f(\omega)$ of the dispersion signal ($H_1 \gg H_{loc}$) will be given by [7]

$$f(\omega) = \int_{-\infty}^{\infty} g_Q(\omega - \Omega) \; S_D(\Omega) \; d\Omega \qquad (26)$$

where g_Q is the shape of the quadrupolar broadened signal and S_D
is given by eq. (23) and figure 4. When the dipolar contribution
S_D is much narrower than g_Q as when $H'_L \sqrt{T_1/T_{1D}}$ is small, the
normal (absorption) signal g_Q determined by quadrupole effects
is observed. On the other hand, dipolar effects predominate for
large H'_L or large T_1/T_{1D} ratios. Figure 7 shows an application
of this technique.

Although, in principle, wide-line NMR signals and transient signals obtained by pulse techniques are correlated by a Fourier transform and provide the same type of information, there are situations where spin echo methods should be applied rather than c.w. methods. In particular, the limitation that the satellites are wiped out and the central line unaffected, may be overcome to a certain extend [11]. "Quadrupole echoes" provide information on the distribution of electrical field gradients.

ANISOTROPIC CHEMICAL SHIFT EFFECTS

If the shift tensor is transformed into the coordinate system (ξ, η, ζ) in which it is diagonal, the first order resonance shift produced by the Hamiltonian of eq. (5) turns out to be approximately

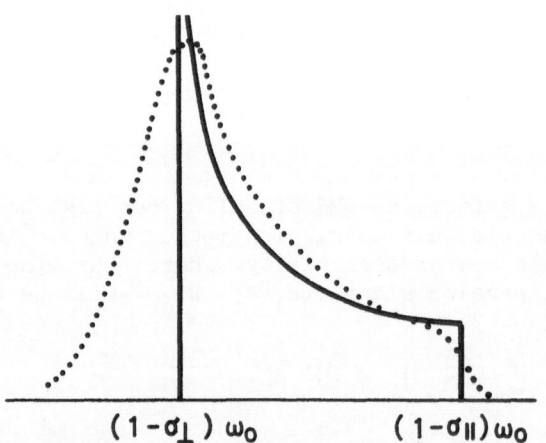

$$(1-\sigma_\perp)\,\omega_0 \qquad\qquad (1-\sigma_{\|})\,\omega_0$$

Fig. 8. The theoretical line shape in a polycrystalline sample with an axially symmetric chemical shift tensor. The dotted line gives the experimental curve, if dipolar interaction and other causes of symmetric line broadening are superimposed.

$$\omega = \omega_o(1-\sigma_{zz})$$

$$(27)$$

$$= \omega_o[(1-\sigma_{\xi\xi})\sin^2\theta\sin^2\phi + (1-\sigma_{\eta\eta})\sin^2\theta\cos^2\phi + (1-\sigma_{\zeta\zeta})\cos^2\theta]$$

θ is again the angle between the axes z and ζ, and ϕ is the angle between x and ξ. In case of an axially symmetric shift tensor, $\sigma_{\xi\xi} = \sigma_{\eta\eta} = \sigma_{\perp}$, and $\sigma_{\zeta\zeta} = \sigma_{//}$. The anisotropy of the shift is $(\sigma_{//}-\sigma_{\perp})$. There is no longer a dependence upon ϕ, and (27) becomes

$$\omega = \omega_o(1 - \sigma_{\perp}\sin^2\theta-\sigma_{//}\cos^2\theta)$$

$$= \omega_o[1 - \frac{1}{3}(\sigma_{//} + 2\sigma_{\perp}) - \frac{1}{3}(\sigma_{//} - \sigma_{\perp})(3\cos^2\theta - 1)] \qquad (28)$$

This expression has the same functional form as eq. (24) for $\eta = 0$. The powder pattern can be obtained using eq. (10) (Fig. 8).

REFERENCES

1. Books
   - E.R. ANDREW, Nuclear Magnetic Resonance, Cambridge University Press (1955)
   - A. LÖSCHE, Kerninduktion, Verlag der Wissenschaften, Berlin (1957)
   - A. ABRAGAM, Nuclear Magnetism, Clarendon Press, Oxford (1961)
   - Ch.P. SLICHTER, Principles of Magnetic Resonance, Harper and Row, New York (1963)
   - J. EBERT, G. SEIFERT, Kernresonanz im Festkörper, Akademische Verlagsgesellschaft, Leipzig (1966)

2. E.R. ANDREW and V.T. WYNN, Proc.Roy.Soc. 291, 257 (1966)

3. F.K. KNEUBÜHL, J.Chem.Phys. 33, 1074 (1960)

4. M. GOLDMAN, Spin Temperature and NMR in Solids, Clarendon Press, Oxford (1970)

5. B.D. MOSEL, W. MÜLLER-WARMUTH and H. DUTZ, Phys.and Chem.of Glasses (1974)

6. I. SOLOMON and J. EZRATTY, Phys.Rev. 127, 78 (1962)

7. F. KRÄMER, W. MÜLLER-WARMUTH, J. SCHEERER and H. DUTZ, Z.Naturforschg. 28a, 1338 (1973)

8. R.A. WIND, B.A. VAN BAREN, S. EMID, J.A. KROONENBURG and J. SMIDT, Physica 65, 522 (1973)

9. Review Articles
  - M.H. COHEN and F. REIF in Seitz-Turnbull, Solid State Physics 5, 321, Academic Press, New York (1957)
  - O. KANERT and M. MEHRING in Diehl-Fluck-Kosfeld, NMR, Basic Principles and Progress 3, 3,Springer, Berlin (1971)

10. J. SCHEERER, Diplomarbeit, Mainz, Unpublished (1966)

11. A. ABRAGAM, Nuclear Magnetism, Clarendon Press, Oxford, 241 (196

THE EFFECT OF MOLECULAR MOTION ON LINE WIDTHS AND RELAXATION TIMES

S. Clough

University of Nottingham

Nottingham, Great Britain

SUMMARY

The effect of molecular motion on n.m.r. line widths and relaxation times is due to the modulation by the motion of nuclear interactions. Analysis of the effects is greatly simplified by a preliminary rearrangement of the more usual expression for the interactions to obtain terms whose time dependence due to the molecular motion is as simple as possible. This procedure enables experimental parameters to be associated directly with lattice sums over specific terms of the hamiltonian. It is illustrated by considering two hypothetical kinds of motion of ethylene molecules:(a) rotation of the whole molecule about the CC bond through 180°, (b) rotation of half of the molecule about the CC bond through 180°.

INTRODUCTION

Molecular motion in solids can be studied by n.m.r. because it modulates those interactions, containing both nuclear spin and space coordinates, which determine line widths and relaxation times. Most important is the nuclear dipole-dipole interaction to which this paper is confined. At low temperatures motional effects can often be neglected (quantum tunnelling motions will be discussed further), but as the temperature is increased and the rate constant τ_c^{-1} increases, its influence on the measured parameters becomes important. Narrowing of the n.m.r. line occurs when τ_c^{-1} is of the order of the low temperature line width and

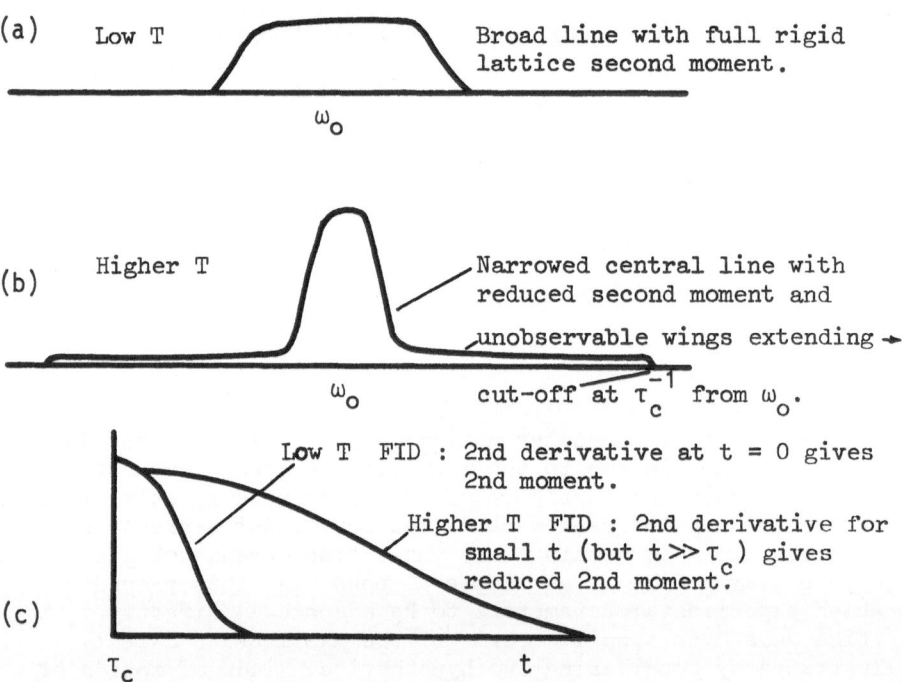

(a)   Low T   Broad line with full rigid
                     lattice second moment.

                     ω_o

(b)   Higher T   Narrowed central line with
                     reduced second moment and
                     unobservable wings extending →
                     ω_o cut-off at τ_c^{-1} from ω_o.

      Low T FID : 2nd derivative at t = 0 gives
                   2nd moment.

      Higher T FID : 2nd derivative for
                   small t (but $t \gg \tau_c$) gives
                   reduced 2nd moment.

(c)

      τ_c t

Fig. 1. Motional Narrowing.
        The effect of molecular motion is to modulate some part
        of the interactions which are responsible at low temperatu
        for the broadening of the n.m.r. line (a). If the motional
        rate greatly exceeds this line width the contribution of
        the modulated terms is limited to unobservable wings (b)
        and the residual broadening of the central line is due
        to unmodulated terms. Equivalent effects are observed in
        the free induction decays (c).

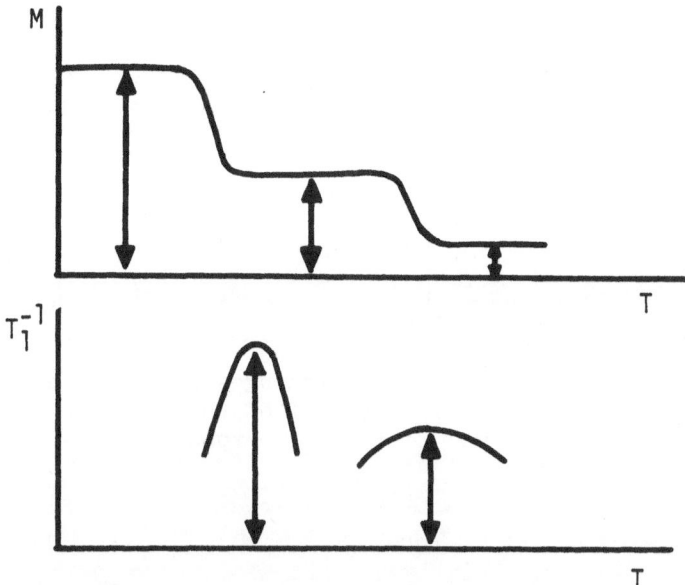

Fig. 2. A schematic representation of the temperature dependence of
second moment M and spin lattice relaxation rate T_1^{-1} due
to the onset of two kinds of molecular motion. Each
of the arrowed lines corresponds to a calculable lattice
sum and may be used to test models for the motion.

along with this goes the lengthening of the free induction decay.
The main features of this motional narrowing are illustrated
schematically in figure 1. When the rate constant is of the order
of the nuclear Larmor frequency the motion is most effective
in causing nuclear spin lattice relaxation and a minimum occurs
in the temperature dependence of T_1. This is illustrated in
figure 2. Two parameters which give information on the type
of molecular motion are changes in second moment and minimum
values of T_1.   (Similar remarks apply to $T_{1\rho}$ but we shall not
consider it here). If a model for the motion is proposed, then
these parameters may be calculated. These notes describe a
systematic approach to this calculation. For the sake of
simplicity a single example will be considered which illustrates
the application of general techniques.

THE EXAMPLE

We consider the molecular motion which might occur in solid
ethylene. We suppose that the molecules undergo two kinds of

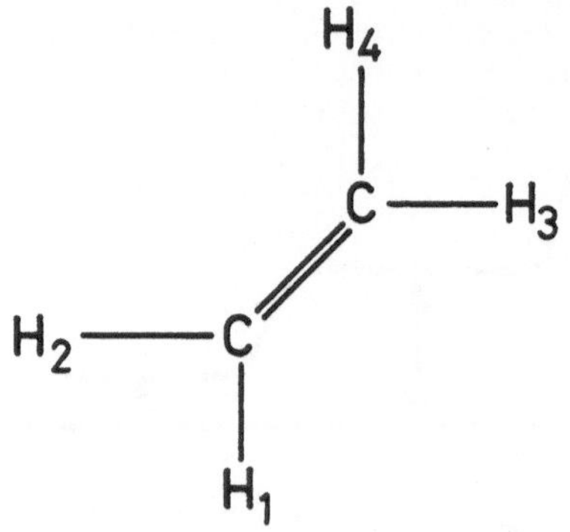

Fig. 3. Rotation of the molecule interchanges H_1 and H_2 and also H_3 and H_4. Rotation of only half of the molecule interchanges only one pair of hydrogen atoms.

motion:(a) the rotation of the whole molecule through 180° about the CC axis,(b) the rotation of one half of the molecule through 180° about the CC axis. Motion of type (b) is assumed to occur much more slowly at a given temperature than motion of type (a). A transition of type (a) has the effect of interchanging the proton pairs 1, 2 and 3, 4 while a transition of type (b) interchanges only one of these pairs (see Fig. 3).

INTERNUCLEAR INTERACTIONS

The internuclear dipole-dipole interaction is usually decompos into 6 terms commonly labelled A to F. Each of these terms has matrix elements between nuclear Zeeman states which differ in total nuclear magnetic spin quantum number $\Delta m = \alpha$ where α is zero for terms A and B, 1 for terms C and D, 2 for terms E and F. Furthermore, each of these terms is a simple product between a spatial function R^α which depends on the length and orientation relative to the external field of the internuclear vector, and a spin function I^α. Thus the dipole-dipole interaction for the whole crystal has the form

$$H_d = \sum_\alpha \sum_{i>j} R^\alpha_{ij} I^\alpha_{ij} \tag{1}$$

where the sum over α signifies a sum over terms A to F. This decomposition permits the $\alpha = 0$ terms to be associated with the broadening of the n.m.r. line and the $\alpha = 1$ and 2 terms with spin lattice relaxation. In the presence of motion the R_{ij}^{α} become time dependent in ways which depend on the geometry and rate of the motion and the calculation of experimental parameters in terms of the R_{ij}^{α} becomes very complicated. Our object is to rewrite (1) in a way which will clearly show the role of each new term in contributing to line narrowing or spin lattice relaxation due to molecular motion.

MOTIONALLY ADAPTED DIPOLAR HAMILTONIAN

a. Intra-Molecular Interactions

For a single molecule (1) has the form

$$H_d = \sum_{\alpha} (R_{12}^{\alpha}I_{12}^{\alpha} + R_{13}^{\alpha}I_{13}^{\alpha} + R_{14}^{\alpha}I_{14}^{\alpha} + R_{23}^{\alpha}I_{23}^{\alpha} + R_{24}^{\alpha}I_{24}^{\alpha} + R_{34}^{\alpha}I_{34}^{\alpha})$$

$$(2)$$

By rewriting (2)

$$H_d = \sum_{\alpha} [\tfrac{1}{2}(R_{12}^{\alpha} + R_{34}^{\alpha})(I_{12}^{\alpha} + I_{34}^{\alpha}) + \tfrac{1}{2}(R_{13}^{\alpha} + R_{24}^{\alpha})(I_{13}^{\alpha} + I_{24}^{\alpha})$$

$$+ \tfrac{1}{2}(R_{14}^{\alpha} + R_{23}^{\alpha})(I_{14}^{\alpha} + I_{23}^{\alpha}) + \tfrac{1}{2}(R_{14}^{\alpha} - R_{23}^{\alpha})(I_{14}^{\alpha} - I_{23}^{\alpha})] \quad (3)$$

sothat each new term is either symmetric or antisymmetric under simultaneous interchange of labels 1 and 2 and of labels 3 and 4, we find three terms in (3) which are not affected by rotation and one term which alternates in sign. (The symmetry of the molecule means that $R_{12}^{\alpha} = R_{34}^{\alpha}$ and $R_{13}^{\alpha} = R_{24}^{\alpha}$ and explains the absence of terms in $R_{12}^{\alpha} - R_{34}^{\alpha}$ and $R_{13}^{\alpha} - R_{24}^{\alpha}$).

If the rotational motion of the molecules is very slow ($\hbar\tau_c^{-1} \ll R$) then the intra-molecular contribution to the second moment of the n.m.r. line is calculated from the $\alpha = 0$ terms in either (2) or (3), but if the motion is fast ($\hbar\tau_c^{-1} \gg R$) the n.m.r. line is narrowed and the residual second moment is calculated from (3) with the final terms $\tfrac{1}{2}(R_{14}^{0} - R_{23}^{0})(I_{14}^{0} - I_{23}^{0})$ omitted. The omitted part can therefore be associated with the intra-molecular contribution to the reduction in second moment due to rotation of the molecule as a whole.

b. Inter-Molecular Interactions

For the interactions between a pair of molecules (1) is

$$H_d = \sum_\alpha (R^\alpha_{11'}\, I^\alpha_{11'} + R^\alpha_{12'}\, I^\alpha_{12'} + R^\alpha_{13'} I^\alpha_{13'} \cdots 16 \text{ terms}) \qquad (4)$$

where the protons of one molecule are labelled 1, 2, 3, 4 and of the other molecule 1', 2', 3,' 4'. The four terms of (4) which connect protons 1 and 2 with 1' and 2' can be rewritten

$$\sum_\alpha\ [\tfrac{1}{4}(R^\alpha_{11'} + R^\alpha_{12'} + R^\alpha_{21'} + R^\alpha_{22'})(I^\alpha_{11'} + I^\alpha_{12'} + I^\alpha_{21'} + I^\alpha_{22'})$$

$$+ \tfrac{1}{4}(R^\alpha_{11'} - R^\alpha_{12'} + R^\alpha_{21'} - R^\alpha_{22'})(I^\alpha_{11'} - I^\alpha_{12'} + I^\alpha_{21'} - I^\alpha_{22'})$$

$$+ \tfrac{1}{4}(R^\alpha_{11'} + R^\alpha_{12'} - R^\alpha_{21'} - R^\alpha_{22'})(I^\alpha_{11'} + I^\alpha_{12'} - I^\alpha_{21'} - I^\alpha_{22'})$$

$$+ \tfrac{1}{4}(R^\alpha_{11'} - R^\alpha_{12'} - R^\alpha_{21'} + R^\alpha_{22'})(I^\alpha_{11'} - I^\alpha_{12'} - I^\alpha_{21'} + I^\alpha_{22'})]$$

$$(5)$$

The first term in (5) is unaffected by rotation of either molecule. The second is changed in sign only if the primed molecule rotates, the third only if the unprimed molecule rotates and the fourth if either rotates. If we suppose that the molecules undergo random uncorrelated rotation with rate constant τ_c^{-1}, then the first term in (5) is time independent, the second and third terms fluctuate with correlation times τ_c and the fourth term fluctuates with correlation time $\tau_c/2$. If on the other hand we suppose that the rotations of the two molecules are completely correlated, then the fourth term is time independent like the first. If the motion is uncorrelated and fast, then only the first term in (5) with $\alpha = 0$ is retained in calculating the reduced second moment.

In calculating T_1 it is the $\alpha = 1$ and 2 terms which are important. With usual assumptions (see refs. 1 and 2) and supposing that the dipolar hamiltonian is written as a sum of products of spatial terms and spin terms $H_d = \sum_\alpha\sum_\beta R^\alpha_\beta I^\alpha_\beta$, as for example in (5)

$$T_1^{-1} = -\frac{1}{2\hbar^2} \sum_\alpha \sum_{\beta,\beta'} J^\alpha_{\beta\beta'}(\alpha\tau)\ \mathrm{Tr}([H_o, I^\alpha_\beta][H_o, I^\alpha_{\beta'}])/\mathrm{Tr}H_o^2 \qquad (6)$$

where $H_o = -\hbar\omega_o \sum_j I_{jz}$ is the Zeeman hamiltonian.

The trace in the numerator of (6) introduces lattice sums over terms in dipole-dipole interactions weighted with appropriate spectral density terms $J(\alpha,\tau)$

$$J(\alpha,\tau) = \tau/[1 + (\alpha\omega_o\tau)^2] \qquad (7)$$

where τ is the correlation time describing the fluctuation of the interaction. When for example the three time dependent terms in (5) are inserted into (6), there are no cross terms; so each makes a separate contribution to T_1^{-1} which therefore takes the form [2]

$$T_1^{-1} = \frac{U\tau_c}{1 + (\omega_o^2\tau_c^2/4)} + \frac{V\tau_c}{1 + \omega_o^2\tau_c^2} + \frac{W\tau_c}{1 + 4\omega_o^2\tau_c^2} \tag{8}$$

The number U derives from the last term in (5) with $\alpha = 1$ while V comes partly from this same term in (5) with $\alpha = 2$ and partly from the second and third terms of (5) with $\alpha = 1$. The second and third terms of (5) with $\alpha = 2$ give rise to W.

A SECOND TYPE OF MOTION

We now suppose that in addition to rapid rotation of the molecules as a whole, at a suitably high temperature the slower rotation of only half of the molecule becomes important. The new effect is that terms in the dipole-dipole interactions which are time independent when the molecules rotate as a whole become time dependent under the new motion.

a. Intra-Molecular Interactions

The first three terms of (3) are now rewritten

$$[\tfrac{1}{2}(R_{12}^\alpha + R_{34}^\alpha)(I_{12}^\alpha + I_{34}^\alpha)$$

$$+ \tfrac{1}{4}(R_{13}^\alpha + R_{24}^\alpha + R_{14}^\alpha + R_{23}^\alpha)(I_{13}^\alpha + I_{24}^\alpha + I_{14}^\alpha + I_{23}^\alpha)$$

$$+ \tfrac{1}{4}(R_{13}^\alpha + R_{24}^\alpha - R_{14}^\alpha - R_{23}^\alpha)(I_{13}^\alpha + I_{24}^\alpha - I_{14}^\alpha - I_{23}^\alpha)] \tag{9}$$

The last term in (9) becomes time dependent under the new motion. The $\alpha = 0$ term ceases to contribute to the second moment and the $\alpha = 1$ and 2 terms begin to contribute to T_1^{-1}.

b. Inter-Molecular Interactions

As the first term in (5) is unaffected by interchange of 1 and 2, the new motion has almost no effect on intermolecular contributions to the second moment and T_1^{-1}. Terms already fluctuating with rate constant τ_c^{-1} are not significantly affected by the additional relatively slow fluctuations introduced by the new motion.

Eastman Kodak Company
Research Library
Kodak Research Labs
Rochester, New York 14650

COLLECTIVE INTERACTION CONSTANTS

The treatment of molecular motion has introduced collective interaction parameters like

$$R^{\alpha} = R^{\alpha}_{13} + R^{\alpha}_{24} - R^{\alpha}_{14} - R^{\alpha}_{23} \tag{10}$$

This is an expression in terms of individual inter-nuclear vectors and their orientation with respect to the external magnetic field. This is certainly not the most convenient form for R^{α}. It may be rewritten in a way which separates internal molecular parameters from the external parameters giving the orientation of the molecule as a whole. A general way of doing this uses Wigner rotation matrices (see for example ref. 3) but the transformation may also be made using cartesian coordinates with perhaps greater conceptual simplicity. We illustrate for R^{o}.

$$R^{o}_{13} = C \, r^{-3}_{13} (1 - 3 \cos^2 \theta_{13}) \tag{11}$$

The vector $\underline{r}_{13}$ has magnitude r_{13} and components r_{13m} ($m = x, y, z$). In cartesian coordinates

$$R^{o}_{13} = C \, r^{-3}_{13} (1 - 3 r^2_{13z} / r^2_{13}) = C \, \underline{n} \cdot \underline{\underline{R}}^{o}_{13} \cdot n \tag{12}$$

where $\underline{n}$ is a unit vector along the direction of H_o, the external field, and $\underline{\underline{R}}^{o}_{13}$ is a second rank tensor.

$$\underline{\underline{R}}^{o}_{13} = r^{-5}_{13} (r^2_{13} \underline{\underline{1}} - 3 \underline{\underline{r}}^{(13)}) \tag{13}$$

In (13) $\underline{\underline{1}}$ is the unit tensor and the elements of $\underline{\underline{r}}^{(13)}$ are just products of components of the vector $\underline{r}_{13}$.

$$\underline{\underline{r}}^{(13)}_{mn} = r_{13m} \, r_{13n} \qquad (m, n = x, y, z) \tag{14}$$

The collective interaction tensor $\underline{\underline{R}}^{o}$ is from (10)

$$\underline{\underline{R}}^{o} = (r^{-3}_{13} + r^{-3}_{24} - r^{-3}_{14} - r^{-3}_{23}) \underline{\underline{1}} - 3(r^{-5}_{13} \underline{\underline{r}}^{(13)} +$$
$$r^{-5}_{24} \underline{\underline{r}}^{(24)} - r^{-5}_{14} \underline{\underline{r}}^{(14)} - r^{-5}_{23} \underline{\underline{r}}^{(23)}) \tag{15}$$

and the required parameter R^{o} is

$$R^{o} = C \, \underline{n} \cdot \underline{\underline{R}}^{o} \cdot \underline{n} \tag{16}$$

The point of recasting (10) in this way is that one may write down the elements of $\underline{\underline{R}}^O$ very simply in a molecule-fixed reference frame and then transform the tensor into a reference frame whose z axis is parallel to $\underline{n}$.

In a molecule-fixed reference frame chosen to have the z axis perpendicular to the molecular plane and the x axis parallel to $\underline{r}_{13}$, the only non-zero tensor elements are

$$r_{xx}^{(13)} = r_{xx}^{(24)} = r_{xx}^{(14)} = r_{xx}^{(23)} = r_{13}^2$$

$$r_{yy}^{(14)} = r_{yy}^{(23)} = r_{12}^2 \qquad\qquad r_{xy}^{(14)} = -r_{xy}^{(23)} = r_{13}r_{12}$$

Thus $\underline{\underline{R}}^O$ is diagonal in this reference frame with

$$R_{xx}^O = -4r_{13}^{-3} - 2r_{14}^{-3} + 6r_{14}^{-5}r_{13}^2$$

$$R_{yy}^O = 2r_{13}^{-3} - 2r_{14}^{-3} + 6r_{12}^2 r_{14}^{-5} \qquad\qquad (17)$$

$$R_{zz}^O = -R_{xx}^O - R_{yy}^O$$

A rotation to a new reference frame related to the molecular frame by the Euler angles θ, ϕ and η transforms to $R_{mn}^{O\prime}$

$$R_{mn}^{O\prime} = \sum_p \sum_s R_{ps}^O \, Q_{mp} \, Q_{ns}$$

with the transformation matrix

$$Q = \begin{pmatrix} (\cos\phi\ \cos\eta & (\sin\phi\ \cos\eta & (\sin\theta\ \sin\eta) \\ \quad -\sin\phi\ \cos\theta\ \sin\eta) & \quad +\cos\phi\ \cos\theta\ \sin\eta) & \\ & & \\ (-\cos\phi\ \sin\eta & (-\sin\phi\ \sin\eta & (\sin\theta\ \cos\eta) \\ \quad -\sin\phi\ \cos\theta\ \cos\eta) & \quad +\cos\phi\ \cos\theta\ \cos\eta) & \\ & & \\ (\sin\phi\ \sin\theta) & (-\cos\phi\ \sin\theta) & (\cos\theta) \end{pmatrix}$$

we require $R^O = C\ z.\ R^{O\prime}.z = C\ R_{zz}^{O}{}'$ (18)

$$R^O = C(R_{xx}^O\ Q_{zx}^2 + R_{yy}^O\ Q_{zy}^2 + R_{zz}^O\ Q_{zz}^2)$$

Thus from (17) and (18) we find the required form of R^O.

$$R^{O} = \frac{1}{2}(R^{O}_{xx} + R^{O}_{yy})(1 - 3\cos^2\theta) + \frac{1}{2}(R^{O}_{yy} - R^{O}_{xx})\sin^2\theta\cos2\phi$$

$$(19)$$

θ is the angle between the normal to the molecular plane and the magnetic field; ϕ gives the orientation of the CC bond in the molecular plane. The reduction in second moment which occurs when the second type of motion becomes important is proportional to $(R^{O})^2$; so (19) shows how this experimental parameter is related to the orientation of the molecular axes.

EFFECT OF MOTION ON SPIN DIFFUSION

In the presence of motion it is necessary to examine critically the assumption of the existence of T_1, i.e. of an exponential recovery of magnetisation after saturation. For example in liquids this assumption is inappropriate and different relaxation times for various lines of high resolution spectra are commonly observed. In rigid solids the thermo-dynamic viewpoint leads to the conclusion that there are only two thermodynamic invariants corresponding to the Zeeman energy and dipolar energy (as described in these theories) with relaxation times T_1 and T_{1D}. The process which brings this system into thermodynamic equilibrium is the nuclear spin flip-flop which is also responsible for spin diffusion. In the presence of molecular motion a situation exists which is intermediate between the rigid solid and the liquid. It is necessary to examine carefully the spin diffusion processes since some are quenched by the motion. The result is the appearance of new quasi-invariants of the motion. Attention has been drawn to this important point by Emid and Wind who have called it symmetry restricted spin diffusion.

We may illustrate the consequences in form of ethylene. Let us suppose that the molecules are rotating rapidly ($\omega_o\tau \ll 1$) about their symmetry axes and that the second type of motion discussed previously does not occur. It is apparent that for an isolated molecule the eigenfunctions must be symmetric or antisymmetric under rotation through π, since any alternative combination would have a lifetime only of under τ. Interactions between molecules, as we have seen, can be divided into those which are symmetric under rotation of either molecule and are in consequence not modulated by the motion and other terms which are modulated. Only the former are able to cause rapid flip-flop transitions and, being symmetric, they do not change the symmetry of either molecule though they may change the magnetisation of both. Thus a new quasi-invariant, which is the difference in number of symmetric and antisymmetric molecular rotational states, is introduced.

In the similar case of solids containing methyl groups two new quasi-invariants must be introduced. One of these corresponds to the difference between clockwise and anticlockwise rotating states and can be called rotational polarization. Following nuclear magnetisation saturation the recovery is complicated by the coupling of the differential equations governing magnetic polarization and rotational polarization. The magnetisation recovery is non-exponential, which is associated with a transient rotational polarization simultaneously induced. The non exponential magnetisation recovery is thus a source of additional information on the type of molecular motion responsible for relaxation.

REFERENCES

1. L.C. HEBEL and C.P. SLICHTER, Phys.Rev. 113, 1504 (1959)

2. D.C. LOOK and I.J. LOWE, J.Chem.Phys. 44, 2995 (1966)

3. V.J. McBRIERTY and D.C. DOUGLASS, J.Mag.Res. 2, 352 (1970)

A COMPARISON BETWEEN CLASSICAL THEORY OF MOTIONAL NARROWING

AND NARROWING DUE TO QUANTUM MECHANICAL TUNNELLING MOTION

S. Clough

University of Nottingham

Nottingham, Great Britain

SUMMARY

At low temperature hindered molecular motion must be
treated quantum mechanically introducing the concept of a
tunnelling frequency which characterises the motional rate in
place of the correlation time τ_c which is a feature of classical
theory. To illustrate the relationship between the two
theories the n.m.r. spectrum is calculated for a proton moving
between two sites having different resonance frequencies (a)
when the motion is described as random hopping (b) when the
proton tunnels between the sites. The central part of the
spectrum is narrowed in each case but spectrum (a) has broad
wings while spectrum (b) has weak tunnelling sidebands. The
interaction of tunnelling groups with nearby paramagnetic centres
is shown to offer the best means of studying the tunnelling
motion.

INTRODUCTION

Motional narrowing of n.m.r. lines is one of the most
important ways of studying molecular motion. At sufficiently
high temperatures rather simple classical pictures of the
motion are perfectly adequate. At sufficiently low temperatures
if motion is important it must be described quantum mechanically
since the classical model then leads to quite wrong predictions.
The essential difference is that tunnelling is coherent and
described by a frequency while classical motion is random and
described by a correlation time. A second more subtle distinction

arises as a result of the exclusion principle where the motion
permutes indistinguishable particles, as for example the
rotation of a methyl group through 120° permutes the proton
coordinates.

THE RELATIONSHIP BETWEEN CLASSICAL AND QUANTUM MOTION

We consider the motion of a particle in a box which is
divided into two halves by a narrow potential barrier as shown
in figure 1. A typical classical approach would be to assume
that there exists a transition probability for penetration of
the barrier τ^{-1}, and to define probabilities p_L and $p_R = 1 - p_L$
for the particle being in the left or right halves of the box.
The rate equation

$$dp_L/dt = (p_R - p_L)\tau^{-1}$$

is easily solved with the initial condition $p_L(0) = 1$:

$$p_L(t) = \frac{1}{2}[1 + \exp(-2t/\tau)] \tag{1}$$

indicating that the random nature of the motion leads to a
loss of information about the location of the particle and after
a time of the order of a few times τ the particle is equally
likely to be in either half of the box.

At first sight the quantum mechanical picture of the
motion appears very different. Eigenfunctions for the particle
extend over both halves of the box and are either symmetrical
or antisymmetrical with respect to reflection in the plane
of the partition. To a first approximation they are of the form

$$\psi_s = \frac{1}{\sqrt{2}}(\psi_L + \psi_R) \quad \psi_a = \frac{1}{\sqrt{2}}(\psi_L - \psi_R) \tag{2}$$

where ψ_L and ψ_R are corresponding eigenfunctions for a particle
confined to only one half of the box, but because the potential
barrier is finite they penetrate into the space occupied by
the barrier. As a result the eigenvalues E_s and E_a differ because
the term arising from the overlap of ψ_L and ψ_R comes in with a
different sign in the two cases. The energy splitting $\hbar\omega_t$
defines the tunnelling frequency.

To illustrate the connection with the classical picture
we must imagine the system prepared in a mixture of eigenstates,
for example $(\psi_s + \psi_a)/\sqrt{2} = \psi_L$. The particle is then clearly
in the left half of the box. This state is time dependent:

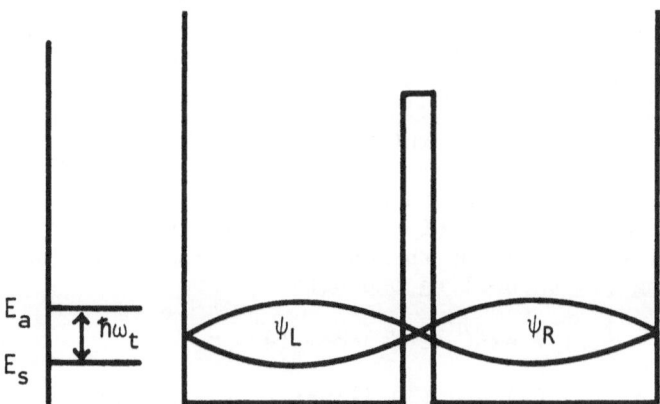

Fig. 1. The potential function for a particle moving in a one
dimensional box divided into two halves by a thin partition
(i.e.: a finite potential barrier). The lowest pair of
levels E_s and E_a differ only a little from the lowest
level for a particle confined to one half of the box. The
energy splitting is $\hbar\omega_t$ where ω_t is the tunnelling frequency.

$$\psi(t) = \frac{1}{\sqrt{2}} [\psi_s \exp(iE_s t/h) + \psi_a \exp(iE_a t/h)]$$

$$= \frac{1}{2} \{\psi_L [\exp(iE_s t/h) + \exp(iE_a t/h)]$$

$$+ \psi_R [\exp(iE_s t/h) - \exp(iE_a t/h)]\}$$

The probability p_L can be identified with the square modulus
of the coefficient of ψ_L

$$p_L = \frac{1}{2}[1 + \cos((E_a - E_s)t/h)] \tag{3}$$

Thus p_L oscillates with angular frequency ω_t.

In order to consider the possibility of both tunnelling
and hopping motion occurring simultaneously, we express the
hopping assumptions in terms of the states ψ_s and ψ_a. A hop
simply changes the coordinates of the particle from one half
of the box to an equivalent state in the other half, i.e. it
interchanges ψ_L and ψ_R. This has no effect on ψ_s but changes
the sign of ψ_a. Thus $\psi(t)$ becomes

$$\psi(t) = \frac{1}{2} [\psi_s \exp(iE_s t/h) + \psi_a k \exp(iE_a t/h)]$$

where k = $\pm$ 1 and changes sign each time a hop occurs, being 1 at t = 0. Expressing this in terms of ψ_L and ψ_R as before, the coefficient of ψ_L is

$$\frac{1}{2} [\exp(iE_s t/h) + k \exp(iE_a t/h)] \tag{4}$$

In evaluating the square modulus of this one notes that $k^2 = 1$ and the ensemble average of k is just $p_L(t) - p_R(t)$ where $p_L(t)$ is given by (1) for the case of hopping only. Thus for both hopping and tunnelling one obtains from (4)

$$p_L(t) = \frac{1}{2} [1 + \exp(-2t/\tau)\cos\omega_t t] \tag{5}$$

Comparison of (1) (3) and (5) illustrates the coherence of tunnelling motion as opposed to the random nature assumed by classical theories. The reason for this difference is that a hop describes a process in which besides the movement of the particle, a change occurs if the state of the surrounding lattice (one or more phonons are scattered). Our ignorance of the detailed state of the lattice means that we can not predict the time of a hop. Tunnelling occurs without an associated change in the lattice.

MOTIONAL NARROWING

    The effects of motion on n.m.r. spectra can be illustrated in the context of the particle in the divided box if we suppose that the particle is a proton and that the magnetic field in the two halves is H_L and H_R with $H_L \neq H_R$. We define the mean resonance frequency as $\omega_o = \gamma(H_L + H_R)/2$ and the difference $\omega_a = \gamma(H_L - H_R)/2$

    The free induction decay is the Fourier transform of the n.m.r. spectrum. The classical theory obtains the spectrum by way of the free induction decay while the tunnelling theory derives the spectrum directly.

CLASSICAL MODEL

    The free induction decay is given by

$$F(t) = R < \exp(i\omega_o t + i\omega_a \int_0^t k \, dt) > \tag{6}$$

where R signifies the real part and the angular bracket indicates an ensemble average. Once again k = $\pm$ 1 and fluctuates between these values with transition probability τ^{-1}. The spectrum line shape is the Fourier transform

$$G(\omega) = \pi^{-1} \int_0^\infty F(t) \exp(-i\omega t)\, dt \tag{7}$$

By a statistical argument [1] given in the appendix

$$G(\omega) = -R\pi^{-1} \begin{bmatrix} \frac{1}{2} & \frac{1}{2} \end{bmatrix} \begin{bmatrix} [i(\omega_o + \omega_a - \omega) - \tau^{-1}] & \tau^{-1} \\ \tau^{-1} & [i(\omega_o - \omega_a - \omega) - \tau^{-1}] \end{bmatrix}^{-1} \begin{pmatrix} 1 \\ 1 \end{pmatrix}$$

$$\tag{8}$$

This is a key result in the theory of motional effects on line shapes and has been extended to deal with many more complicated problems [2] than the present one. The central matrix contains the hopping rate τ^{-1} and the two frequencies $\omega_o + \omega_a$ and $\omega_o - \omega_a$. After inversion of the central matrix and multiplification of the matrix product one has

$$G(\omega) = 2\tau^{-1}\pi^{-1}\omega_a^2 / [(\omega_o - \omega)^4 - 2(\omega_o - \omega)^2(\omega_a^2 - 2\tau^{-2}) + \omega_a^4] \tag{9}$$

Equation (9) is the simplest example of a spectrum showing motional narrowing. When $\tau^{-1} = 0$ the denominator of (9) reduces to $(\omega_o - \omega_a - \omega)^2 (\omega_o + \omega_a - \omega)^2$ and $G(\omega)$ consists of two δ functions (spectrum a in figure 2). When τ^{-1} is very large the denominator of (9) is large except for $\omega \sim \omega_o$ and then it approximates to $4\tau^{-2}(\omega_o - \omega)^2 + \omega_a^4$. This approximation makes $G(\omega)$ a single Lorentzian line at the centre of the original pair of δ functions (spectrum d in figure 2). The change in the spectrum is illustrated in figure 2. As the motional rate increases, the central line continues to narrow, approaching zero width at infinite τ^{-1}. In many solids the molecular motion modulates only a part of the interactions responsible for line width so that the unmodulated interactions give rise to a residual width unaffected by the motion.

TUNNELLING MODEL

A quantum mechanical treatment of the motion starts from the Hamiltonian which consists of a purely spatial part H_s and a magnetic part H_m

$$H_s = -(\hbar^2/2m)\, \partial^2/\partial x^2 + V(x) \tag{10}$$

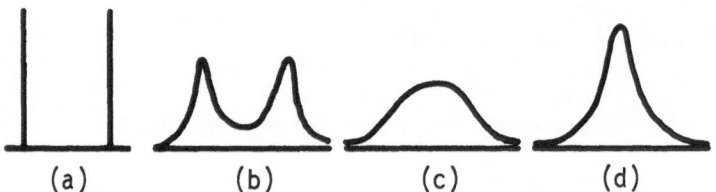

(a)   (b)   (c)   (d)

Fig. 2. The n.m.r. spectrum $G(\omega)$ for a proton jumping randomly
between two sites with resonance frequencies $\omega_o \pm \omega_a$.
The jumping rate τ_c^{-1} is (a) zero, (b) $\omega_a/2$,
(c) $\sqrt{2}\,\omega_a$, (d) $2\sqrt{2}\,\omega_a$.

$$H_m = -\gamma\hbar[H_s + H_a(x)]\,I_z \tag{11}$$

The eigenfunctions of H_s are ψ_s and ψ_a but since we now wish to
introduce the spin state as well as the motional state we have
four basis functions $\psi_s\alpha$, $\psi_a\beta$, $\psi_a\alpha$, $\psi_s\beta$, where α and β are
eigenfunctions of I_z. We note that the field H_a is a function of
x since it has different values in the two halves of the box.
The hamiltonian matrix is

$$\begin{pmatrix} E_s - \hbar\omega_o/2 & -\hbar\omega_a/2 & 0 & 0 \\ -\hbar\omega_a/2 & E_a - \hbar\omega_o/2 & 0 & 0 \\ 0 & 0 & E_s + \hbar\omega_o/2 & \hbar\omega_a/2 \\ 0 & 0 & \hbar\omega_a/2 & E_a + \hbar\omega_o/2 \end{pmatrix}$$

Diagonalization leads to the eigenfunctions

$$\begin{aligned} \psi_1 &= \cos\theta\,\psi_s\alpha \;+\; \sin\theta\,\psi_a\alpha \\ \psi_2 &= \cos\theta\,\psi_a\alpha \;-\; \sin\theta\,\psi_s\alpha \\ \psi_3 &= \cos\theta\,\psi_s\beta \;-\; \sin\theta\,\psi_a\beta \\ \psi_4 &= \cos\theta\,\psi_a\beta \;+\; \sin\theta\,\psi_s\beta \end{aligned} \tag{12}$$

and eigenvalues

$$E_1 = \frac{1}{2}\{E_a + E_s - \hbar\omega_o - [(E_a - E_s)^2 + \hbar^2\omega_a^2]^{1/2}\}$$

$$E_2 = \frac{1}{2}\{E_a + E_s - \hbar\omega_o + [(E_a - E_s)^2 + \hbar^2\omega_a^2]^{1/2}\}$$ (13)

$$E_3 = \frac{1}{2}\{E_a + E_s + \hbar\omega_o - [(E_a - E_s)^2 + \hbar^2\omega_a^2]^{1/2}\}$$

$$E_4 = \frac{1}{2}\{E_a + E_s + \hbar\omega_o + [(E_a - E_s)^2 + \hbar^2\omega_a^2]^{1/2}\}$$

with

$$\tan 2\theta = \hbar\omega_a/(E_a - E_s) = \omega_a/\omega_t$$

The n.m.r. transition probabilities are easily calculated from the eigenfunctions

$$T_{13} = T_{24} = \cos^2 2\theta$$

$$T_{14} = T_{23} = \sin^2 2\theta$$ (14)

The eigenvalues are shown in the upper part of figure 3 as a function of ω_t/ω_a. Spectra are also shown for small ω_t (spectrum a) and for $\omega_t/\omega_a = \sqrt{3}$ when the effect of motional narrowing becomes apparent. When ω_t becomes large the weak side bands occur at approximately ω_t from the central line and correspond roughly to a nuclear spin flip plus a change in the motional wavefunction from ψ_s to ψ_a. Comparison of figures 2 and 3 shows the similarity of the averaging effect in that a pair of lines (spectrum (a) in each case) is replaced by a narrowed spectrum consisting of a single central line. The classical theory (Fig. 2 (d)) always gives a finite width for the central line but in most practical cases this would be unobservable for small τ_c due to the residual broadening caused by interactions not modulated by the motion. It is therefore normally not possible to tell from the central observable part of the spectrum whether the motion should be described classically or as tunneling motion. Only when the tunnelling frequency is comparable with the dipolar line width can distinctive features be seen.

PRACTICAL EXAMPLES OF TUNNELLING MOTIONS IN MOLECULAR SOLIDS

The observation of tunnelling phenomena by n.m.r. normally requires that ω_t should be at least comparable with the dipolar line width. Molecular motions satisfying this criterion are usually rotations of small symmetric groups, for example CH_3 and NH_4. The tunnelling frequency depends in a sensitive

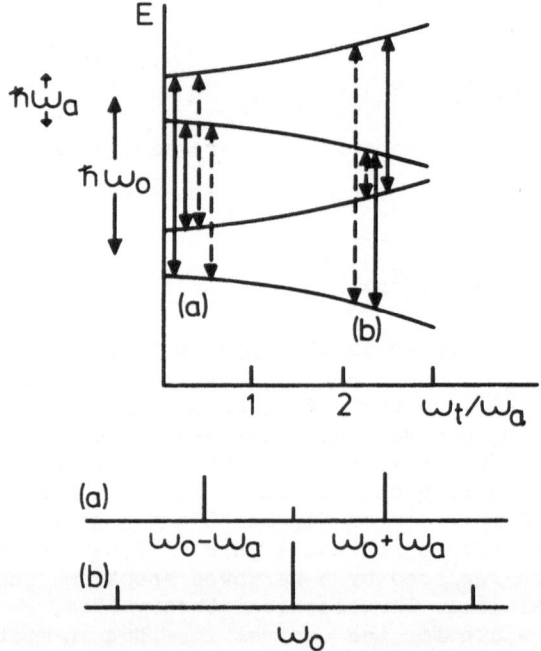

Fig. 3. The upper diagram shows the energy levels for a proton
        tunnelling at frequency ω_t between two sites with
        resonance frequencies $\omega_o \pm \omega_a$. Strong transitions in the
        n.m.r. spectrum are shown by continuous arrowed lines,
        weak ones by broken lines. Spectra for $\omega_t \ll \omega_a$ (a) and
        for $\omega_t/\omega_a = \sqrt{3}$ (b) are shown in the lower diagram.
        Spectrum (b) illustrates the appearance of tunnelling
        sidebands. The second moments of spectra (a) and (b)
        are equal.

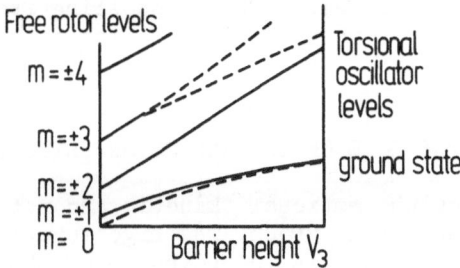

Fig. 4. The energy levels of a hindered methyl group. The
eigenstates have a simple interpretation in the two
extremes of zero barrier height and large barrier
height, being free rotor states in the former case and
triply degenerate harmonic oscillator states in the
latter case. Between these extremes the levels occur
as singlets and doublets (shown as broken and continuous
lines) and the splitting may be regarded as giving the
frequency of tunnelling between harmonic oscillator
states.

way on the height of the hindering potential barrier. Figure 4
shows how the energy levels for a methyl group depend on the
barrier height. At the left of the diagram the barrier height
is zero and the energies are just those of a free rotor $m^2h/2I$
where I is the moment of inertia about the symmetry axis. The
potential energy is assumed to have the form

$$V(\phi) = \frac{1}{2} V_3 \ (1 - \cos 3\phi)$$

where ϕ is the rotational coordinate. As the barrier height V_3
increases towards the right of figure 4 the splitting between
the lowest singlet and the higher doublet progressively decreases
becoming zero in the limit of the infinite barrier when one has

three degenerate harmonic oscillator states corresponding
to the three possible orientations. This splitting corresponds to
the tunnelling frequency for tunnelling between these three
equivalent localized states. If the tunnelling frequency exceeds
the dipolar line width, then the n.m.r. line is narrow even at
very low temperature.

SPIN SPECIES AND THE ROLE OF THE EXCLUSION PRINCIPLE

Figure 4 gives the motional (kinetic and potential) energy
associated with the methyl group. The eigenfunctions have the
property that when the spatial coordinates of the three protons
are cyclically permuted the eigenfunction is multiplied by one
of the three cube roots of 1, i.e. 1, $\exp(2\pi i/3)$, $\exp(-2\pi i/3)$.
The indistinguishability of the protons then requires that the
nuclear spin functions associated with these spatial functions
have the property that cyclic permutation of the three proton
spin coordinates multiplies the spin function by the conjugate
number, namely 1, $\exp(-2\pi i/3)$, $\exp(2\pi i/3)$ so that cyclic number,
of both spin and space coordinates leaves all eigenfunctions
unchanged. The effect is that transitions between the three
levels of the torsional oscillator ground state (figure 4) are
often slow because they involve a change in the spin state as
well as the motional state and require the mediation of an
interaction which contains both spin and space coordinates.
These long lived states are often called spin symmetry species
and transitions between them spin symmetry conversion. Ortho
and para hydrogen are the best known examples.

MEASUREMENT OF TUNNELLING FREQUENCY BY N.M.R.

Except in very favourable cases [3, 4] the features
characteristic of tunnelling are unobservable in the n.m.r.
spectrum. Inspection of (12) suggests an alternative approach.
If we had a radiofrequency field which was asymmetric (i.e.
had opposite phase in the two halves of the box) then the
intensities of the lines in figure 3, spectrum (b) would be
transposed. It would be the tunnelling sidebands which were
most intense. This suggests that we should look for an r.f.
field with a strong spatial dependence, for example the field
of a precessing electron. Inspection of the dipole–dipole
coupling between a tunnelling methyl group and a distant
electron spin shows that transitions can occur which involve an
electron spin flip plus spin symmetry conversion of the methyl
group. These transitions become rapid when the electron Larmor
frequency and the methyl group tunnelling frequency coincide.
They then provide a resonant means of transfering energy between

methyl group and electron, and thence to the lattice and so
appear as resonant anomalies [5] in the magnetic field dependence
of nuclear spin lattice relaxation time in samples containing
both tunnelling groups and paramagnetic impurities. The tunnelling
frequency is easily found from the magnetic field at which
relaxation is anomalous.

APPENDIX A

We wish to evaluate the average (6). Consider first

$$\exp(i\omega_a \sum_1^n k_n t/n) \tag{A1}$$

where the integral in (6) is replaced by a sum, the time t
being divided into n intervals each of duration t/n. In each
interval k has the value 1 or -1. There are thus 2^n possible
sequencies. Each sequence has a probability of occurrence. For
example the probability that $k_n = 1$ for all n is
$(1/2)(1 - \tau^{-1}t/n)^{(n-1)}$ because the probability that $k_1 = 1$
is 1/2 and the probability that k does not change in each
successive interval is $(1 - \tau^{-1}t/n)$. For this same sequence (A1)
is $\exp(i\omega_a t)$. The average is to be found by taking the value
of the exponential function weighted by its probability and
summing for all possible 2^n sequences of k values. This ap-
parently formidable sum is identical with the expansion of the
matrix product (A2)

$$\langle\exp(i\omega_a \sum_1^n k_n t/n)\rangle$$

$$= \begin{bmatrix} \frac{1}{2} & \frac{1}{2} \end{bmatrix} \left[\begin{bmatrix} e^{(i\omega_a t/n)} & 0 \\ 0 & e^{(-i\omega_a t/n)} \end{bmatrix} \begin{bmatrix} (1 - t\tau^{-1}/n) & t\tau^{-1}/n \\ t\tau^{-1}/n & (1 - t\tau^{-1}/n) \end{bmatrix} \right]^{n-1} \begin{bmatrix} 1 \\ 1 \end{bmatrix} \tag{A2}$$

Now we let $n \to \infty$. We may expand $\exp(i\omega_a t/n) = 1 + i\omega_a t/n$,
evaluate the inner matrix product retaining only terms up to
first order in t/n and then use Limit $(1 + x/n)^n = e^x$ to
rewrite (A2)

$$\langle\exp(i\omega_a \int_0^t k \, dt)\rangle$$

$$= \begin{bmatrix} \frac{1}{2} & \frac{1}{2} \end{bmatrix} \begin{bmatrix} \exp[t(i\underline{\underline{\omega}}_a + \underline{\underline{\pi}})] \end{bmatrix} \begin{bmatrix} 1 \\ 1 \end{bmatrix} \tag{A3}$$

where the matrices $\underline{\underline{\omega}}_a$ and $\underline{\underline{\pi}}$ are

$$\underline{\underline{\omega}}_a = \begin{pmatrix} \omega_a & 0 \\ 0 & -\omega_a \end{pmatrix} \qquad \underline{\underline{\pi}} = \begin{pmatrix} -\tau^{-1} & \tau^{-1} \\ \tau^{-1} & -\tau^{-1} \end{pmatrix} \qquad (A4)$$

Using (A3) in (6) and Fourier transforming according to (7) leads to (8).

APPENDIX B

It is of interest to see the connection between the above calculation and the equivalent treatment using the formalism of the density matrix. When an appropriate term is introduced to describe the classical jumping process between the two sites, the evolution of the density matrix is described by the equation

$$d\rho/dt = -i\hbar^{-1}[H_m,\rho] + (P\rho P^{-1} - \rho)\tau^{-1} \qquad (B1)$$

where H_m is given by (11) and the operator P is defined by

$$P\psi_L = \psi_R \qquad\qquad P\psi_R = \psi_L \qquad (B2)$$

Spin lattice relaxation terms have been ignored in (B1). Following a 90 degree pulse the density matrix $\rho(0)$ has the form

$$\rho(0) = 1 - aI_x \qquad (B3)$$

and the free induction decay is described by

$$F(t) \propto \text{Tr}[I_x\rho(t)] \qquad (B4)$$

With basic functions $\psi_L\alpha$, $\psi_L\beta$, $\psi_R\alpha$, $\psi_R\beta$, numbered 1 to 4 respectively, we have

$$F(t) \propto R\,(\rho_{12} + \rho_{34}) \qquad (B5)$$

R indicating the real part. From (B2) we find the non zero matrix elements of P to be $P_{13} = P_{24} = P_{31} = P_{42} = 1$ and inserting these in (B1) gives

$$\frac{d}{dt} \begin{bmatrix} \rho_{12} \\ \rho_{34} \end{bmatrix} = \begin{bmatrix} (i\omega_{12} - \tau^{-1}) & \tau^{-1} \\ \tau^{-1} & (i\omega_{34} - \tau^{-1}) \end{bmatrix} \begin{bmatrix} \rho_{12} \\ \rho_{34} \end{bmatrix} \tag{B6}$$

where $\hbar\omega_{ij} = (H_m)_{22} - (H_m)_{ij}$ (B7)

After integrating (B6) and inserting it into (B5) one has a result entirely equivalent to (A3).

APPENDIX C

The density matrix formalism permits the treatment of more complex motional averaging problems in which the random jumping process leads to changes in eigenfunctions as well as eigenvalues. A simple example of this would be jumping between two sites where the fields were not parallel. We shall illustrate the main features of this type of calculation by discussing the line shape which should be expected of a proton which is both tunnelling and jumping between two sites.

The hamiltonian is given by (10) and (11) and we now choose to work in terms of the eigenfunctions (12). In this representation the non zero matrix elements of P are

$$P_{11} = -P_{22} = P_{33} = -P_{44} = \cos 2\theta$$
$$P_{34} = P_{43} = -P_{12} = -P_{21} = \sin 2\theta \tag{C1}$$

and I_x has matrix elements

$$(I_x)_{13} = (I_x)_{31} = (I_x)_{24} = (I_x)_{42} = \tfrac{1}{2}\cos 2\theta$$
$$(I_x)_{14} = (I_x)_{41} = -(I_x)_{23} = -(I_x)_{32} = \tfrac{1}{2}\sin 2\theta \tag{C2}$$

The free induction decay is

$$F(t) \propto R\left[(\rho_{13} + \rho_{24})\cos 2\theta + (\rho_{14} - \rho_{23})\sin 2\theta\right] \tag{C3}$$

From (B1) (with H_m replaced by $H_m + H_s$) we find differential equations for the matrix elements of the density matrix

$$\frac{d}{dt}\begin{pmatrix} \rho_{13} \\ \rho_{24} \\ \rho_{14} \\ \rho_{23} \end{pmatrix} = \begin{pmatrix} (i\omega_{13} - x) & -x & y & -y \\ -x & (i\omega_{24} - x) & y & -y \\ y & y & (i\omega_{14} - z) & -x \\ -y & -y & -x & (i\omega_{23} - z) \end{pmatrix} \begin{pmatrix} \rho_{13} \\ \rho_{24} \\ \rho_{14} \\ \rho_{23} \end{pmatrix}$$

$$\tag{C4}$$

where $x = \tau^{-1}\sin^2 2\theta$, $y = \tau^{-1}(1/2)\sin 4\theta$, $z = \tau^{-1}(1 + \cos^2 2\theta)$.
(C4) is integrated to give

$$\underline{\rho}(t) = \exp(\underline{A}\, t).\underline{\rho}(0) \tag{C5}$$

where $\underline{\rho}$ and $\underline{A}$ are the column vector and matrix occurring
in (C4). With (C5) this now gives the free induction decay.
After Fourier transforming we have the spectrum line shape

$$G(\omega) \propto R\, \overset{\sim}{\underline{\rho}}(0).(\underline{A} - i\omega\underline{1})^{-1}.\underline{\rho}(0) \tag{C6}$$

where $\overset{\sim}{\underline{\rho}}(0)$ is a row vector and $\underline{1}$ is the unit matrix.

$$\overset{\sim}{\underline{\rho}}(0) = [\cos 2\theta \quad \cos 2\theta \quad \sin 2\theta \quad -\sin 2\theta]$$

The problem is therefore solved by the inversion of a complex
matrix.

   Our previously obtained results for the simple cases of
classical jumping alone or tunnelling alone are easily recovered
from (C6). For no classical jumping, i.e. $\tau^{-1} = 0$, $x = y = z = 0$.
The matrix to be inverted in (C6) is diagonal and one easily
obtains

$$G(\omega) = \cos^2 2\theta[\delta(\omega_{13} - \omega) + \delta(\omega_{24} - \omega)]$$
$$+ \sin^2 2\theta\,[\delta(\omega_{14} - \omega) + \delta(\omega_{23} - \omega)]$$

which is the spectrum of figure 3.

   If $\omega_t = 0$ then $x = z = \tau^{-1}$ and $y = 0$; $\overset{\sim}{\underline{\rho}}(0) = [0 \quad 0 \quad 1 \quad -1]$
and (C6) reduces to

$$G(\omega) \propto R \; [1 \quad -1] \begin{bmatrix} i(\omega_{13} - \omega) - \tau^{-1} & -\tau^{-1} \\ -\tau^{-1} & i(\omega_{23} - \omega) - \tau^{-1} \end{bmatrix} \begin{bmatrix} 1 \\ -1 \end{bmatrix}$$

which is equivalent to (8).

REFERENCES

1. P.W. ANDERSON, J.Phys.Soc.Japan 9, 316 (1954)

2. T.B. COBB and C.S. JOHNSON, J.Chem.Phys. 52, 6224 (1970)

3. C. MOTTLEY, T.B. COBB and C.S. JOHNSON, J. Chem.Phys. 55, 5823 (1971)

4. R. IKEDA and C.A. McDOWELL, Molecular Phys. 25, 161 (1973)

5. S. CLOUGH and B.J. MULADY, Phys.Rev.Letters 30, 161 (1974); H. GLATTLI, A. SENTZ and M. EISENKREMER, Phys.Rev.Letters 28, 871 (1972)

MAGNETIC RESONANCE AND RELAXATION: A PROBE OF THE PHONON SPECTRUM[*]

R. Orbach[**]

University of California

Los Angeles, California 90024, U.S.A.

ABSTRACT

The resonance relaxation process is sensitive to the occupation of phonon states with energies much greater than kT. There is no method currently known to the author for the determination of the lifetime of these vibrational states. It is shown how the phonon bottleneck of the resonance relaxation process can be used to determine the anharmonic (frequency shifting) lifetime of these high energy phonons.

[*] Supported in part by the National Science Foundation and the U.S. Office of Naval Research, Contract No. N00014-69-0200-4032.

[**] Part of this work was performed at the Physics Department, University of Tel Aviv, Ramat Aviv (Tel Aviv), Israel.

   We wish to discuss spin-lattice relaxation of dilute
paramagnetic salts in the so-called "resonance relaxation"
process regime. In order to acquaint you with the various
relaxation processes, imagine that the energy levels of a
paramagnetic ion are as in figure 1. Here, δ represents the
Zeeman splitting of the ground doublet (or, for that matter,
any splitting such that the magnetic resonance is performed at
the frequency $\delta/\hbar$), Δ is the (crystal field) splitting between
the ground doublet and the first excited state (or doublet,
depending on the number of magnetic electrons), and τ_c the
lifetime of the excited state $|c\rangle$. The double arrow is supposed
to represent the magnetic resonance transition between the
ground doublet states $|a\rangle$ and $|b\rangle$. The finite lifetime of the
excited state is caused by phonon emission to the ground states,
with phonons of energies $\Delta \pm \delta/2$.

   When magnetic resonance is performed as indicated, the
populations of the ground doublet are changed. Under steady state
conditions, the populations are driven to equality, and thermal
equilibrium restored as a result of contact between the phonon
bath and the spin system. In general, the phonon bath is assumed
to be at the temperature of the sample (or He bath in which it
may be immersed). Thus, the spins relax to the bath temperature
such that the ratio of populations of states $|a\rangle$ and $|b\rangle$ is given

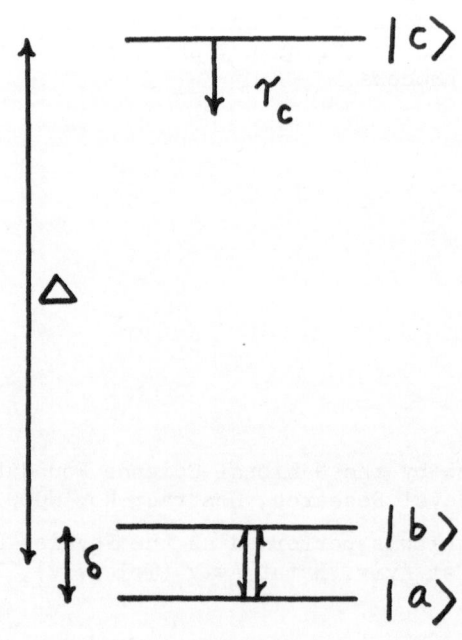

Fig. 1

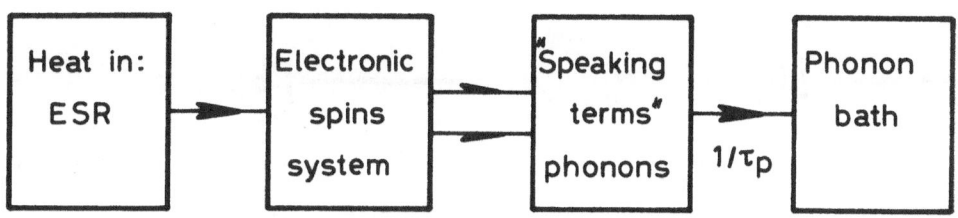

Fig. 2

by

$$P_b/P_a = \exp(-\delta/kT) \tag{1}$$

This conclusion is valid for very dilute spin systems, or for phonon systems where anharmonic processes (phonon-phonon interactions) are very strong. It is now generally recognized that this is seldom the case, and that one must consider the details of the phonon system when computing the electronic relaxation process [1]. The result of energy transfer between the paramagnetic spins and the phonon system is a heating up of the latter, but only at the "speaking terms" [2] energy, i.e. that energy characterizing the transitions of the paramagnetic ion. A schematic is given in figure 2. We denote the temperature of the "speaking terms" phonons by T_p. Then, depending on the relaxation process, the phonon distribution is altered from its thermal equilibrium value. In the schematic above, $1/\tau_p$ represents the phonon-phonon relaxation rate. If this rate is slower than the "feeding" rate (denoted by the double arrow), then the phonons will heat up as a result of the relaxation between spin and phonon.

The first process which was investigated extensively exhibiting this effect was the "direct" or one-phonon relaxation process (Fig. 3).

The cross denotes occupancy of the electronic state, the wavy line the phonon. If the spin system is heated to a temperature greater than T (denotes the true bath temperature), then the "speaking terms" phonons will be heated to a temperature greater than T. This new temperature, T_p, can be made very great

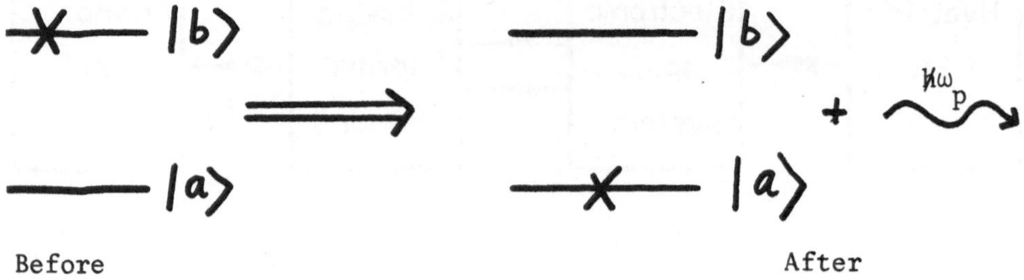

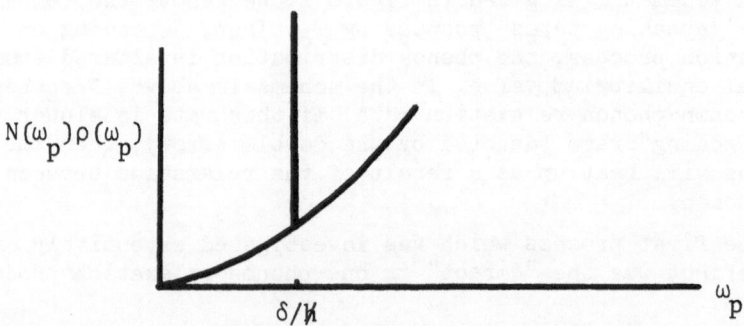

Before   After

Fig. 3

because the frequency shifting anharmonic processes are very
weak at low frequencies, i.e. those of the magnitude $\delta/\hbar$,
where usual magnetic resonance frequencies are involved. If we
denote the phonon density of states by $\rho(\omega_p)$, and the phonon
occupation number by $N(\omega_p)$, then the phonon "number" curve is
altered in the following manner (Fig. 4).

$$N(\omega_p)\rho(\omega_p)$$

$\delta/\hbar$ ω_p

Fig. 4

The heated "spike" is placed at $\delta/\hbar$ in the phonon frequency distribution because of the requirement of conservation of energy: the change in spin energy must equal the energy of the emitted phonon.

In general, relaxation involving substantial heating of the "speaking terms" phonons does not result in a simple exponential decay of the spin populations after the microwave field has been suddenly reduced. The coupled rate equations for the spins and phonons must be solved simultaneously [3]. In the limit of only slight phonon heating, the "effective" spin relaxation rate (that which would be observed in an experiment) has the characteristic form:

$$1/T_1 \propto \coth^2(\delta/2kT) \tag{2}$$

It should be clear that what is actually relaxing is not the spin, but rather the "speaking terms" phonons. One can derive the result (2) by using Casimir's heat flow argument. Refering to figure 2, we see that the spins relax "in series" to the bath. Thus, the time for relaxation is the sum of the times to go from box to box, weighted by the relative specific heats. Thus,

$$T_1 = T_1^0 + \tau_p \left(C_{spin}/C_{\text{"speaking terms phonons"}}\right) \tag{3}$$

where T_1^0 is the (short) time for contact between the spin and the speaking terms phonons, and C the specific heat of the respective species. At temperatures large compared to δ/k, the spin specific heat equals $(1/4)Nk(\delta/kT)^2$ while the phonon specific heat equals $\rho(\delta)k\Gamma$, where Γ is the linewidth for phonon emission in the direct process. That is, it is the bandwidth of the heated phonon "spike". Inserting these expressions into (3), the high temperature form of (2) is immediately obtained.

It is customary to refer to the case of phonon heating as the "magnetic resonance bottleneck" though that phrase is extensively used for quite a different effect in dilute magnetic alloys.

In the absence of a "bottleneck", the direct process rate can be easily shown to be proportional to

$$1/T_1 \propto \coth(\delta/2kT) \tag{4}$$

Thus, the presence or absence of a "bottleneck" can be rather quickly assessed on the basis of the temperature dependence of the relaxation rate. This was done with great clarity in the work of Scott and Jeffries [1].

The effect of phonon heating at selective frequencies is
absent in the usual two-phonon Raman (non-resonant) relaxation
process. This is because relaxation occurs via the absorption and
emission of two phonons, the difference in phonon energy being
restricted to δ, but there being no restriction (other than
density of states and thermal occupation) on the absolute phonon
energies. This process is important normally at the upper end of
the He^4 temperature range, and higher temperatures. The rapid
increase in phonon density of states with increasing phonon energy,
and the generally small value of δ/k relative to T, is responsible
for the dominance of the Raman process over the direct process.
As the latter involves essentially all of the phonons, only an
overall heating of the crystal could result from the energy
transfer depicted in figure 2. With any sensible contact between
the sample and the surrounding He bath, such heating is negligible.

There is another, sometimes rather important, two phonon spin-
lattice relaxation process which does selectively interact with the
phonon system. This is the resonant relaxation process, proceeding
according to the steps pictured in figure 5.

It should be emphasized that these "steps" are not uncorrelated.
They should be pictured as two parts of a "one step" second-order
perturbation theory calculation. The resonance character of figure
5 arises from the great increase in scattering amplitude when the
energy of the "incoming" phonon equals the splitting $\Delta - (\delta/2)$.
Overall energy conservation then requires that the emitted
phonon ("outgoing") have energy $\Delta + (\delta/2)$. Such a process is the
spin-phonon analogue of the resonance-fluorescence effect well
known in optical spectroscopy [4].

The resonance relaxation process thereby "heats up" a
band of phonons at energy $\Delta + (\delta/2)$, and "cools off" a band of
phonons at energy $\Delta - (\delta/2)$, if the initial population of spins
in the state $|b>$ exceeds that in $|a>$, with the reverse true if
the spin system is initially at negative temperature. This causes
the phonon number, as a function of frequency, to be altered
as shown in figure 6.

The width of the heated/cooled bands of phonons is determined
by a variety of factors. The most important is usually the
width of the excited state $|c>$ caused by lifetime effects.
These can be due to spin-spin interactions, but most likely are
caused by phonon emission to either $|b>$ or $|a>$ from $|c>$. Since
this matrix element is also contained in the relaxation process
itself, there is an intimate connection between the excited
state linewidth and the spin-lattice relaxation rate [5].Another
complication (in fact the point of this paper) is the lifetime
of the phonons involved in the resonance relaxation process.
As we shall see below, what is important is the frequency

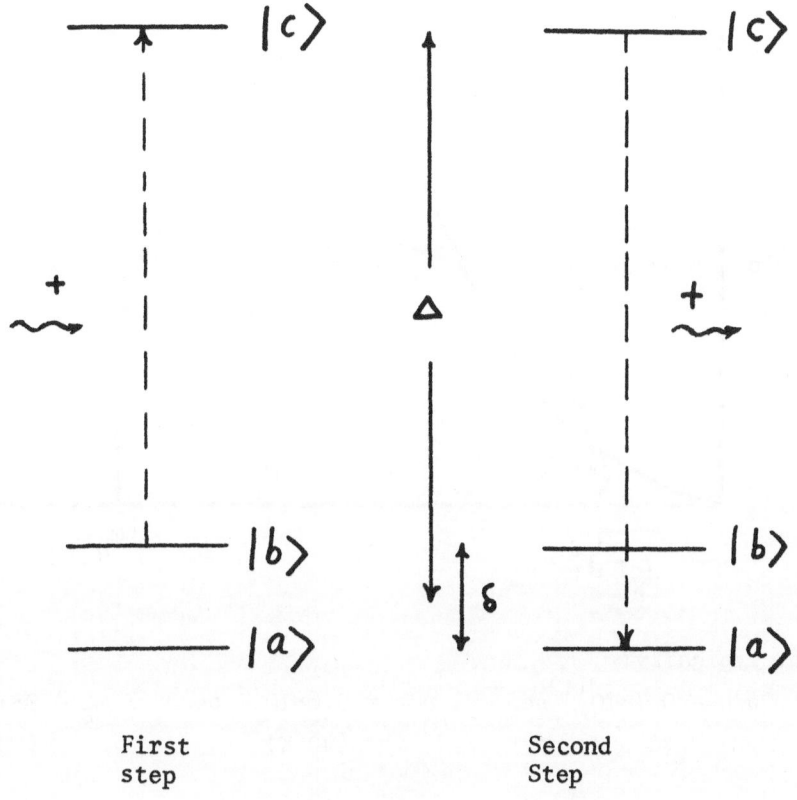

Fig. 5

shifting ability of the anharmonic phonon-phonon scattering process.

It is instructive to plot the change in phonon number as a function of frequency in the presence of resonant phonon heating (and cooling) (Fig. 7).

One sees that though the heated and cooled phonons possess high energies, the actual difference in energy transfer rate is proportional to the spin-lattice relaxation rate times the Zeeman energy difference δ. One also sees that if the anharmonic

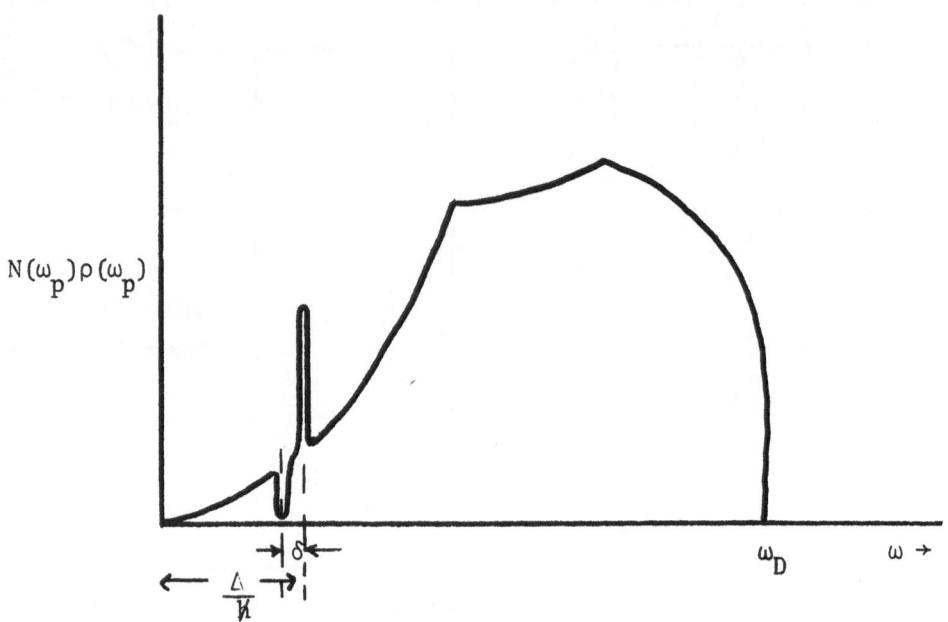

Fig. 6. Typically:

$\delta/\hbar \cong 0.7 \times 10^{11}$ sec^{-1} ; $\Delta/\hbar \cong 5.6 \times 10^{12}$ sec^{-1} ; $\omega_D \cong 3 \times 10^{13}$ se

(0.5 K)   (40 K)   (200 K)

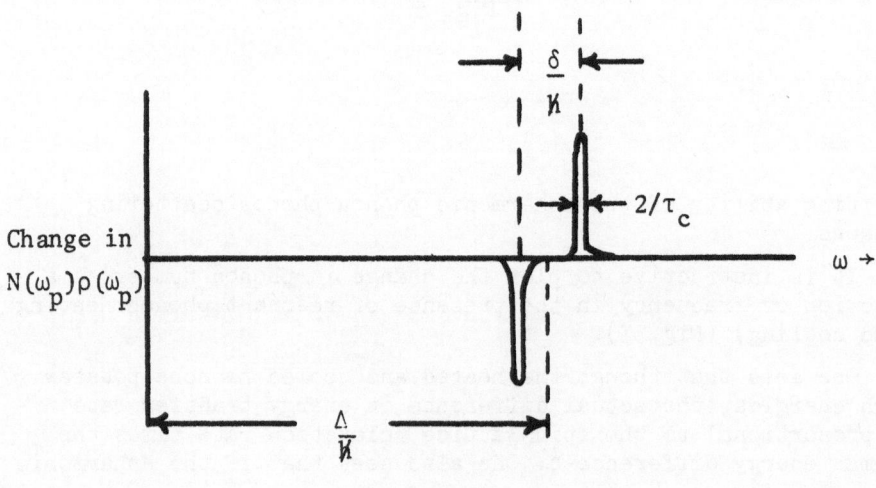

Fig. 7

phonon-phonon lifetime is sufficiently short, the breadth of the heated and cooled spike in figure 7 will exceed the spin lifetime width $1/\tau_c$. In fact, if the phonon anharmonic rate $1/\tau_p$ greatly exceeds the spin lifetime width $1/\tau_c$, then the breadth of the spikes in figure 7 is determined by the former. For the Zeeman splitting $\delta/\hbar$ much larger than $1/\tau_c$, the picture remains as in figure 7. For the reverse, however, the two spikes "overlap", and only a much more modest heating will result. The relevant quantity is therefore the ratio (defined by Gill [3])

$$z = \delta\tau_c/2\hbar \qquad (5)$$

for the former limit, with τ_p replacing τ_c for the latter limit. In general, one finds that the former is the more usually encountered, but the latter should in principal also occur in practice. Another parameter which plays a role in the phonon bottleneck is the quantity (correct to within quantities of order unity)

$$Q = n\tau_p/\rho(\Delta/\hbar) \qquad (6)$$

The ratios (5) and (6) refer, respectively, to the Zeeman splitting versus excited state(s) level width, and to the total number of spins in the crystal, n, versus the number of phonon states in a bandwidth $1/\tau_p$.

It is clear that if z is much greater than unity, the resonance spin-lattice relaxation rate will be sharply reduced. This follows because the separation of the heated and cooled spike (see figures 6 and 7) is increased relative to the width of the spikes the larger z. The relaxation rate also decreases as Q increases, because this ratio represents the rate at which spins make relaxation transitions divided by the rate at which phonons leave or enter a band of width $\pi\hbar/\tau_c$. One has assumed in (6) that the rates just referred to represent thermal equilibrium values, and that $\delta \ll kT$. Gill [3] has calculated the relative increase in relaxation time as a function of z and Q over a physically interesting region (Fig. 8).

In figure 8, the broken lines are in accordance with the equations of Gill [3] while the solid lines were obtained by direct computation. Lichti and Culvahouse [3] used Gill's solutions to rather completely analyze data on the resonant relaxation process currently available in the literature. The major features of Gill's theory, displayed in figure 8, are that the relaxation time decreases as the magnetic field, H, to the one-half power, and as the one-fourth power of the paramagnetic ion concentration, c. Hence,

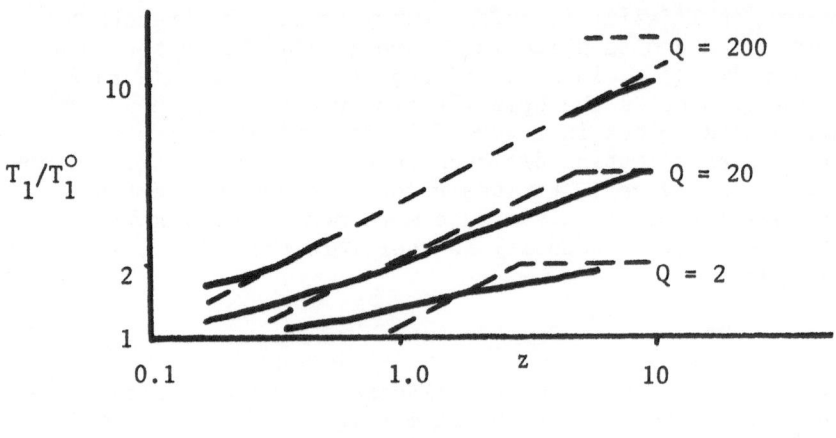

Fig. 8

$$1/T_1 \propto H^{1/2} c^{1/4} \tag{7}$$

In figure 8, T_1 represents the observed relaxation time, T_1^0 the time which would have been observed in the absence of the bottleneck.

Probably the most puzzling aspect of Lichti and Culvahouse's analysis is the relatively long phonon lifetimes they extracted from the experimental data. For example, in $La_2Mg_3(NO_3)_{12} \cdot 24H_2O$ doped with Ce^{3+}, the excited state splitting $\Delta = 36$ K, which must be at nearly 50 % of the acoustic zone-boundary frequency. Yet, they found that τ_p was (in the author's view) remarkably large, of the order of 10^{-9} sec at 4.2 K. This time is much shorter than expected from boundary scattering, yet by no means as short as one would expect from anharmonic phonon calculations of the conventional sort. Such calculations [6] predict phonon lifetimes varying as τ_p^5. To have a feeling for the magnitudes involved, at microwave frequencies, in "soft" materials, at temperatures greater than $\hbar\omega/k$, $1/\tau_p \simeq 10^{-7} \omega T^4$. At 10 K, and 10 Gcs, this results in $1/\tau_p \simeq 10^{+7}$ sec^{-1}, or a mean free path of approximately 10^{-2} cm. To find the relaxation rate for phonons of energy 40 K, at temperatures much less than Δ/k, replace T with $\hbar\omega/k$, where $\omega = \Delta/\hbar$. One finds

$$1/\tau_p = 10^{-7} \omega(\hbar\omega/k)^4 = 10^{-7}(k/\hbar)(\Delta/k)^5 = 10^4(\Delta/k)^5$$

Taking $\Delta = 40$ K, we find

$$1/\tau_p = 10^4 (40)^5 = 10^{12} \text{ sec}^{-1}$$

This value is much shorter than that obtained by Lichti and Culvahouse [3], and certainly exceeds $1/\tau_c$ and $\delta/\hbar$.

In fact, there is reason to believe that the rate of 10^9 sec^{-1} extracted by Lichti and Culvahouse is reasonable physically. To see why, it is necessary to consider the frequency shifting mechanisms for phonons whose energies greatly exceed kT. Under such conditions, the principal phonon decay mechanism caused by anharmonic interactions is one of "splitting". The high energy phonon splits into two lower energy phonons, such that energy and wave vector are conserved. These two conditions severely limit the relaxation channels. The only allowed processes are:

$$t_2 + t_2 \longleftrightarrow t_1$$

$$t_2 + t_1 \longleftrightarrow t_1 \qquad\qquad t_1 + t_1 \longleftrightarrow \ell$$

$$t_2 + \ell \longleftrightarrow \ell \qquad\qquad t_1 + \ell \longleftrightarrow \ell \qquad\qquad (8)$$

$$t_2 + t_1 \longleftrightarrow \ell \qquad\qquad t_1 + t_1 \longleftrightarrow t_1$$

For the splitting process, the transition goes from right to left. The symbols represent phonons of difference polarization surfaces. Thus, ℓ represents a phonon on the highest branch, t_1 a phonon on the intermediate branch, and t_2 a phonon on the lowest branch. It is not necessary that the branches be pure; only that they order in some sensible manner. Then, energy and wave vector conservation can be assured by drawing the energy versus wave vector curves for all three branches. One then shifts the origin up along the lowest branch (t_2), and notes where the unshifted and shifted dispersion curves cross. The crossing points denote the allowed processes (Fig. 9).

It is immediately seen that the lowest energy phonon branch, t_2, never can split into two branches. It can only combine with a phonon on branch t_1 or on ℓ. However, when kT is much smaller than the phonon energy $\hbar\omega_p(t_2)$, then the phase space available for such a combination is very small. This is because the combining phonon only has an energy kT, and phase space per phonon varies as $(\omega)^2$. Said another way, at temperatures low compared to the phonon energy in question (in this case, a phonon of branch t_2, with an energy $\Delta \pm \delta/2$), splitting would involve phonons of energies $\Delta/2$, whereas combining would involve phonons of energies Δ and kT. The quadratic dependence of the density of states on energy makes the latter process negligible compared to

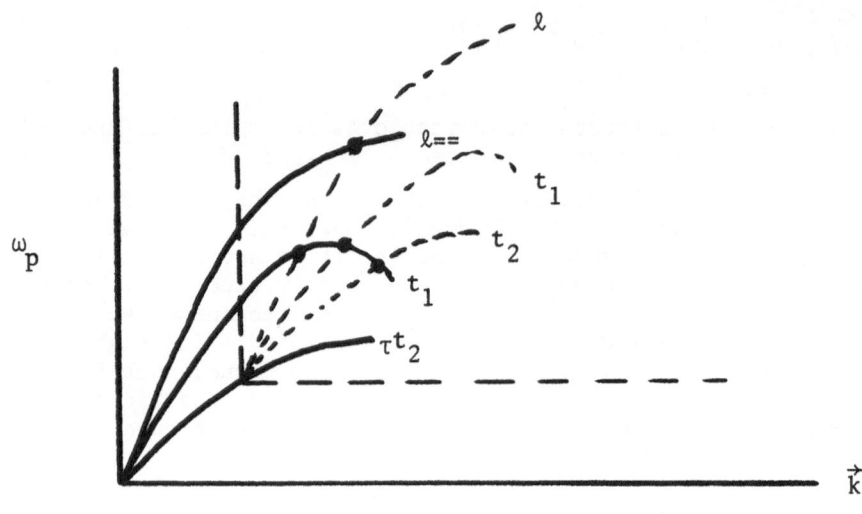

Fig. 9

the former. The only way to obtain significant phase space is to absorb a phonon of energy of order Δ, but this is exponentially improbable at low temperatures. Orbach and Vredevoe [6] predict that

$$1/\tau_p \simeq 10^4 (\Delta/k)^5 \exp(-\Delta/2kT) \tag{9}$$

at temperatures low compared to Δ/k, for the lowest energy phonon branch. For numbers appropriate to the system examined in detail by Lichti and Culvanhouse [3], (9) results in an increase by four orders for τ_p over what one would have expected from direct extrapolation from lower frequencies. Curiously enough, this is close to that value extracted by them for Ce^{3+} in the double nitrate.

Finally, one should comment on the significance of the dependence of the measured spin-lattice relaxation time T_1 on the resonant phonon lifetime τ_p. As can be seen from figure 8, and as calculated explicitly by Gill [3], T_1 increases as τ_p increases. Explicitly (with a numerical correction provided by Lichti and Culvahouse [3]), for

$$zQ^{1/2} \gg Q, \quad T_1 = T_1^0 (1 + Q)^{1/2} \operatorname{sech}(\delta/2kT) \tag{10a}$$

and for

$$Q \gg zQ^{1/2} \gg 1, \quad T_1 = T_1^0 \; (0.804) \; Q^{1/4} z^{1/2} \tag{10b}$$

Because the resonance relaxation process requires the absorption of a phonon of energy $\simeq \Delta$, at temperatures low compared to Δ/k,

$$T_1^0 \propto \exp(\Delta/kT)$$

so that at temperatures greater than δ/k,

$$T_1 \propto \exp(\Delta/kT)\tau_p^{1/2}$$

or $\hspace{9cm}$ (11)

$$T_1 \propto \exp (\Delta/kT)\tau_p^{1/4}$$

according to whichever limit of (10) is appropriate. Noting (9), we see that the bottleneck of the resonant phonon will change the temperature dependence of the observed spin-lattice relaxation time. Indeed, measurement of T_1 will yield, in this limit, a direct measure of the phonon lifetime at an energy Δ which is much greater than kT. In general, there is no method currently available to study the lifetime or dynamics of such high energy phonons. The careful analysis of spin-lattice relaxation at low temperatures and conventional microwave frequencies therefore offers an opportunity to probe a part of the phonon spectrum with a vastly different scale of energy. It is an example of the use of electron spin resonance to study the dynamics of the host crystal.

REFERENCES

1. J.H. VAN VLECK, Phys.Rev. 59, 724 (1941)
   J.H. VAN VLECK, Phys.Rev. 59, 730 (1941)
   B.W. FAUGHNAN and W.W.P. STRANDBERG, Phys.Chem.Solids 19, 155 (1961)
   P.L. SCOTT and C.D. JEFFRIES, Phys.Rev. 127, 32 (1962)
   W.J. BRYA and P.E. WAGNER, Phys.Rev.Lett. 14, 431 (1965)
   W.J. BRYA and S. GESCHWIND and G.E. DEVLIN, Phys.Rev.Lett. 21, 1800 (1968)
   A.M. STONEHAM, Proc.Phys.Soc.Lond. 147, 239 (1966)
   R. ADDE, S. GESCHWIND and L.R. WALKER, "Proceedings of the Fifteenth Colloque AMPERE", 460, North-Holland, Amsterdam (1969)

2. J.H. VAN VLECK, Phys.Rev. 57, 426 (1940); and [1]

3. J.C. GILL, J.Phys. C6, 109 (1973)
   R.L. LICHTI and J.W. CULVAHOUSE, Phys.Rev. B9, 4816 (1974)

4. W. HEITLER, "The Quantum Theory of Radiation", 196, Oxford, Clarendon (1957)

5. R. ORBACH, Proc.R.Soc. A264, 485 (1961)
   J.W. CULVAHOUSE and P.M. RICHARDS, Phys.Rev. 178, 485 (1969)

6. R. ORBACH and L.A. VREDEVOE, Physics 1, 91 (1964)

MAGNETIC RESONANCE AND STRUCTURAL PHASE TRANSITIONS

R. Blinc

University of Ljubljana, Institut Jozef Stefan

Ljubljana, Jugoslavija

INTRODUCTION

Structural phase transitions in condensed matter are characterized by:

1) The appearance of a static order parameter (like the spontaneous polarization in ferroelectrics, the sublattice polarization in antiferroelectrics, or the long range orientational ordering in nematic liquid crystals) which breaks the symmetry of the high temperature phase and characterizes the new phase, and by

2) Critical order parameter fluctuations, which increase in amplitude and which relax back to equilibrium ever more slowly as the critical point is approached.

At the transition temperature T_0 itself, the displaced atoms no longer return to their original equilibrium positions, and the structure of the low temperature phase represents the superposition of the "frozen-in" soft mode displacements on the structure of the high temperature phase.

An illuminating description of this situation is given by the Landau theory. In the simplest possible case of a homogeneous system we expand the non-equilibrium free energy density in powers of the order parameter

$$\Delta F = F(\eta,T) - F(0,T) = \frac{1}{2}a(T)\eta^2 + \frac{1}{4}b\eta^4 + \ldots \qquad (1.1)$$

For sake of simplicity let us further restrict our discussion to the case of II. order phase transitions, where $b = \text{const} > 0$

and where there are no third order terms in η. We shall further assume that we can expand $a(T)$ in a Taylor series in powers of $(T-T_O)$ and keep only the first order term:

$$a(T) = a'(T-T_O) \tag{1.2}$$

The equilibrium value of the order parameter η at the temperature T is this value of η which minimizes the free energy. The condition that ΔF is stationary with respect to an infinitesimal change in η results in:

$$[\partial(\Delta F)/\partial\eta]_{\eta_O} = 0 \tag{1.3}$$

or, using (1.1), in

$$\eta_O[a(T) + b\eta_O^2] = 0 \tag{1.4}$$

One solution of eq. (1.3) is always

$$\eta_O = 0$$

corresponding to the high symmetry phase. Another solution,

$$\eta_O = \pm \sqrt{-a/b} = \pm \sqrt{-a'(T-T_O)/b} \tag{1.5}$$

which breaks the symmetry of the high temperature phase, exists only for $T < T_O$.

In a real system, the system is not homogeneous and fluctuations in the order parameter must be taken into account $[\eta = \eta(r)]$:

$$\Delta F = F(\eta,T) - F(0,T) = \frac{1}{2}a(T)\eta^2 + \frac{1}{2}c(\nabla\eta)^2 + \ldots \tag{1.6}$$

Here we have for sake of simplicity neglected the fourth order term in η.

Expanding $\eta(r)$ in a Fourier series

$$\eta(r) = \sum_q \eta_q e^{iqr} \tag{1.7}$$

we find

$$\Delta F = \frac{1}{2} \sum_q a(T) \ |\eta_q|^2 \ [1 + q^2\xi^2] \tag{1.8}$$

where the correlation length ξ is introduced as

$$\xi^2 = c/a(T) = c/a'(T-T_o) \tag{1.9}$$

The mean square fluctuation in the Fourier component of the order parameter with wave vector $\vec{q}$ $<|\eta_q|^2>$ is now obtained as

$$\overline{<|\eta_q|^2>} = A \int_{-\infty}^{+\infty} |\eta_q|^2 \ e^{-\Delta F/kT} \ d\eta_q = kT/a'(T-T_o)(1+q^2\xi^2) \tag{1.10}$$

For $q \neq 0$, $\overline{|\eta_q|^2}$ approaches a finite value for $T \to T_o$, whereas it diverges for $q = 0$ as $(T-T_o)^{-1}$.

The dynamics of the order parameter fluctuations can be as well described by the Landau theory. The regression of an order parameter fluctuation $\delta\eta = \eta - \eta_o$ towards equilibrium is given in the Fourier representation by

$$\partial|\eta_q|/\partial t = - L \ \partial(\Delta F)/\partial(\eta_q) = -L \ a(T) \ |\eta_q| \ (1 + q^2\xi^2) \tag{1.11}$$

where L is a non-critical kinetic coefficient.
Expression (1.11) can be written as

$$d \ |\eta_q|/dt = -(1/\tau_q) \ |\eta_q| \tag{1.12}$$

where the order parameter relaxation time τ_q is given by

$$\tau_q = L^{-1}/a'(T-T_o)(1+q^2\xi^2) \tag{1.13}$$

This expression shows that for II. order phase transitions

the macroscopic order parameter relaxation time $\tau_{q=0}$ diverges at T_O:

$$\tau_{\underset{q=0}{\longrightarrow}} \to \infty \qquad \text{as} \quad |T-T_O|^{-1} \tag{1.14}$$

The critical slowing down of the order fluctuations results from the vanishing of the termodynamic restoring force

$$\partial(\Delta F)/\partial\eta \quad \text{at} \quad T_O.$$

ORDER PARAMETER DETERMINATION BY MAGNETIC RESONANCE

Hydrogen Bonded Solids

A classical example of order parameter determination by magnetic resonance is provided by the protonic order-disorder transitions in cooperative O-H--O hydrogen bonded systems like KH_2PO_4 and KD_2PO_4. In these systems the protons respectively deuterons move in double minimum potentials. The two equilibrium sites O-H--O and O---H-O of a proton or deuteron in such a potential can be described by two different values of the local order parameter p_i:

$$p_i = +1 \quad \text{and} \quad p_i = -1 \tag{2.1}$$

The electric field gradient (EFG) tensor at the deuteron ($I=1$) site changes when the deuteron is transferred from one potential well into another. The quadrupole perturbed deuteron magnetic resonance spectrum can be hence used for a determination of the order in deuterated systems. The actual calculation proceeds as follows: when the deuteron moves between the two off-centre sites it spends a fraction $\frac{1}{2}(1-p_i)$ of its time at the "left" and $\frac{1}{2}(1+p_i)$ at the "right" equilibrium site in the O-H--O bond. The time dependence of the deuteron EFG tensor for a given O-D--O bond can be hence expressed as:

$$\underline{T}_i(t) = \frac{1}{2}[1+p_i(t)]\,\underline{T}(1) + \frac{1}{2}[1-p_i(t)]\,\underline{T}(2) \tag{2.2}$$

where $\underline{T}(1)$ refers to the "left" and $\underline{T}(2)$ to the "right" equilibrium site.

Expression (2.2) can be rewritten as:

$$\underline{T}_i(t) = \underline{T}_o + \underline{A} \, p_i(t) \tag{2.3}$$

where

$$\underline{T}_o = \frac{1}{2} \left[\underline{T}(1) + \underline{T}(2) \right] \tag{2.4}$$

represents the static, and

$$\underline{A} = \frac{1}{2} \left[\underline{T}(1) - \underline{T}(2) \right] \tag{2.5}$$

the fluctuating part of the deuteron EFG tensor. Since the deuterons move between the two off centre sites at a rate which is much faster than the quadrupole splitting frequency between the two sites, the NMR spectrum is determined by the time averaged value of the EFG tensor. As the static value of the order parameter

$$\langle p_i \rangle = \langle p_j \rangle = p \tag{2.6}$$

equals zero in the paraelectric phase, we get for:

$$T > T_c : \qquad \langle \underline{T} \rangle = \underline{T}_o \tag{2.7}$$

$$T < T_c : \qquad \langle \underline{T} \rangle = \underline{T}_o + \underline{A}p \tag{2.8}$$

The deuteron resonance lines are thus temperature independent for $T > T_c$ and are proportional to the order parameter below T_c (Fig. 1). For ferroelectric systems, p is proportional to the spontaneous polarization and for antiferroelectric systems it is proportional to the sublattice polarization.

The matrix $\underline{A}$ can be determined by measuring the angular dependence of the deuteron NMR spectrum above and below T_c. In KD_2PO_4 type crystals, the matrix $\underline{A}$ expressed in the crystal fixed coordinate system has just one non-zero matrix element $T_{YZ} = T_{ZX}$.

Since the proton does not have a quadrupole moment — and since the long range magnetic dipolar interactions are much less sensitive to small nuclear displacements than quadrupole interactions — the order-disorder transition in undeuterated

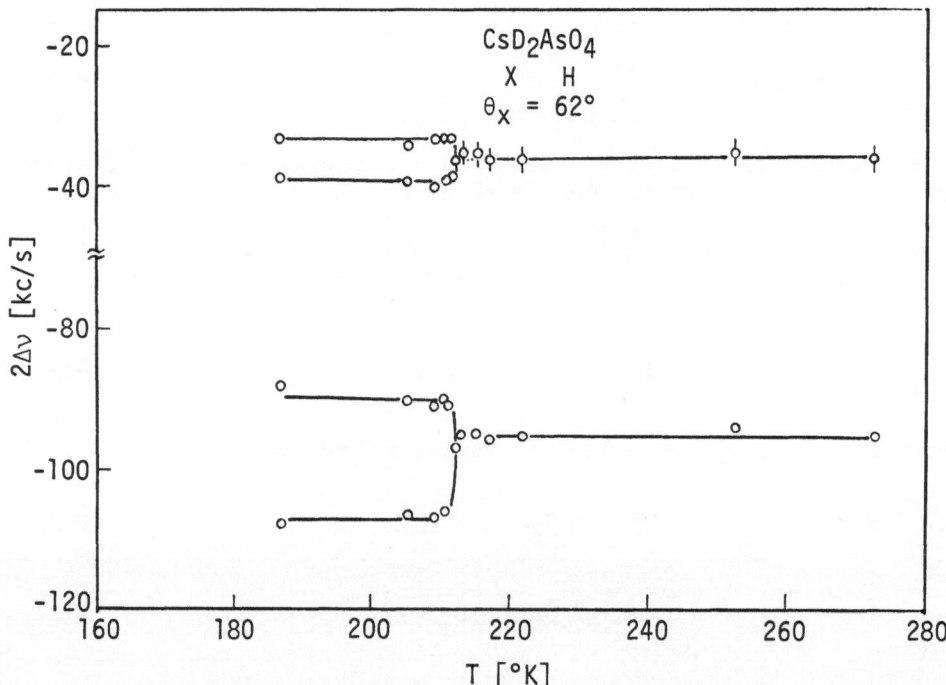

Fig. 1. Temperature dependence of the quadrupole splitting of the deuteron magnetic resonance lines in CsD_2AsO_4.

H-bonded systems is best studied by O^{17} (I = 5/2) nuclear quadrupole resonance (NQR) spectroscopy.

The results of an O^{17} NQR study in a 10 % O^{17} enriched sample of KH_2PO_4 are shown in figure 2. Below T_c we have two non-equivalent O^{17} sites, resulting in six NQR lines. One of the two sites represents the O^{17} nucleus to which the proton is directly attached (O^{17}-H---O), whereas the second represents the oxygen 17 to which the proton is only hydrogen bonded (O^{17}---H-O). The difference in the quadrupole coupling between these two sites with the proton in a "close" and a "far" position is quite large.

Above T_c all oxygen sites are chemically equivalent and we see only three O^{17} lines. This can be readily understood if the O-H--O hydrogen bond potential is of the double minimum type. The O^{17} quadrupole coupling constant above T_c is the average of the two O^{17}-H--O and O^{17}--H-O quadrupole coupling constants observed below T_c:

$$\underline{T}_i(t) = \frac{1}{2} \left[1+p_i(t)\right] \underline{T}(1) + \frac{1}{2} \left[1-p_i(t)\right] \underline{T}(2) \tag{2.9}$$

$$\langle \underline{T}_i(t) \rangle = \frac{1}{2}[\underline{T}(1) + \underline{T}(2)] = \underline{T}_o \qquad (T > T_c) \tag{2.10}$$

$$\langle \underline{T}_i(t) \rangle = \underline{T}(1) \text{ or } \underline{T}(2) \qquad (p = \pm 1,\ T < T_c) \tag{2.11}$$

In the above two cases the deuteron and O^{17} EFG tensors were functions of only one local order parameter $p_i(t)$. In the general case, however, the EFG tensor at a given site will depend on the state of order of the whole crystal:

$$\underline{T} = \underline{T}\ (p_1,\ p_2\ \cdots\ p_N) \tag{2.12}$$

and the higher terms in the expansion of $\underline{T}_i$ in powers of p_i cannot be neglected. In such a case it is important to know the relation between the point symmetry of a given lattice site and the form of the EFG tensor at this site.

Changes in the symmetry of a given lattice site produced by a phase transition are thus easily detected by a measurement of the EFG tensor, provided that a nucleus with $I>1/2$ is located at this site and thus serves as a probe for the local electric field distribution.

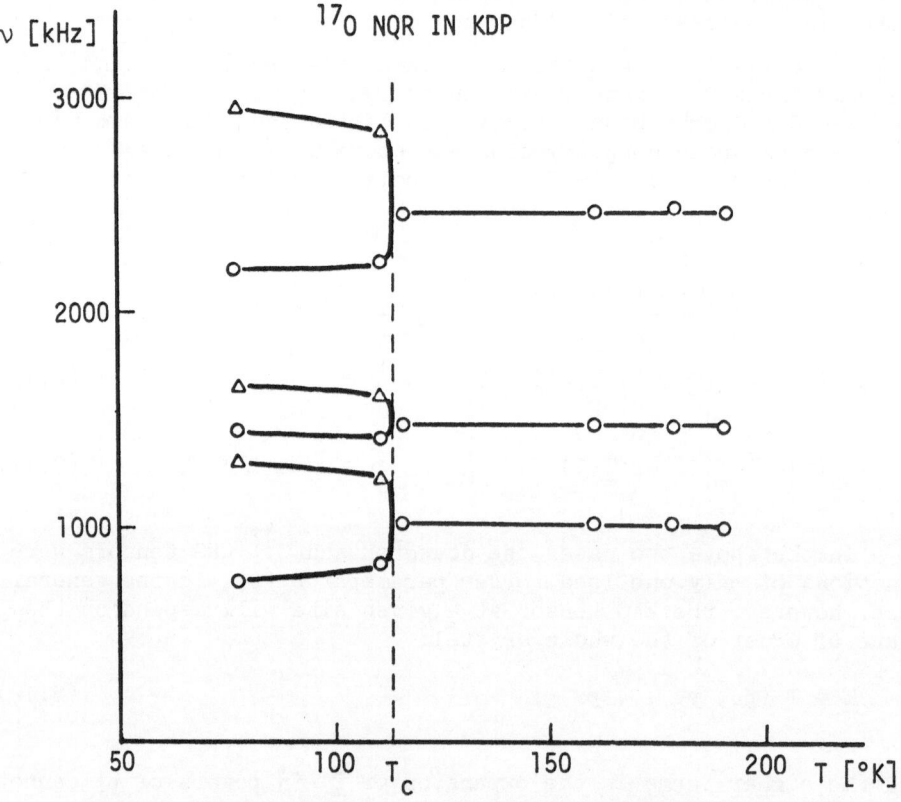

Fig. 2. O^{17} NQR spectrum of KH_2PO_4.

Table 1. Relation between the point symmetry of a given lattice
site and the form of the EFG tensor at this site

Point symmetry of lattice site	EFG tensor
1 or $\bar{1}$	No restrictions
2 or $\bar{2}$	One of the principal axes is parallel to the 2-fold axis
m	One of the principal axes is perpendicular to the mirror plane, whereas the other two lie in the plane
222, mmm, mm2	EFG tensor axes parallel to crystal axes
3, $\bar{3}$, 4, $\bar{4}$, 6, $\bar{6}$	EFG tensor is axially symmetric and the largest principal axis is parallel to the rotation axis
cubic	All EFG tensor elements vanish

Nematic Liquid Crystal Systems

    Liquid crystals are systems which exhibit in addition to
the solid and the liquid an intermediate anisotropic liquid
phase. Nematic liquid crystals, in particular, are characterized
by the existence of long range order in the orientation of the
long molecular axis whereas the centres of gravity of the molecules
are not fixed.

    For such systems the natural order parameter is

$$S = \langle \frac{1}{2} (3 \cos^2\theta - 1) \rangle \tag{2.13}$$

where θ is the angle between the long molecular axis and the
direction of nematic ordering.

    The above orientational order parameter can be obtained from
the splitting of the NMR proton line due to the dipolar interaction
between the two close ortho-protons in the phenyl ring of a
liquid crystal molecule.

The significant term in the magnetic dipole interaction Hamiltonian between two nuclear spins at a fixed separation distance is

$$\langle \tfrac{1}{2}(3 \cos^2\Theta_o - 1)\rangle$$

where Θ_o is the angle between the internuclear vector, $\vec{r}$, and the magnetic field direction $\vec{H}_o$. Transforming from the laboratory frame to the molecular frame with the z-axis parallel to the molecular axis we find:

$$\tfrac{1}{2}\langle 3 \cos^2\Theta_o - 1\rangle = \langle \tfrac{3}{2} \cos^2\theta - \tfrac{1}{2}\rangle \tfrac{1}{2}(3n^2-1)$$

$$-3 \sin^2\theta \;\; [\langle e^{2i\phi}\rangle(\tfrac{1 + im}{2})^2 + \langle e^{-2i\phi}\rangle(\tfrac{1 - im}{2})^2]$$

$$-6i \sin\theta\cos\theta \;\; [\langle e^{i\phi}\rangle(\tfrac{1 + in}{2})n - \langle e^{-i\phi}\rangle(\tfrac{1 - im}{2})n] \qquad (2.14)$$

Here θ is the angle between the long molecular axis and the magnetic field direction which is also the direction of nematic ordering. ϕ is the azimuthal angle for rotations around the long molecular axis, and l, m, n are direction cosines describing the orientation of the internuclear vector with respect to the molecular frame. The brackets <> designate a time average.

If there is a complete rotation about the long molecular axis

$$\langle e^{\pm in\phi}\rangle = 0 \qquad (2.15)$$

and the line splitting is given by

$$\Delta H = \tfrac{3}{2} \frac{\gamma \hbar}{r^3} \cdot S \cdot \tfrac{1}{2}(3n^2-1) \qquad (2.16)$$

where S is the nematic order parameter. For $r = 2.45$ Å we get from the observed splitting for nematic systems $0.3 \leq S \leq 0.7$. In the isotropic phase, S is of course zero.

If there is no longer complete rotation about the molecular axis - as this is claimed for some smectic C systems - , and the x-axis is in the plane of the phenyl ring,

$$m = 0, \; \langle e^{\pm in\phi}\rangle = \langle \cos n\phi\rangle \neq 0$$

and expression (2.14) becomes:

$$\Delta H \propto \frac{1}{2}\langle 3\cos^2\Theta_o - \frac{2}{\lambda}\rangle = S\frac{1}{2}(3n^2-1) + (1-S)\langle\cos2\phi\rangle(1-n^2) \quad (2.17)$$

ORDER PARAMETER DYNAMICS VIA T_1

Quadrupolar Relaxation in Hydrogen Bonded Systems

Information about deuteron dynamics has been obtained from deuteron spin-lattice relaxation studies which show an anomalous increase in the spin-lattice relaxation rate T_1^{-1} on approaching T_o (Fig. 3).

The spin lattice relaxation rate depends on the spectral density of the autocorrelation function of the matrix elements of the fluctuating part $H_1(t)$ of the spin Hamiltonian. Relating with the help of expression (2.2) the spectral density of the EFG tensor fluctuations to the spectral density of the order parameter fluctuations, we find the spin transition probabilities P as:

$$P_\mu = P_{m,m'} = \frac{1}{\hbar^2}\int_{-\infty}^{+\infty}\overline{\langle m|H_1(0)|m'\rangle\langle m'|H_1(t)|m\rangle}\,e^{-i\omega_{m,m'}t}\,dt =$$

$$\frac{1}{\hbar^2}|Q_{\mu m}|^2\int_{-\infty}^{+\infty}\overline{\underline{T}_{1,\mu}(0)\,\underline{T}_{1,\mu}^*(t)}\,e^{-i\omega_{m,m'}t}\,dt =$$

$$(\frac{eQ}{4I(2I-I)}|A_\mu|/\hbar)^2(|\mu|/2)[\int_{-\infty}^{+\infty}\overline{\Delta p_i(0)\Delta p_i(t)}\,e^{i\omega_\mu t} + c.c.] \quad (3.1)$$

Here $H_1(t)$ is the fluctuating part of the spin Hamiltonian

$$H_1(t) = H_Q(t) - \overline{H_Q(t)} \quad (3.2)$$

$\underline{T}_1$ is the fluctuating part of the EFG tensor expressed in the frame where $\bar{H}$ is diagonal

$$\underline{T}_1 = \underline{T}_i(t) - \underline{T}_o \quad (3.3)$$

Δp_i is the fluctuation in the local order parameter

$$\Delta p_i = p_i(t) - p \quad (3.4)$$

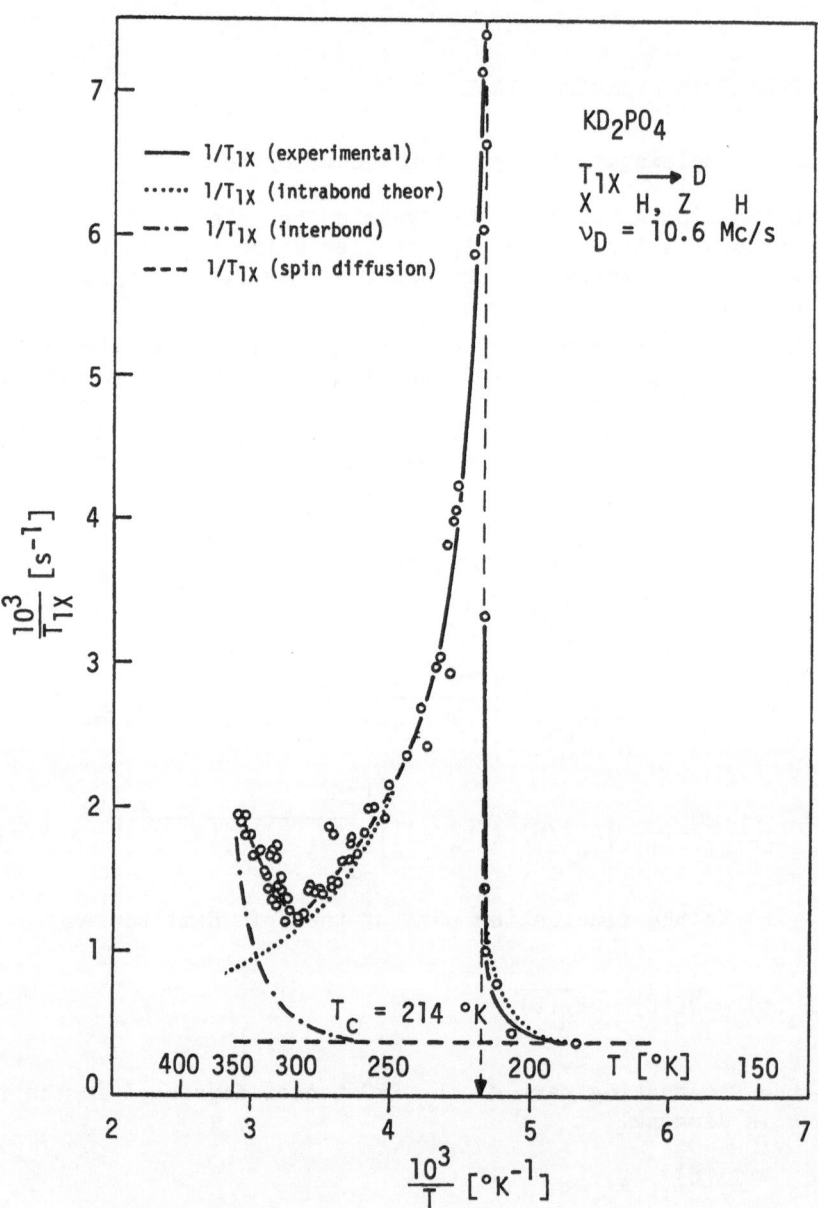

Fig. 3. Temperature dependence of the deuteron spin-lattice relaxation time in KD_2PO_4.

and

$$\mu = \Delta m = \pm 1 \text{ or } \pm 2 \qquad (3.5)$$

Q is the quadrupole moment and I the spin of the investigated nucleus. For $\Delta m = \mu = +1$ we find for KD_2PO_4

$$|A_1|^2 = 4T_{YZ}^2 \cdot \sin^2\Theta_Z \qquad (3.6)$$

where Θ_Z is the angle between H_o and the crystal Z axis.

For $\Theta_Z = 55°$, the deuteron energy levels are equidistant and

$$T_1 = P_1 + 2P_2 \qquad (3.7)$$

A T_1 experiment thus directly measures the spectral density of the local order parameter autocorrelation function $\overline{\Delta p_i(0)\Delta p_i(t)}$ at the nuclear Larmor frequency ω_μ.

For KD_2PO_4, a non-interacting deuteron intra-bond correlation time

$$\tau = 2.7 \times 10^{-13} \text{ sec}$$

was obtained from the T_1 data. In expression (1.14) τ corresponds to L^{-1}.

Using the dynamic Ising model to evaluate $\overline{\Delta p_i(0)\Delta p_i(t)}$, we find that in the fast motion regime ($\omega\tau << 1$):

$$T_1 \propto (T-T_c)^{1/2} , \quad T > T_c \qquad (3.8)$$

or

$$T_1 \propto \{\ln[(T-T_c)/T_c]\}^{-1} + \text{const}, \quad T > T_c \qquad (3.9)$$

depending on the isotropy – expr.(3.8)– or anisotropy – expr. (3.9)– of the Fourier transform of the Ising model interaction constant $J_{\vec{q}}$.

As T_1^{-1} is a local quantity it cannot really diverge. The apparant divergence in expressions (3.8) and (3.9) are a result of the approximation $\omega\tau << 1$. The inclusion of this factor results in a finite value for T_1^{-1} at $T = T_c$.

Below T_c we find

$$T_1^{-1} \propto (1-p^2), \quad T < T_c \tag{3.10}$$

Zeeman Dipolar and Rotating Frame Spin-Lattice Relaxation in the
Case of Fast and Ultraslow Motion

According to Ailion, Slichter and Zumer, the dipolar spin-lattice relaxation rate is, in the slow motion case $\tau \gg T_2 = 1/\gamma\Delta H$ given by

$$T_{1D}^{-1} = \frac{1-p^2}{\tau} \frac{A_1}{A_o + A_1} \tag{3.11.a}$$

where A_o and A_1 are constants characterizing the molecular motion as described by the correlation function

$$G(t) = Tr\{\overline{H_D'(0), H_D'(t)}\} = A_o + A_1 \frac{1}{N} \sum_q \overline{p_q(0)p_q(t)} \tag{3.11.b}$$

of the secular part of the dipolar Hamiltonian.

The rotating frame spin-lattice relaxation rate T_1^{-1} is, on the other hand, given by

$$T_{1\rho}^{-1} = \frac{3}{2}\gamma^4\hbar^2 \, I(I+1)[\tfrac{1}{4} J_o(2\omega_1) + \tfrac{5}{2} J_1(\omega) + \tfrac{1}{4} J_2(2\omega)] \tag{3.12.a}$$

In the limit $\omega\tau \gg \omega_1\tau \gg 1$, $T_{1\rho}^{-1}$ becomes for protons:

$$T_{1\rho}^{-1} = \frac{9}{256} \gamma_H^4\hbar^2 \, |\Delta F|^2 \, (1-p^2)/\omega_1^2\tau \tag{3.12.b}$$

Here $|\Delta F|^2$ is the change in the F_o term of the dipolar Hamiltonian due to proton motion and $\omega_1 = \gamma H_1$ is proportional to the strength of the applied radiofrequency field H_1.

Figure 4 shows the temperature dependence of the proton dipolar and rotating frame spin-lattice relaxation times near the antiferroelectric transition in $Cu(HCOO)_2 \cdot 4H_2O$.

As we are in the slow motion regime, no critical effects are observable. In the antiferroelectric phase, the temperature dependence of T_{1D} is determined by the dependence of the sublattice polarization p on temperature and by the thermally activated nature of τ.

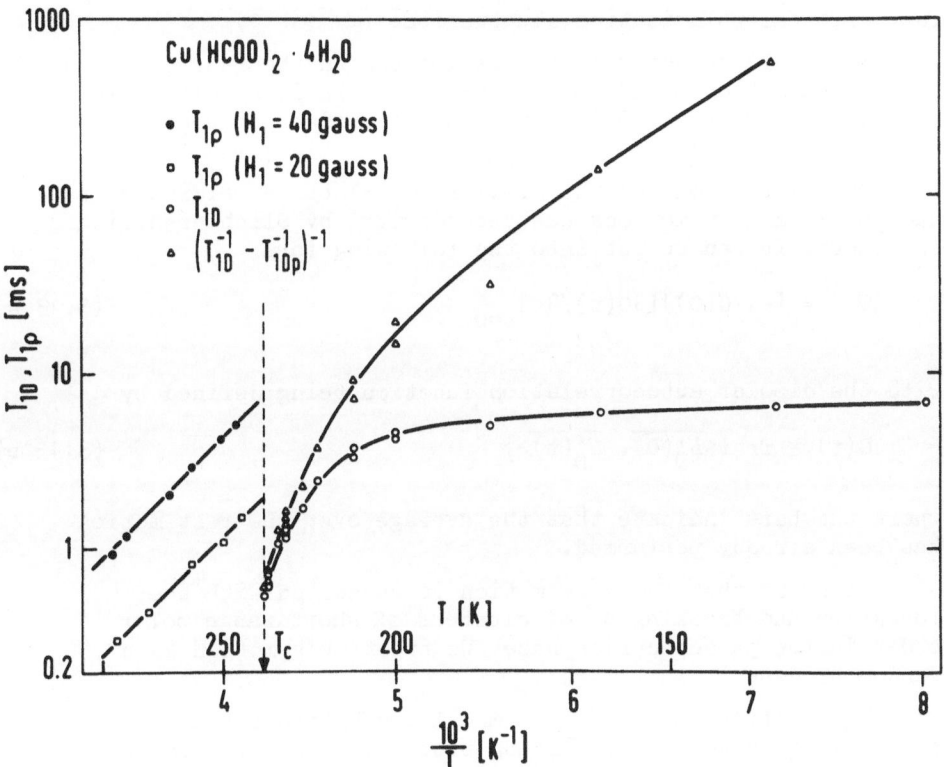

Fig. 4. Temperature dependence of the dipolar and rotating
frame spin-lattice relaxation times in $Cu(HCOO)_2 \cdot 4H_2O$.

In the fast motion regime, on the other hand, critical effects can be seen in the spin-lattice relaxation rate even though T_1^{-1} is dominated by the modulation of the dipolar inter-actions and not by the modulation of the quadrupolar interactions.

One of the first cases where this was demonstrated was the ferroelectric transition in $Ca_2Sr(C_2H_5COO)_6$. The Zeeman proton spin-lattice relaxation time, T_1, exhibits in this crystal an anomalous dip on approaching T_c (Fig. 5), which is due to the modulation of the magnetic dipolar interactions by fast critical polarization fluctuations ($\omega\tau \ll$ 1).

It should be noted that superimposed on the fast deuteron intra-bond motion in KD_2PO_4 and KH_2PO_4 there is an ultra-slow motion [$\tau \gg T_2 \simeq (1/\gamma\Delta H)$] which may determine the proton dipolar spin-lattice relaxation rate T_{1D}^{-1}.

The expression for the dipolar spin-lattice relaxation rate due to ultra-slow motions has been derived by Slichter-Ailion and Zumer. It can be put into the following form

$$T_{1D}^{-1} = [-1/G(0)][\partial G(t)/\partial t]_{t=0} \tag{3.13.a}$$

with the dipolar autocorrelation function being defined by

$$G(t) = \text{Tr } \{<\overline{H_D'(0)} . \overline{H_D'(t)}>\} \tag{3.13.b}$$

where the bars indicate that the average over the fast motion has been already performed.

It seems that the slow motion is connected with the formation and breaking up of clusters of short range polar order in the paraelectric phase. We treat this motion as a random process.

The lattice part of H_D' is modulated by the formation of polar clusters and the resulting proton motion. $G(t)$ can be hence written as

$$G(t) = \text{Tr } \{C_{ij}^2\} [A_o + A_2\bar{g}(t)] \tag{3.13.c}$$

where C_{ij} represents the spin part of the dipolar interactions; A_o and A_2 are structure factors and

$$\bar{g}(t) = <\overline{p_i}(0) \ \overline{p_i}(t)> = f \ e^{-t/\tau_c} \tag{3.13.d}$$

Finally one obtains

$$T_{1D}^{-1} = (A_2/A_o)(f/\tau_c) \tag{3.13.e}$$

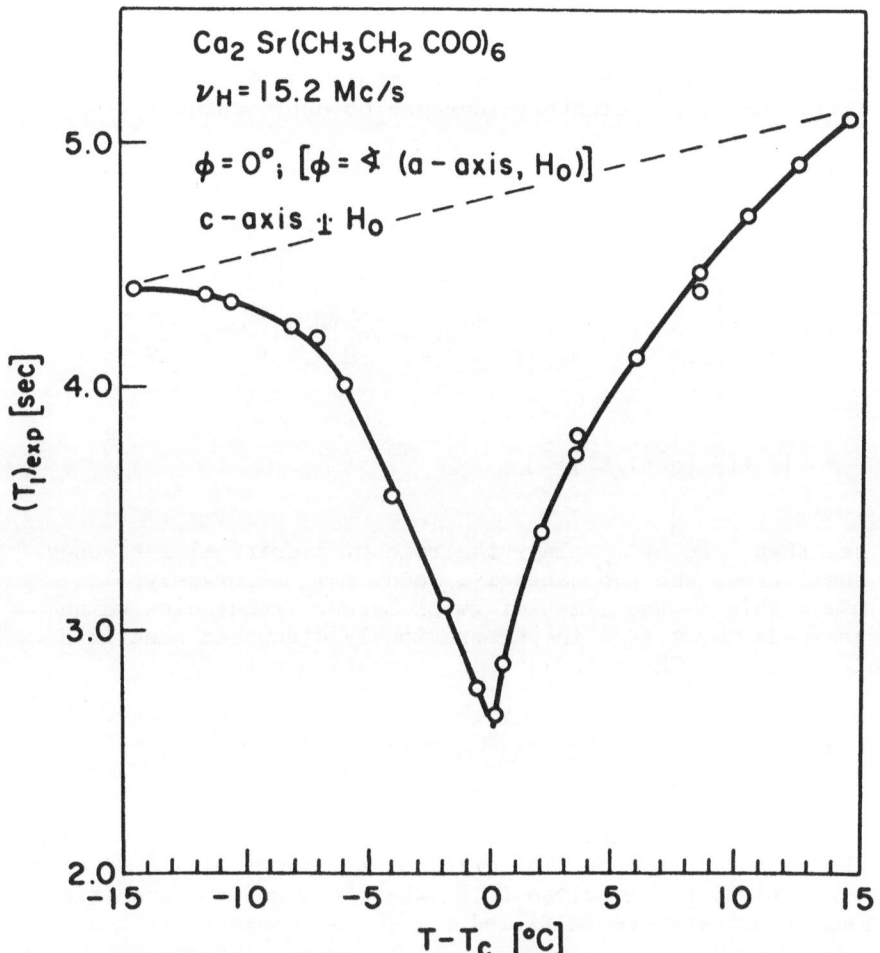

Fig. 5. Temperature dependence of the Zeeman proton spin-lattice
relaxation time in $Ca_2Sr(CH_3CH_2COO)_6$.

where f is the fraction of the protons in the polar short range order clusters and τ_c is the lifetime of these clusters.

When the external field is in the z-direction, we find (Zumer, 1974)

$$\tau_c/f = 0.3 \ 10^{-2} \ T_{1D}$$

(τ_c/f) does not seem to be critical as both the size and the lifetime of the clusters increase on approaching T_c.

Liquid Crystals

Using expressions (1.10) and (1.13) we find for the spectral density of nematic order fluctuations with wave vector q

$$J_q(\omega) = \int_{-\infty}^{+\infty} <\eta_q(0)\eta_q(t)>e^{i\omega t}dt = <\overline{|\eta_q|^2}> 2\tau_q/[1+(\omega\tau_q)^2]$$

$$\propto \quad [\omega^2+\{\frac{C}{L} (q^2+1/\xi^2)\}^2]^{-1}$$

We see that $C/L\xi^2 = \omega_o$ plays the role of a critical frequency. We shall treat the two cases $\omega<\omega_o$ and $\omega>\omega_o$ separately. (i)$\omega<\omega_o$: This is the limit of small Larmor frequencies which corresponds to $\omega\tau << 1$ in the previously discussed case of H-bonded systems. The spin-lattice relaxation rate is

$$1/T_1 \quad \propto \int_0^{q_{max}} J_q(\omega).4\pi q^2 dq \propto \xi \propto (T-T_o)^{-1/2} \quad \neq f(\omega)$$

The spin-lattice relaxation rate is critical and does not depend on the Larmor frequency. The critical behaviour of T_1^{-1} of course does not persist very close to T_o where $\xi\rightarrow\infty$ and the condition $\omega < \omega_o$ is not anymore fulfilled.

(ii)$\omega > \omega_o$: This is the limit of large Larmor frequencies which corresponds to the case $\omega\tau> 1$ in case of H-bonded systems.

Here

$$1/T_1 \propto \int_0^{q_{max}} J_q(\omega)4\pi q^2 dq \propto 1/\sqrt{\omega} \neq f(T-T_o)$$

In this limit T_1^{-1} is not critical but shows a strong ω-dependence.

Both cases have already been observed experimentally.

NMR STUDIES OF MOLECULAR SOLIDS, POLYMERS AND GLASSES

W. Müller-Warmuth

Westfälische Wilhelms-Universität

Münster, Deutschland

INTRODUCTION

The present lecture is based upon the introductory papers: "NMR parameters for studying structure and motion", and "NMR studies of molecular motion in solids". Measurement of dipolar line shapes and moments, of quadrupole interactions, and of relaxation times (Zeeman relaxation, dipolar relaxation, spin-lattice relaxation in the rotating frame of reference) will be examined in order to determine the types of structural and dynamic information which may be obtained for molecular solids, polymers and glasses. The emphasis here will be on materials rather than on phenomena and theoretical background. However, it is not the purpose of this lecture to present a review of all previous NMR work in this field; rather, typical applications yielding characteristic information will be discussed. Application of high resolution methods to molecular solids where interest is primarily focused on the chemical shift are excluded in this lecture since they will be treated elsewhere.

The materials to be considered are all non-metallic solids. For these systems, the NMR technique only provides information about the short range environment of the nuclei in question since all functions describing the spin-spin interactions decrease rapidly with distance. However, it should be kept in mind that NMR techniques measure collective properties which depend on contributions from all the nulcei in a sample.

MOLECULAR SOLIDS

Dipolar Line Shapes

In the majority of studies, proton resonance has been
observed because of the ease with which protons are detected and
because many solids and polymers contain protons. An early NMR
study of 1,2-dichloroethane and acetonitrile revealed typical
two-spin and three-spin line shapes, as shown in figure 1.
The curves correspond to theory for a crystal powder with proton
separations of 1.70 Å, and 1.10 Å, respectively. Only intra-
molecular broadening was considered, and it was assumed that
the dipole-dipole interaction of the CH_2Cl or CH_3 protons
dominated all others. Simple structures, such as those in
figure 1 are only observed in solids when the nuclear spins
form small groups spaced relatively far apart.

For a single crystal, the separation between the two or
three resonance lines is a function of the angle θ between the
vector which joins the protons within a group and the magnetic
field. In this case, the orientation of the vector may be
obtained, in addition to the distance between spins. These
parameters are important, particularly since the best X-ray
data can yield only approximate positions of protons.

For most molecular crystals where the nuclei are not localized
in small groups, dipolar broadening leads to an unresolved
symmetric line structure. In this case, one is generally limited
to the evaluation of second (and possibly higher) moments of the
spectrum. In applying the two-spin and moment formulae derived
for a rigid lattice, care must be taken to account for molecular

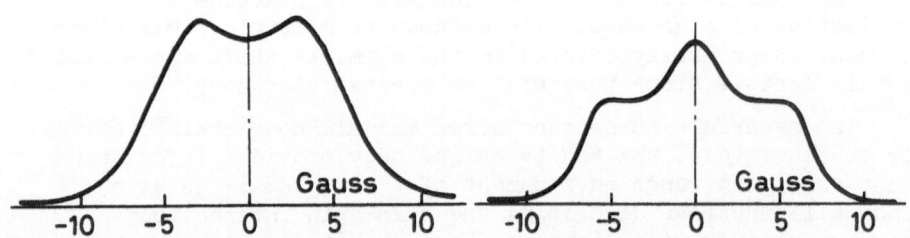

Fig. 1. Resonance line shapes for protons in polycrystalline
        CH_2Cl-CH_2Cl (left) and CH_3CN (right) at 90 K [1].

motions which modulate the spin-spin interactions and which may narrow the spectral lines. At low temperatures, motional effects can often be neglected, but even here quantum mechanical tunnelling motion may occur, particularly with protons.

Benzene and Cyclohexane

Polycrystalline benzene [2] and cyclohexane [3] are typical examples of molecular crystals that have been studied extensively following the early work of Andrew and Eades. Figure 2 shows the variation of temperature of the second moment and of the spin-lattice relaxation times for benzene. The rigid lattice value of M_2 (9.7 Gauss^2) was observed at temperatures below 90 K. Intra- and intermolecular contributions to M_2 were separated by the method of isotopic substitution [2]. Substitution of the 1, 3, 5 protons in C_6H_6 by deuterons affect these contributions in a different manner, and a separation was thus possible. The second moment formula can be applied to the rigid lattice value $(M_2)_{R.L.}$ obtained at a sufficiently low temperature. The change in both M_2 and $1/T_1$ with increasing temperature was attributed to the random rotational motion of benzene molecules about their hexad axes with a frequency which increases with temperature. Below 90 K, the rate of molecular transitions, characterized by the correlation time τ_c, is very low. In this region,

$$1/\tau_c \ll \sqrt{(M_2)}_{R.L.} \tag{1}$$

and the spectrum is not affected. A rapid increase of M_2 shown in figure 2 occurs when $1/\tau_c \simeq \sqrt{(M_2)}_{R.L}$, while for sufficiently rapid rotation,

$$1/\tau_c \gg \sqrt{(M_2)}_{R.L.} \tag{2}$$

Above 120 K, M_2 is reduced to 1.6 Gauss^2, which is approximately a factor of 6 below the rigid lattice value. For this case, it was shown that the intramolecular contribution to M_2 was reduced by a factor 4.

The $1/T_1$ data can be described by the following equation:

$$1/T_1 = C[\tau_c/(1 + \omega_o^2\tau_c^2) + 4\tau_c/(1 + 4\omega_o^2\tau_c^2)] \tag{3}$$

where τ_c is of the form

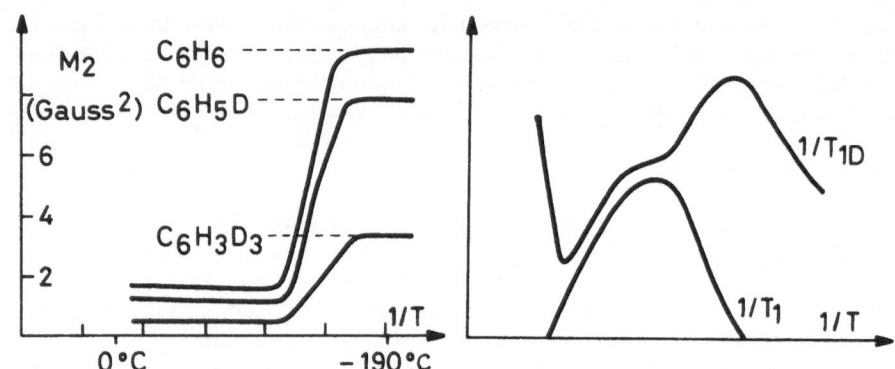

Fig. 2. The second moment (left), and the Zeeman and the dipolar
spin-lattice relaxation times for polycrystalline benzene
as a function of reciprocal temperature [2, 4].

$$\tau_c = \tau_o e^{E_A/RT} \qquad\qquad (4)$$

E_A is the activation energy which corresponds approximately
to the height of the potential barrier between two equivalent
molecular positions; for benzene, E_A was found to be 3.7 kcal/mole.
For maximum Zeeman relaxation, $\omega_o \tau_c = 0.6158$, whereas the maximum
dipolar relaxation rate occurs when $\omega_{loc} \cdot \tau_c \approx 1$ ($\omega_{loc} \approx \sqrt{(M_2)}_{R.L.}$).
The maximum dipolar relaxation rate [4], corresponding to the
thermally activated rotation of the molecules about their hexad
axis, occurs at a lower temperature than the Zeeman maximum.
However, above 230 K it increases again, and this was explained
by the onset of a new type of molecular motion with an activation
energy of 16 kcal/mole.

In studying molecular motion, attempts are often made to
obtain τ_c as a function of temperature from either the line
width or the second moment in the transition region. Such attempts
should be undertaken with caution, for they can give misleading
results. In certain cases, however, the temperature dependence
of the second moment rather than the line width of the
properly truncated spectrum can be used to obtain the

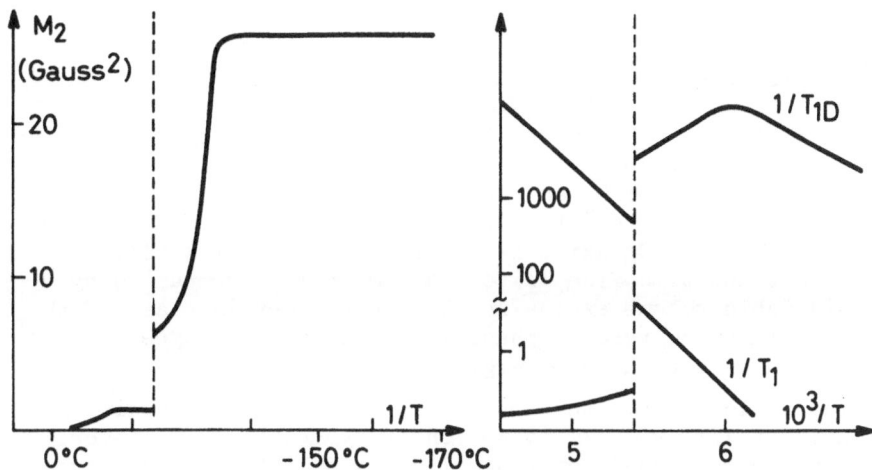

Fig. 3. The second moment and the spin-lattice relaxation times
for polycrystalline cyclohexane as a function of reciprocal
temperature [3, 4].

activiation energy characterizing the motion [5]. For example,
with benzene, agreement between the E_A values obtained from
M_2, $1/T_1$, and $1/T_{1D}$ is quite satisfactory.

Cyclohexane (Fig. 3) shows an analogous behaviour, except
that, at 186 K, a solid phase transition occurs which leads to
discontinuities in the curves. Below the transition temperature
both the reduction in second moment and the change in relaxation
rates are due primarily to reorientation of the molecules about
their triad axis. From the rigid lattice value of $(M_2)_{R.L.}$, the
HCH angle in the equivalent methylene groups was obtained.
The "plastic phase" of cyclohexane is stable above 186 K and
is distinguished by an unusually large amount of molecular motion,
including reorientation about axes other than the triad axes, and
even molecular diffusion. From the straight line behaviour of
$1/T_{1D}$ vs. $1/T$, an activation energy of 9.1 kcal/mole for diffusion
was determined [4]. Just below the melting point, the self-diffusion
coefficient has also been measured directly using pulsed-gradient

spin-echo NMR [6] techniques.

Hexamethylbenzene

 Hexamethylbenzene exists in three different solid phases and shows extraordinary behaviour at very low temperatures (Fig. 4). The sharp decrease of M_2 and the maxima of $1/T_1$ and $1/T_{1D}$ which occur in phase II were attributed to the rotation of the whole molecule about its C_6 axis. E_A was found to be 6.7 kcal/mole. The phase transition from II to I at 383 K is accompanied by a considerable volume expansion. The very steep increase of the dipolar relaxation rate in phase I indicates a new type of motion with an activation energy of 28 kcal/mole.

 In the low temperature phase (III), there is an additional maximum, but $1/T_1$ vs. $1/T$ is no longer symmetrical. Moreover, the second moment does not reach its rigid lattice value, even

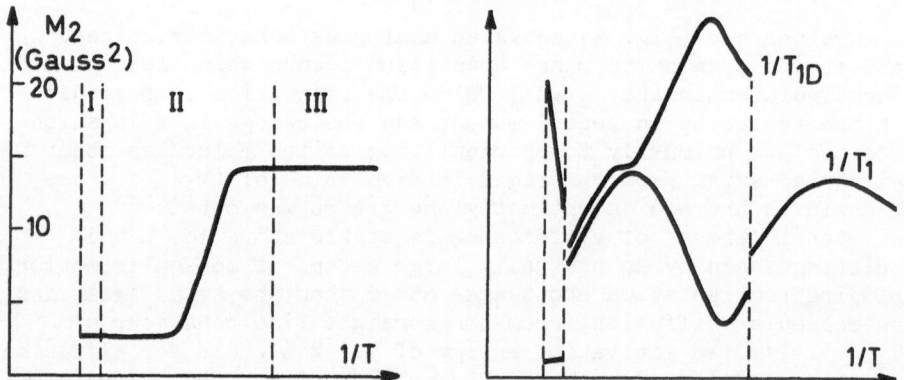

Fig. 4. Variation with reciprocal temperature of the second moment and of the spin-lattice relaxation times for solid hexamethylbenzene.

at extremely low temperatures (2 K). These findings were explained
by the rotational motion of methyl groups about the pseudo-C_3 axes.
At low temperatures, tunnelling effects play an important role and
the methyl group rotation has to be treated quantum mechanically.
Different quantum mechanical treatments have been given by Haupt
[8], by Allen and Clough [9], and by Clough [10].

Toluene [11], Xylenes [12], Methyl-naphtalenes [13]

    The hindered rotation of methyl groups has been studied
in detail at low temperatures using polycrystalline toluene and
some of its derivatives. Motional narrowing of the proton
resonance lines was found to persist to very low temperatures.
The spin-lattice relaxation times, shown as $1/T_1$ vs. $1/T$ in
figure 5, yielded an asymmetric curve with a maximum relaxation
rate smaller than that expected from the strength of the local
field alone. In toluene, two maxima occured rather than one.
The interpretation of these results utilized quantization
of rotational and torsional motion, and the coupling between
spin and rotational degrees of freedom with the phonons of the
lattice. At low temperatures, the motion causing relaxation is
reduced to certain torsional transitions between the ground and
first excited state of the methyl rotator; here tunnelling
transitions are important. The potential barriers hindering the

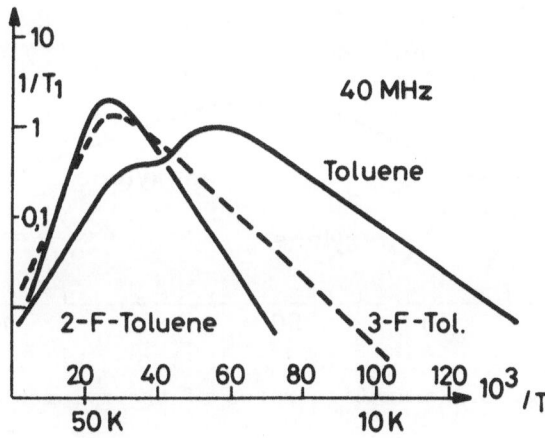

Fig. 5. Variation with reciprocal temperature of the relaxation
        rates of toluene, 2-F-toluene, and 3-F-toluene.

methyl group reorientations in the three compounds were found
to be rather small: toluene 0.1 - 0.2 kcal/mole, 2-fluorotoluene
0.50 kcal/mole, 3-fluorotoluene 0.30 kcal/mole. To a first
approximation, the slope of the high temperature branch of the
curves in figure 5 corresponds to the height of the barrier,
whereas the low temperature branch reflects the energy difference
between the lowest states of the rotator.

Similar results were found earlier for the three xylenes,
and are shown in figure 6. Motionally narrowed second moments
were observed over the whole temperature range. The theoretical
explanation is similar to that for toluene. The potential
function contains sixfold and threefold terms. The absolute
barrier heights were found to be 2.2 kcal/mole (o-xylene),
0.27 kcal/mole (m-xylene), and 1.1 kcal/mole (p-xylene).

Methyl group rotation has also been studied in several methyl
naphtalene crystals which differed in the arrangement and number
of CH_3 groups. Correlation times and activations energies showed
a strong dependence on the methyl group arrangement. Single
crystals of naphtalenes have also been investigated.

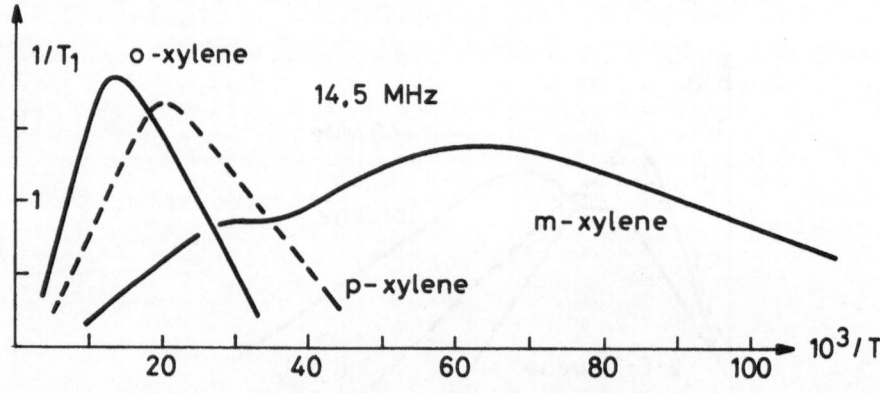

Fig. 6. Relaxation rates of the xylenes.

Further Studies of CH_3-Reorientation

    Hindered rotation of methyl groups in molecular solids has been studied extensively by NMR techniques. In most cases, the potential barrier hindering this motion has been found to be between 2 and 4 kcal/mole. Narrowing of NMR lines and maximum relaxation rates occur in such cases at temperatures not much below 77 K. Only if the barriers are low (which seems often to be the case when CH_3 is attached to unsaturated systems such as benzene) is a quantum mechanical treatment necessary. Otherwise, semiclassical theory, leading to eq. (3), can be applied. In eq. (3), then

$$C_{CH_3} = (9/20)(\gamma^4 \hbar^2/b^6) = 7.8 \cdot 10^9 \ s^{-2} \qquad (5)$$

where $b = 1.79 \cdot 10^{-8}$ cm is the proton-proton distance. The molecular solids in which CH_3 rotation is studied, frequently contain protons in addition to those in the methyl groups. Generally, there exists an unique spin-lattice relaxation time, since spin diffusion will occur within the molecule as a result of the spin-spin interaction between adjacent groups. However, eq. (5) holds only for an isolated CH_3 group and the observed C is related to eq. (5) by [14]

$$C/C_{CH_3} = n_{CH_3}/n_H \qquad (6)$$

where n_{CH_3} is the number of CH_3-protons, and n_H the total number of protons in the molecule. The relationship (6) was verified for solid alkanes [14], ethers [15], ketones [16], methyl-piperidines, and methyl-piperazines [17].

POLYMERS

    Most NMR studies of polymers are conducted with solutions in order to utilize high resolution methods; indeed, in this respect, high resolution NMR has become a major tool for polymer research. However, since polymers occur naturally as solids or as highly viscous liquids, solid state NMR techniques are also of considerable importance, and these will form the basis of the present lecture. The use of NMR for studying solid polymers is similar to that for molecular solids and its greatest merit is the examination of molecular motions. As with small molecules, dipole-dipole interactions are partially averagered and the resonance lines narrow when motion occurs at a sufficient rate and with sufficient amplitude. Maxima occur in the relaxation rates $1/T_1$, $1/T_1$, or $1/T_{1D}$ when the correlation time for a motional process fulfils one of the following conditions: $\omega_o \cdot \tau_c \overset{\sim}{\sim} 1$, $\omega_1 \cdot \tau_c \overset{\sim}{\sim} 1$ or $\omega_{loc} \cdot \tau_c \overset{\sim}{\sim} 1$. Because of this, a large range of motional frequencies can be covered.

Natural Rubber

Data from a representative polymer [18, 19] are shown in figure 7. $1/T_2$ is equivalent to the line width and is a measure

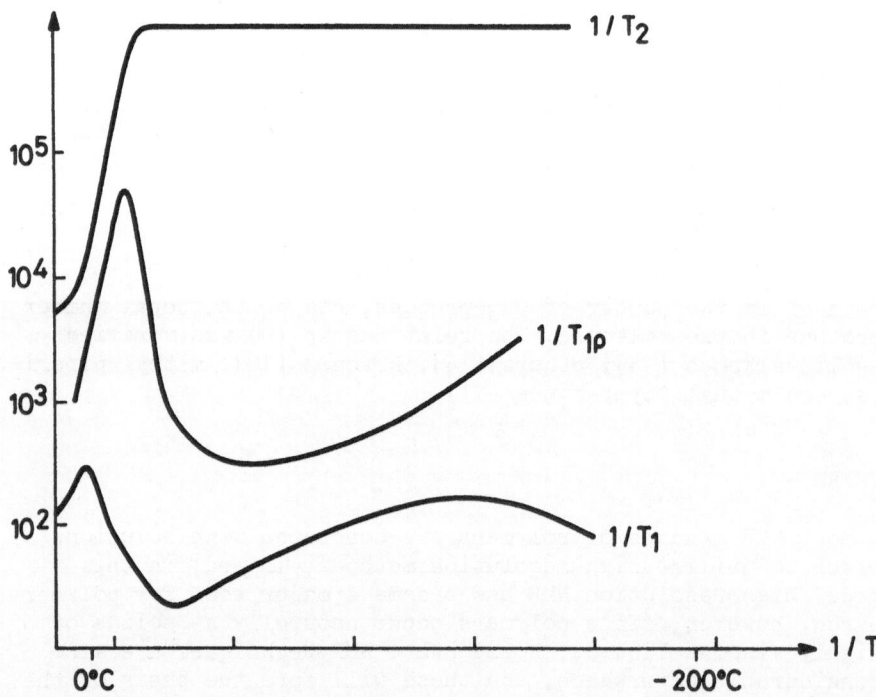

Fig. 7. Spin-lattice relaxation rate, rotating frame relaxation rate, and spin-spin relaxation rate for natural rubber as a function of reciprocal temperature [19].

of the efficiency of dipolar spin-spin interaction. The various relaxation rates and their changes with temperature extend over nearly five orders of magnitude. The low temperature maximum for $1/T_1$ was ascribed to the rotation of substituent methyl groups, while the maximum at higher temperatures was associated with reorientation of segments of the main polymer chain. There are corresponding maxima for $1/T_1$ which occur at lower temperatures and which reflect the shift of the spectrum of molecular motions to lower frequencies. The reduction of $1/T_2$ at high temperature occurs because the motions are fast compared with the rigid lattice line width on a frequency scale. The transition lies in the same temperature range as the major $1/T_1$ maximum. Quantitatively, the observed behaviour cannot be described by single correlation times for the respective motional processes. However, a jump model with a "Cole-Davidson" distribution of correlation times was successfully applied to the interpretation of experimental results for natural rubber and other solid polymers [20].

Further Polymer Studies

Solid state NMR techniques applied to polymers, typically yield qualitative rather than quantitative results. Relaxation rate vs. T^{-1} curves are all similar in principle and additional information is obtained from a comparison of different systems, from relations to other measurements, and from the relationship of the NMR data to other factors which are known to restrict the segmental motions of polymer chains. As a further example of the application of NMR techniques to polymers, the relaxation rates of three elastomers are shown [22] in figure 8. The high temperature maximum in all three cases was associated with reorientation of the main chain segments. The low temperature maxima only occur if the material contains methyl substituents and arise from hindered rotation.

Even when a complete quantitative description of the data is not possible, estimates of correlation times may be obtained from the T_1 measurements. The τ_c value for the high temperature maximum of $1/T_1$ is roughly the time required for a segment to move from one site to another. Comparison of τ_c data obtained from dielectric, mechanical and NMR measurements of a number of polymers showed very good agreement [23].

It would be beyond the scope of this lecture to present all possibilities of employing NMR in polymer research. Most previous studies have involved comparison of data of the type shown in figures 7 and 8 with known properties of the polymer in an attempt to answer questions such as: "How will a proper heat treatment or a variation of the degree of crystallinity influence the results?" The difference between various isomers, isotactic and atactic polymers, has also often been investigated,

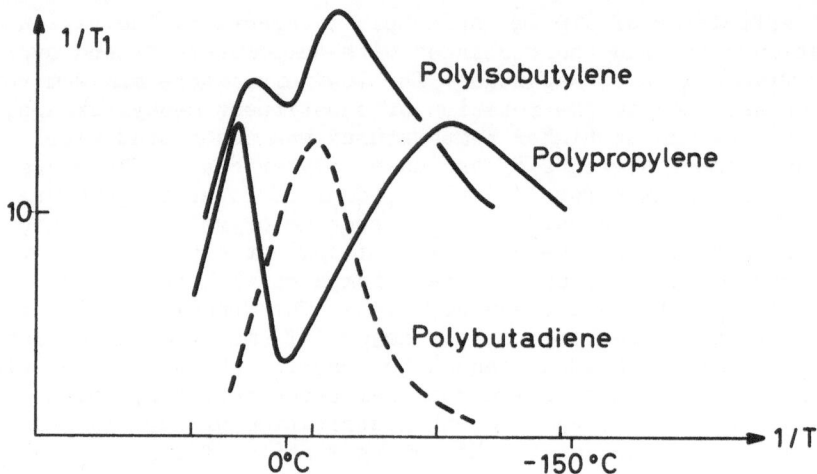

Fig. 8. Spin-lattice relaxation rate at 30 MHz for atactic poly-
       propylene, polyisobutylene, and cis-polybutadiene as a
       function of reciprocal temperature [22].

and it was found that stereoregularity and copolymer composition
restrict the segmental motions of chains. The mobility of chain
segments may also be altered by mechanical stress, by vulcanization
and by other operations upon the sample.

GLASSES

    Nuclear magnetic resonance techniques have been used in
recent years to study atomic or ionic arrangements, short range
structures, and motional processes in glasses. Glass is an
amorphous solid with no long-range order; but, within dimensions
of about 10 Å, it is reminiscent of "crystallites". The glassy
or vitreous state of matter is usually discussed in terms of a
combination of a random network hypothesis, a crystallite model,
and a heterogenic microstructure. Traditional and technical glasses
have been made of inorganic materials, but organic glasses formed
of molecules or polymers also exist. The principal objective of
NMR is to contribute to the general understanding of glass
structure; in this respect, most other techniques have not been

extremely powerful and have led to considerable controversy. However, NMR should not be regarded as a panacea, but rather as one of numerous techniques which may shed light on the various unsolved problems. This is so because, work with glasses suffers frequently from a bad sensitivity and from a smearing out of the spectra and their structure.

The principal difference between a glass and a crystal important for NMR applications is demonstrated by figure 9. Each glass forming ion is surrounded by oxygen, and the latter is shared between two SiO_4 units. In a three-dimensional model, the glass former is in the centre of a polyhedron of oxygen (as far as oxide glasses are concerned), and different polyhedra are connected by their corners only. Cavities may arise in such a network and other ("modifying") ions may enter or even jump from one cavity to another. Nuclei utilized for NMR, to a first approximation, are sensitive to their nearest neighbour environment only, and should give wide line spectra similar to those of crystalline materials. The structure of glass is expected to produce a random orientation of spin connecting vectors and of local principal axes of the electric field gradient, so that the signals are expected to look as those from polycrystalline powders. However, in many cases, the local fields vary from site to site and only a structureless broad distribution curve can be observed. Glasses containing ^1H, ^7Li, ^{11}B, ^{19}F, ^{23}Na, ^{27}Al, ^{29}Si, ^{31}P, ^{51}V, ^{133}Cs, ^{205}Tl, and ^{207}Pb nuclei have been examined by NMR to date, but most work has been done using ^{11}B, ^7Li and ^{27}Al nuclei, all of which have quadrupole moments.

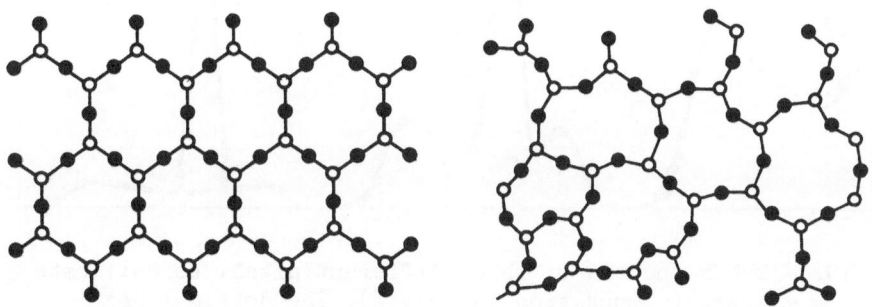

Fig. 9. Two-dimensional quartz crystal (left) and a SiO_2 glass (schematic).

Coordination Number of Boron

The most striking results concern glasses containing boron
oxide. The [11]B spectra look generally like those shown in figure
10 [24]; the detailed appearance of the spectrum depends on
the composition of the glass. The spectra consist of a rather
narrow line superimposed on a large quadrupole broadened
pattern. The broad part of the spectrum is due to planar BO_3
units in the glass, where the low symmetry oxygen arrangement
around boron creates strong electric field gradients at the
nuclear positions. The narrow line arises from symmetrical
tetrahedral BO_4 units which also exist in the glass. To determine
the fraction of four and three coordinated boron atoms, the
spectra were simulated by superimposing two curves, one
characteristic of BO_4 and the other of BO_3. The final result
of this procedure for the ternary system of figure 10 is
presented in figure 11. At low alkali concentration, each added
oxygen (introduced with Na_2O) transforms the surroundings of two
three coordinated boron atoms to four coordinated borons (dotted
line). At larger x values, as in the case of binary alkali
borate glasses [25], the formation of four coordinated boron
no longer proceeds in the same way. The maximum possible fraction

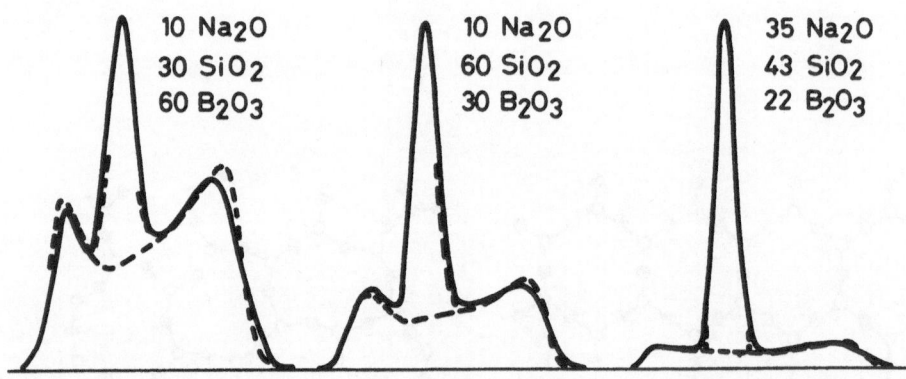

Fig. 10. ^{11}B-NMR spectra of three different alkali borosilicate
glasses (composition in mole %). The dotted lines
represent the theoretical spectra obtained by computer
simulation.

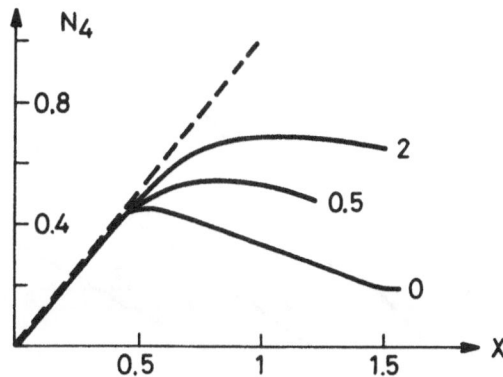

Fig. 11. The relative proportion N_4 of four coordinated boron in
$Na_2O-B_2O_3-SiO_2$ glasses against the mole % ratio of
Na_2O/B_2O_3. Numbers beside the various curves give the
ratio SiO_2/B_2O_3.

of BO_4 groups is, however, greater than in the absence of silica.
The results can be explained by a statistical distribution of
boron and silica polyhedra in the glass and a preferred association
of the alkali oxide with the boron rather than with the silicon
groups.

In $KF-B_2O_3$ glasses, the fraction of four coordinated boron
increases as the percentage of potassium fluoride increases [26].
Somewhat different results have been obtained recently for the
ternary glasses of the "NABAL" system, $Na_2O-B_2O_3-Al_2O_3$ [27].
N_4, obtained from the [11]B spectra using a procedure analogous to
that described above, is a linear function of the ratio R = mole %
Na_2O/mole % B_2O, with a slope that depends on the alumina
contents. This is shown in figure 12. With increasing
concentration of aluminium oxide the rate of conversion of boron
from three to four coordination decreases, while at a given
value of R, addition of Al_2O_3 to the glass reduces the number
of four coordinated boron atoms. This is explained by assuming
that aluminum plays the role of a glass former and displaces boron
from its tetrahedral positions.

Aluminium in Glasses

The ^{27}Al NMR spectra of NABAL glasses, in contrast to
boron, are very broad and structureless; a typical example,
with a half line width of 18 Gauss [27] , is shown in figure 13.

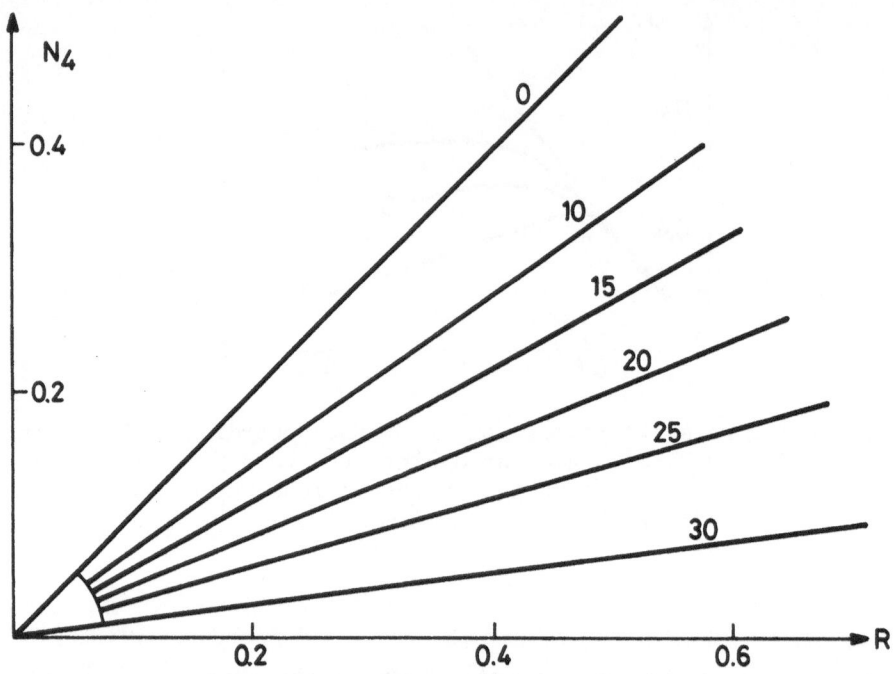

Fig. 12. N_4 values of the NABAL system; numbers beside the curves give the concentration of Al_2O_3 in mole %.

These spectra can be explained by second order quadrupole inter-actions with a random distribution of environments for the aluminium nuclei. To interpret these results aluminium ions are assumed to be centered in distorted tetrahedra of oxygen atoms regardless of the composition of the glass.

In contrast, structural changes are observed in alkali-alumo-silicate glasses. The spectra are similar to those for the NABAL glasses, but show variations in line width [28].

The transitions in figure 14 do not correspond to molecular motions but rather changes in composition. The structure of these glasses are characterized by the K_2O_3/Al_2O_3 ratio.

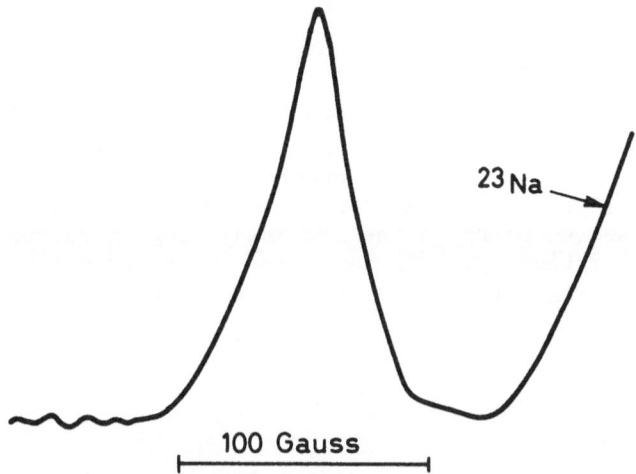

100 Gauss

Fig. 13. ^{27}Al NMR spectrum of a NABAL glass measured at 10 MHz.

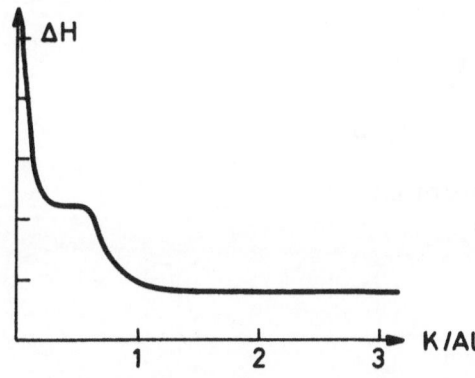

Fig. 14. Line widths of ^{27}Al in K_2O-Al_2O_3-SiO_2 glasses as a
function of composition.

At low alkali concentration, the short range order around the
Al^{3+} ions is substantially reduced.

Alkali Silicate Glasses

NMR studies of binary Na_2O-SiO_2 glasses have revealed the
existence of a microstructure consisting of different regions
or phases in which the alkali contents are much higher or lower
than average [29, 30]. There is some evidence that sodium ions
are distributed as pairs in the low alkali region rather than
evenly or statistically. ^{29}Si data were collected using a
computer of average transients. Adiabatic fast passage,

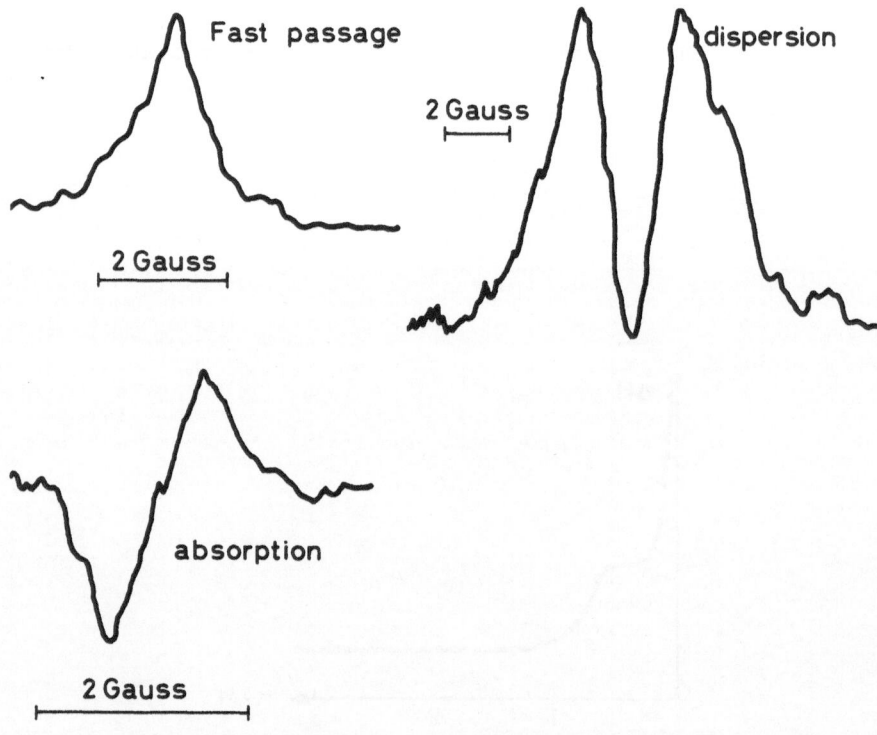

Fig. 15. ^{29}Si NMR signals from glass consisting of 14 mole % Na_2O
and 86 % SiO_2. Measurement times varied from 5 to 20
minutes.

dispersion mode signals at strong H_1 and absorption derivatives were recorded and are shown in figure 15. To shorten the relaxation times, a small percentage of Fe^{3+} ions was added to the glass. ^{23}Na data consisted of absorption spectra which looked very much like the ^{27}Al NMR signals of the NABAL glasses; they were characterized by quadrupole broadening. Thus in this case, information regarding the dipolar interaction between sodium ions must be obtained either from the dispersion signals or indirectly from the ^{29}Si resonance.

^{7}Li has also been examined [30]. Because both the quadrupole moment and the Sternheimer antishielding factor are much smaller than for ^{23}Na, ^{7}Li spectra show only first order quadrupole effects. First order satellites are observable only at low temperatures; at room temperature, motional averaging, due to the diffusion of lithium ions, occurs and the satellites disappear.

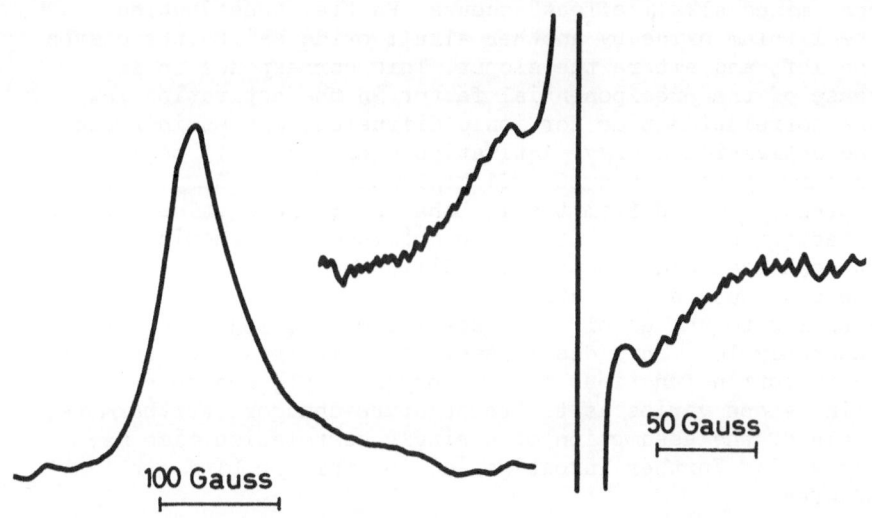

Fig. 16. ^{23}Na (left) and ^{7}Li NMR signals from a silicate glass containing 33 mole % NA_2O or Li_2O, respectively. For the lithium signal, the derivative is shown [30] (peaks of the central resonance omitted).

For sodium, only the dispersion spectra are temperature dependent.
The electric field gradient at the sodium and lithium sites
is approximately the same. Lastly, ^{133}Cs resonances in glasses
have been detected [31]. This nucleus gives broad structureless
spectra which contain contributions from both quadrupole
interactions and chemical shift anisotropy. A study of frequency
dependence in this case led to a partial separation of the two
effects.

Diffusion

By examining the temperature dependence of the NMR parameters, motional effects in glasses can also be studied.
Treatment of these effects is similar to that for molecular
crystals and polymers. There are, however, important experimental
restrictions in measuring relaxation times because of the
broadness of many spectra and because of poor sensitivity. Thus,
most studies have dealt with line narrowing rather than T_1 or
T_{1D}, and this makes certain results somewhat dubious.

As alkali ions move randomly in the glass, local magnetic
and electric fields become fluctuating functions of time and
a reduction of the second moment or increasing relaxation
may result. In figure 17, the effect of ionic diffusion upon
the lithium resonance is shown. As seen from this data, a
strong "mixed alkali effect" occurs. Partial substitution
of the lithium oxide by another alkali oxide shifts the curves
to the left and alters the slopes. This corresponds to an
increase of the preexponential factor in the activation law
of the correlation time for ionic diffusion, and an increase
in the activation energy. Activation energies of 15 to 16 kcal/mole
for binary systems, and 20 - 21 kcal/mole for ternary systems
have been estimated from the T_1 behaviour; these values are in
satisfactory agreement with those obtained from chemical,
electrochemical and conductivity diffusion data.
Attempts to obtain activation energies from line narrowing
effects led to values of 3 - 5 kcal/mole [32, 33] which are
unreasonably low. This discrepancy most likely arises because
the correlation functions are not exponential, and because
the line shape varies as the temperature changes. Furthermore,
in a glass, the assumption of a single correlation time may
not be valid. Further investigations of this problem are
in progress.

In alkali borate glasses, similar effects were observed.
Figure 18 differs from figure 17 in that the spectra from
mixed alkali glasses contain a second maximum superimposed on
the low temperature decay of the first broad maximum. The low
temperature maximum corresponds to a mechanism of motion whose

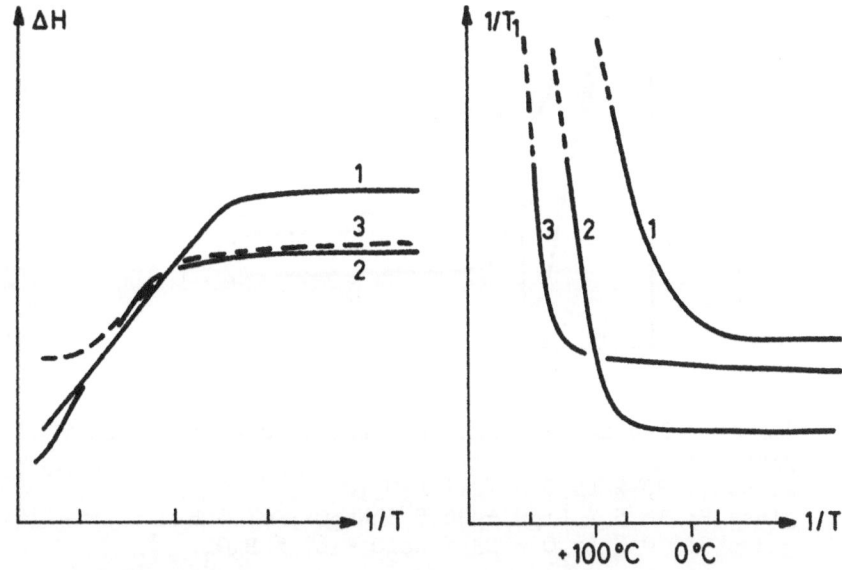

Fig. 17. ^{7}Li half line width (left) and spin-lattice relaxation
rates of three alkali silicate glasses as a function of
reciprocal temperature [30];
glass 1: 33 % Li_2O - 67 % SiO_2;
glass 2: 16.5 % Li_2O - 16.5 % Na_2O - 67 % SiO_2;
glass 3: 16.5 % Li_2O - 16.5 % Cs_2O - 67 % SiO_2.

correlation frequency and activation energy is much higher than
that for cation diffusion. Here, one is tempted to postulate
some kind of defect diffusion as occurs in ionic crystals and
which modulates the quadrupole interaction. Similar peaks have
also been observed in systems with partial crystallization.

Glasses in the Transformation Range

    In a glass, there is no abrupt phase transition between the
liquid and solid state. However, as a glass forming liquid is

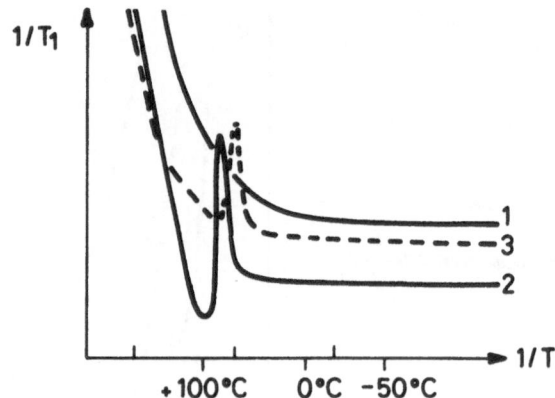

Fig. 18. ^7Li spin-lattice relaxation rates of alkali borate
glasses;
glass 1: 30 % Li_2O - 70 % B_2O_3;
glass 2: 16.5 % Li_2O - 16.5 % Na_2O - 67 % B_2O_3;
glass 3: 10 % Li_2O - 23 % Na_2O - 67 % B_2O_3.

cooled some of its properties such as viscosity, heat capacity
and electrical conductivity change sharply in a narrow temperature
range. The glass transition or transformation temperature is
associated with the slowing down of atomic rearrangements in
the glass structure [34]. When these rearrangements occur rapidly
during an experimental measurement, the glass has the properties
of a liquid; when they are slow, the structure is frozen. It is
interesting to relate NMR line widths and relaxation rates
to the structural behaviour of glasses in the transition range.
As a first example, the behaviour of a molecular solid (toluene)
is presented in figure 19. Toluene solidifies either as a
crystal or as a glass (supercooled liquid). In the glass transition
range, $1/T_2$ or the line width changes steadily and varies over
about four orders of magnitude. At the same time, a maximum in
$1/T_1$ appears. In the rigid lattice, below the transition temperature,
the relaxation rate of the glassy state is larger than that of
the crystalline state, as expected. Upon cooling the vitreous
solid, the liquid phase is only obtained after the system passes
through a crystalline state. At the glass-crystal transition
temperature, motion occurs which narrows the resonance signals.
In figure 20, NMR data for an inorganic glass in the transition
range are shown [35]. The line width decreases as the temperature
increases, and a sharp bend occurs near the glass transition
temperature. The curves can be interpreted semi-quantitatively
using the following relations:

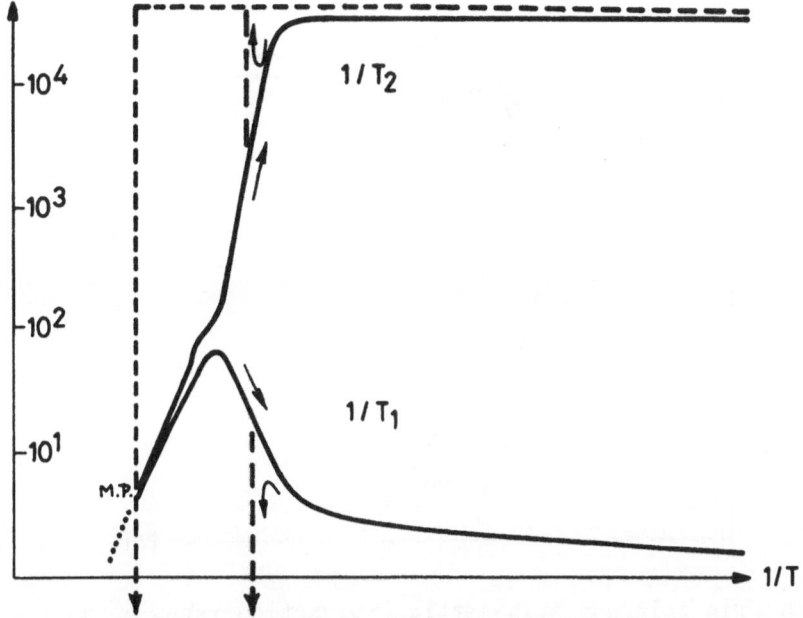

Fig. 19. ^{1}H spin-spin and spin-lattice relaxation rates of
toluene as a function of reciprocal temperature
(schematic); solid lines: vitreous state; dotted lines:
crystalline state.

$$1/T_1 = C_I \ [F(\omega_I) + 4F(2\omega_I)] + C_s \ [\tfrac{1}{3} F(\omega_I - \omega_s)$$

$$+ \ F(\omega_I) + 2F(\omega_I + \omega_s)]$$

$$\text{(7)}$$

$$1/T_2 = C_I \ [\tfrac{3}{2} F(0) + \tfrac{5}{2} F(\omega_I) + F(2\omega_I)]$$

$$+ \ C_s \ [\tfrac{2}{3}F(0) + \tfrac{1}{6} F(\omega_I - \omega_s) + \tfrac{1}{2} F(\omega_I) + F(\omega_s) + F(\omega_I + \omega_s)]$$

In eq. (7) , C_I and C_s, defined in a similar way as C in eq. (4),

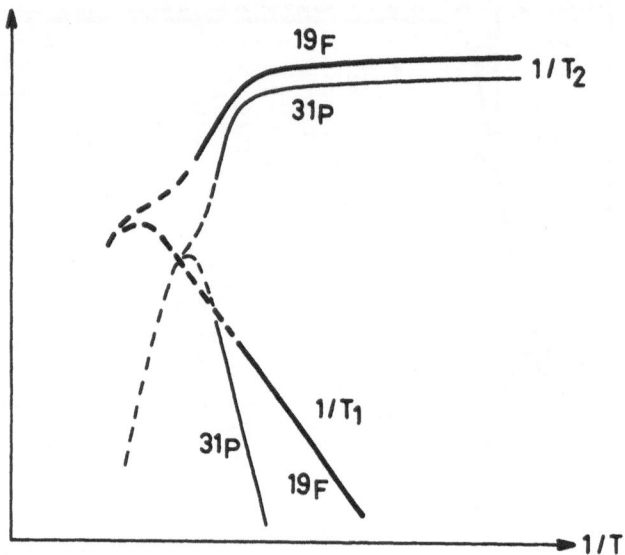

Fig. 20. Spin-spin and spin-lattice relaxation rates as a
         function of reciprocal temperature for the glass 27 %
         AlF_3, 23 % BaF_2, 50 % $NaPO_3$.

provide a measure of the local field at the nuclei I produced
either by other nuclei of the same type, or by unlike nuclei S.
The various spectral density functions $F(\omega)$ depend on the
resonance frequencies of either the nuclei I or S, or on
$\omega_I \pm \omega_s$. For the sake of simplicity, exponential correlation
functions were assumed leading to

$$F(\omega) = \tau_c / (1 + \omega^2 \tau_c^2)$$

The motion of the smaller ionic components containing fluorine
is characterized by an activation energy of 12 kcal/mole. From
the phosphorus curve, 36 kcal/mole is derived, which demonstrates
that in the same glass considerably more energy is required to
cause the metaphosphate chains to move than the fluorine containing
groups. This result compares favorably with other activation energy
measurements for $NaPO_3$ glasses, and for the system AlF_3-Na_2SO_4.
In the transformation range, correlation times are of the order
of 10^{-4} to 10^{-6}s.

REFERENCES

1. H.S. GUTOWSKY, G.B. KISTIAKOWSKY, G.E. PAKE and E.M. PURCELL, J.Chem.Phys. 17, 972 (1949)
   H.S. GUTOWSKY and G.E. PAKE, J.Chem.Phys. 18, 162 (1950)

2. E.R. ANDREW and R.G. EADES, Proc.Roy.Soc. A218, 537 (1953)

3. E.R. ANDREW and R.G. EADES, Proc.Roy.Soc. A216, 398 (1953)

4. R. VAN STEENWINKEL, Z. Naturforsch. 24a, 1526, (1969)

5. E.R. ANDREW and J. LIPOFSKY, J.Magn.Res. 8, 217 (1972)

6. J.E. TANNER, J.Chem.Phys. 56, 3850 (1972)

7. P.S. ALLEN and A. COWKING, J.Chem.Phys. 47, 4286 (1967)

8. J. HAUPT, Z. Naturforsch. 26a, 1578 (1971)

9. P.S. ALLEN, J.Chem.Phys. 48, 3031 (1968)
   P.S. ALLEN and S. CLOUGH, Phys.Rev.Let. 22, 1351 (1969)

10. S. CLOUGH, J.Phys. C4, 1075 and 2180 (1971)

11. J. HAUPT and W. MÜLLER-WARMUTH, Z.Naturforsch. 24a, 1066 (1969)

12. J. HAUPT and W. MÜLLER-WARMUTH, Z.Naturforsch. 23a, 208 (1968)

13. J.U. VON SCHÜTZ and H.C. WOLF, Z. Naturforsch. 27a, 42 (1972)
    J.U. VON SCHÜTZ and F. NOACK, Z.Naturforsch. 27a, 645 (1972)

14. J.E. ANDERSON and W.P. SLICHTER, J.Phys.Chem. 69, 3099 (1965)

15. K. GRUDE, J. HAUPT and W.MÜLLER-WARMUTH, Z.Naturforsch. 21a, 1231 (1966)

16. K.P.A.M. VAN PUTTE, Trans.Faraday Soc. 65, 1709 (1969)

17. R. SCHÜLER, Diplomarbeit, Universität Münster (1974)

18. W.P. SLICHTER, "NMR Basic Principles and Progress", 209, Springer-Verlag, Berlin (1971)

19. D.W. Mc CALL and D.R. FALCONE, Trans.Faraday Soc. 66, 262 (1970)

20. M. STOHRER, F. NOACK and J. V. SCHÜTZ, Kolloid-Z. and Z. Polymere 241, 937 (1970)

21. U. KIENZLE, F. NOACK and J. V. SCHÜTZ, Kolloid-Z. and Z. Polymere 236, 129 (1970)

22. W.P. SLICHTER, J.Polymer Sci. C14, 33 (1966)

23. D.W. Mc CALL, "Molecular Dynamics and Structure of Solids", 475, NBS Publication (1969)

24. J. SCHEERER, W. MÜLLER-WARMUTH and H. DUTZ, Glastechn. Ber. 46, 109 (1973)

25. S.E. SVANSON, E. FORSLIND and J. KROGH-MOE, J.Phys.Chem. 66, 174 (1962)

P.J. BRAY and J.G. O'KEEFE, Phys.Chem.Glasses 4, 37 (1963)

26. W. MÜLLER-WARMUTH, W. POCH and G. SIELAFF, Glastechn.Ber. 43, 5 (1970)

27. R. GRESCH, Diplomarbeit, Universität Münster (1974)

28. J.W. SCHULZ, W. MÜLLER-WARMUTH, W. POCH and J. SCHEERER, Glas-techn.Ber. 41, 435 (1968)

29. B.D. MOSEL, W. MÜLLER-WARMUTH and H. DUTZ, Phys.Chem.Glasses, in press (1974)

30. F. KRÄMER, W. MÜLLER-WARMUTH, J. SCHEERER and H. DUTZ, Z.Natur-forsch. 28a, 1338 (1973)

31. C. RHEE and P.J. BRAY, Phys.Chem.Glasses 12, 165 (1971)

32. S.G. BISHOP and P.J. BRAY, J.Chem.Phys. 48, 1709 (1968)

33. S.E. SVANSON and R. JOHANSSON, Act.Chem.Scand. 24, 755 (1970)

34. R.H. DOREMUS, Glass Science, John Wiley, New York (1973)

35. F. KRÄMER, W.MÜLLER-WARMUTH and H. DUTZ, Glastechn.Ber. 46, 191 (1973)

NMR IN METALS AND ALLOYS

H. Alloul

Université Paris Sud

91405-Centre d'Orsay, France

INTRODUCTION

Among the great number of NMR studies which have been performed in the condensed matter, a quite important part has been devoted to investigate the properties of metals and alloys, since the discovery of the Knight shift, which showed that NMR experiments were quite sensitive to the electronic structure of metals. The increasing number of experimental results did quickly raise difficult problems of interpretation as the NMR data required very refined details of the electronic structure, even in simple elemental alkali metals. It became quite evident that results from other experimental techniques such as susceptibility, specific heat, Mossbäuer effect, etc..., as well as developments of some specific theoretical aspects were required in order to achieve some understanding of the NMR properties of more complicated systems such as those involving transition metals. The NMR technique revealed to be even a more powerful tool for investigating compounds or alloys as it permits to study the particular electronic properties of a given constituent, and even allows to distinguish atoms which have a given environment. Finally it should be emphasized that NMR is not only sensitive to static electronic properties through the positions, splittings, widths of the NMR spectra, but also to the dynamic properties of the electronic system through the nuclear spin T_1.

Since the pioneer works of Knight [1], Bloembergen and Rowland [2], the developments of technical aspects, through the use of careful metallurgy (high purity samples, single crystals), advanced electronics (pulse techniques, signal averaging), and cryogenics (high fields, up to 120 kgauss, temperatures down to a few mK) allowed a multiplication of experiment data during this last decade. It is certainly out of the scope of the present lectures to provide a complete description of the work which has been performed. It shall rather be assumed that the reader is somewhat acquainted with NMR techniques, which will allow then to focus on the specific properties of NMR in metallic systems. A choice of results in cases where direct comparisons with theory, band structure or other experimental data would be performed will illustrate the main physical ideas, the present level of understanding of NMR experiments in metals and alloys, and point out the type of information which can be obtained.

The main theoretical results for the Knight shift and spin lattice relaxation time in metals will be recalled first (sec. II), the derivation of which might be found in great details in various monographs which deal with the subject [3, 4, 5]. The experimental results for these quantities in some classes of pure elemental metals will be discussed briefly in sec. III. More details and a selected bibliography are available from several review articles [5 - 9]. The nature of the spin-spin interactions and their influence on the line-broadening in metals will be reviewed in sec. IV. The specific field of dilute alloys, where the impurities perturb locally the properties of the host metal, in which important developments have been achieved these last few years, will be considered then in sec. V separately from the studies of more concentrated alloys (sec. VI) in which more pronounced modifications of the electronic structure, both of local and non local origin, may occur with composition.

THEORY OF THE KNIGHT SHIFT AND SPIN LATTICE RELAXATION IN METALS

1. Hyperfine Interactions

The magnetic interactions between the electron and nuclear magnetic moments are at the origin of the most important specific properties of   NMR in metals and alloys. Three hyperfine interactions between a given electron and a nuclear spin $\underset{\sim}{I}$ can be distinguished.

  - The contact interaction

$$H_c = \frac{8\pi}{3} \gamma_e \gamma_n \hbar^2 \underset{\sim}{I} \underset{\sim}{S} \, \delta(\underset{\sim}{r}) \tag{1}$$

where γ_n and γ_e are the nuclear and electronic gyromagnetic ratios, $\underset{\sim}{S}$ is the electron spin, whose position is denoted by the vector $\underset{\sim}{r}$ with the origin taken at the nuclear site.

- The dipolar interaction

$$H_D = -(\gamma_e \gamma_n \hbar^2/r^3) \ \{\underset{\sim}{I}\cdot\underset{\sim}{s} - [3(\underset{\sim}{I}\cdot\underset{\sim}{r})(\underset{\sim}{s}\cdot\underset{\sim}{r})]/r^2\} \tag{2}$$

- The interaction with the electronic orbital moment ℓ

$$H_{orb} = -\gamma_e \gamma_n \hbar^2 \ (\underset{\sim}{I}\cdot\underset{\sim}{\ell}/r^3) \tag{3}$$

The total hyperfine interaction for a given electron can then be written:

$$H_{hyf} = - \ \gamma_n \hbar \ \underset{\sim}{I}\cdot\underset{\sim}{H}_{eff} \tag{4}$$

from which it can be considered that each electron induces an effective field $\underset{\sim}{H}_{eff}$ at the nuclear site. As the fluctuations of the electronic moments are usually very rapid as compared with the nuclear Larmor period, the position of the NMR line in a given material is then determined by the time average of the total effective field due to all electrons in the material. It can be easily seen that the hyperfine interaction vanishes for filled electron shells as they have zero total orbital moment and spin. Without any applied external field, $\underset{\sim}{H}_{eff}$ might be non-zero only for materials in which the electronic system possess net spin or electronic orbital moments, which is only the case for magnetic (ferro- or antiferromagnetic) substances In para- or diamagnetic substances the static part of the hyperfine coupling will be non zero only when an external magnetic field H_0 is applied, and is usually proportional to H_0, yielding then a shift of the NMR with respect to the free atom resonance. Such a shift might be due to the orbital part of the coupling (chemical shift), but the spin hyperfine interaction might be important when unpaired electron spins do exist, as in metals, where H_0 unpairs the conduction electrons at the Fermi level, yielding the electronic paramagnetic Pauli susceptibility. In most metals the Fermi contact interaction with the conduction electrons will be the most important and deserves special attention.

2. Knight Shift and Spin Lattice Relaxation due to the Contact
   Interactions

a) Simple case of a s electron band

    The contact interaction is non zero only for electrons of s character as they have a non zero wave function at the nuclear site. The polarized s electrons (Fig. 1) then yield the well

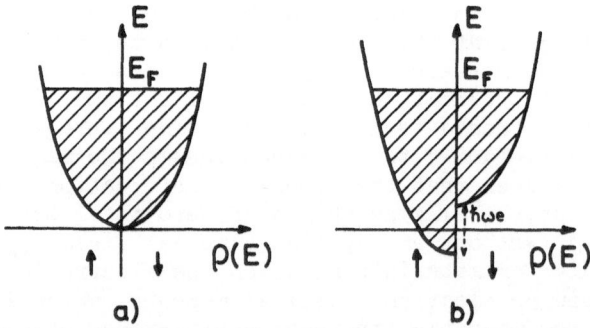

Fig. 1. Density of states $\rho(E)$ of a free electron band without (a)
or with (b) an external applied field H_0 such as $g\mu_B H_0 = \hbar\omega_e$
The excess of ↑ spins in figure (b) corresponds to the
Pauli paramagnetic susceptibility   of the electron gas.

known positive Knight shift of the resonance frequency in metals, which is given by:

$$K_{(s)} = (8\pi/3) < |\Psi_{\underset{\sim}{k}}(0)|^2 >_F \chi_s \qquad (5)$$

where χ_s is the susceptibility per atom of the s electrons and $< |\Psi_{\underset{\sim}{k}}(0)|^2 >_F$ the probability density of electrons of wave vector $\underset{\sim}{k}$ averaged over all electronic states at the Fermi level. Equation 5 is also often written:

$$K_{(s)} = \mu_B^{-1} H_{hf}^{(s)} \chi_s \qquad (6)$$

where $H_{hf}^{(s)}$ is defined as the s hyperfine field per Bohr magneton μ_B. The contact interaction is also at the origin of a spin lattice relaxation mechanism for the nuclear spins through the coupling with the fluctuating part of $H_{\underset{\sim}{eff}}$.

This can be seen as this exchange interaction allows a change of the nuclear spin state as the result of a scattering of a conduction electron from a state $\underset{\sim}{k}\uparrow$ to a state $\underset{\sim}{k}'\downarrow$, through the I^+s^- term of equation 1. The conservation of energy in such a scattering process requires that the change of kinetic energy of the electron balances the difference between the electron and nuclear spin Zeeman energies. As a result of the Fermi Dirac statistics, only electrons at the Fermi level can be scattered to a nearby unoccupied state (Fig. 2). The resulting relaxation rate, for a free non interacting electron band is given by:

$$(1/T_1)_{(s)} = \pi\hbar^3 k_B T [\gamma_e \gamma_n H_{hf}^s \mu_B^{-1} \rho(E_F)]^2 \qquad (7)$$

where $\rho(E_F)$ is the density of states at the Fermi surface per atom and for one spin direction. The factor $k_B T$ is associated with the fact that only electrons within $k_B T$ of the Fermi level contribute to the relaxation. An important difference between equations 5 and 7 appears, as the Knight shift involves the spin susceptibility of the electron band, while the relaxation rate involves directly the density of states at the Fermi surface. For a free electron band of non interacting s electrons:

$$\chi_s = \frac{1}{2}(\gamma_e \hbar)^2 \rho(E_F) \qquad (8)$$

and in this particular case a simple relation between K and T_1 is obtained by combining equations 6, 7 and 8:

$$K_{(s)}^2 (T_1 T)_{(s)} = (\hbar/4\pi k_B)(\gamma_e/\gamma_n)^2 = \zeta \qquad (9)$$

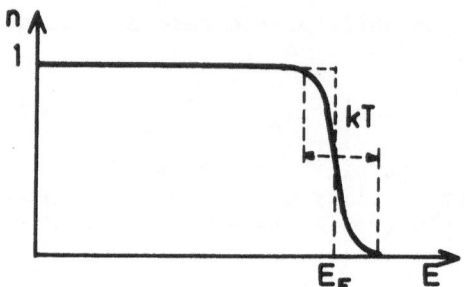

Fig. 2. Thermal population n of the electronic states, for one
        spin direction, as a function of energy. The only
        electrons which can be excited from a state of energy
        E_0 to a nearby state of energy $E_0 + \hbar\omega$, are within kT
        of the Fermi level (for $\hbar\omega \ll k_B T$).

where ζ is called the Korringa constant [10].

b) Non s electron bands: core polarization effects

When the conduction electrons have a non s character,
especially for transition elements for which the d band density
of states may dominate at the Fermi level, they are not expected
to be coupled with the nuclear spins through the contact interaction. In fact these electrons are coupled through an electro-
static exchange interaction with the inner s electron shells
of the atoms. These core electrons are then polarized when the
non s conduction electrons are polarized yielding an indirect
contact interaction for p, d .. f electrons, which behaves
like if the p, d, f electrons were coupled with the nucleus
through genuine contact interactions. This core polarization
effect yields a contribution to the Knight shift, which can be
written, for instance for d electrons:

$$K_d = \mu_B^{-1} H_{hf}^{(d)} \chi_d \tag{10}$$

where $H_{hf}^{(d)}$ is the corresponding hyperfine field and χ_d the susceptibility of the d electron band. The corresponding relaxation rate follows a slightly modified Korringa relation:

$$K_d^2 (T_1 T)_{(d)} = p\zeta \qquad (11)$$

where p is a coefficient which takes into account the orbital degeneracy of the d electron band ($2 < p < 5$ for cubic symmetry). This comes in as $(T_1)^{-1}$ involves matrix elements between electrons in different states, which vanish for electron states with different orbital wave functions and results in a reduction of the relaxation rate.

3. Effect of the Other Hyperfine Interactions

Though in most cases the s contact interaction has been found to be dominant, the dipolar and orbital hyperfine couplings may also contribute to the Knight shift and relaxation of the nuclear spins.

It can be shown easily that the Knight shift due to the dipolar coupling vanishes for electrons with an s character, owing to their spherical symmetry. It also vanishes for nuclear spins in position of cubic symmetry. It will then yield an anisotropy of the Knight shift in non cubic metals linked with the amplitude of the non s character of the wave functions at the Fermi level.

The orbital coupling with the conduction electrons is also expected to yield an orbital susceptibility $\chi_{(orb)}$ and a corresponding Knight shift:

$$K_{(orb)} = \mu_B^{-1} H_{hf}^{(orb)} \chi_{(orb)} \qquad (12)$$

Such an orbital susceptibility arises as H_0 admixes the wave functions of the occupied states, for which the orbital moment is quenched, with higher energy states, yielding an average total orbital moment. It will only be important when this admixture can take place between states which are not far apart, which is the case for transition metals. These interactions also yield orbital and dipolar relaxation rates which in any case can be written:

$$(1/T_1)_{(i)} \propto [H_{hf}^{(i)} \rho(E_F)]^2 k_B T \qquad (13)$$

as the energy conservation restricts the relaxation processes to electron states at the Fermi level [(i) stands for (dip) or (orb), and $\rho(E_F)$ is the density of states at the Fermi level of the band which contributes to $H_{hf}^{(i)}$]. It can be seen that no general

Korringa relation can be written for these relaxation processes
as, for instance, K_{dip} vanishes for cubic symmetry while $(T_1T)^{-1}_{(dip)}$
do not vanish. Similarly the orbital shift is related with
a susceptibility which involves all the electrons of the conduction
band while $(T_1T)^{-1}_{orb}$ only involves the density of states at the
Fermi level.

KNIGHT SHIFT AND T_1 IN PURE METALS: EXPERIMENTAL RESULTS

In most cases the experimentally measured shift of the
resonance line ($\sim 10^{-2}$ to 10^{-3}) are found much greater than the
chemical shifts obtained in insulators. Similarly the relaxation
rates are extremely high ($T_1T \sim 10$ to 10^{-2} sec K), and overwhelm
for instance the quadrupolar relaxation associated with the
modulation of the electric field gradients by the phonons, which
are found to dominate the relaxation of nuclei in pure insulators
(for spins $I > 1/2$). The effect of atomic motions on T_1 and $T_{1\rho}$
have only been detected by NMR at high temperatures, mainly for
light metals for which atomic diffusion is efficient, while the
contact relaxation rate is rather small [11]. In any experimental
case, at low temperatures, a Korrigna like relaxation holds
between K and T_1, but the measured Korringa constant may
substantially differ from the theoretical one of equation 9. This
is not surprising as the theoretical results presented in sec. II
indicate that this would only be true for the s contact interaction,
while other hyperfine interactions might contribute to the
measured shift and relaxation rate. In order to distinguish
the information which can be drawn from these important experimental
parameters, specific metals for which simplifying assumptions can
be made have to be considered separately.

1. Alkali Metals

These metals have a simple band structure, with a nearly
spherical Fermi surface of mainly s character, which does not
intersect the zone boundaries. They are then expected to be the
best candidates for a direct quantitative comparison with the
theory of the Knight shift and relaxation through the contact
interaction, as the other hyperfine interactions should be
unimportant. Even in this case, it can be seen that the Knight
shift involves the product of two quantities $H_{hf}^{(s)}$ and χ_s which
are difficult to obtain independently. In the case of Li and Na,
χ_s could be measured directly as a narrow electron spin resonance
(ESR) of the conduction electrons could be observed. As both NMR
and ESR signals could be detected at the same fequency, in the
same experimental conditions, the ratio χ_s/χ_n, where the nuclear
susceptibility χ_n is well known, could be measured [12].

Table 1. Experimental values for K, T_1T, Δ and χ_s in pure lithium
         and sodium metal. Theoretical value for the electronic
         susceptibility for non interacting electrons. The
         experimental values for χ_s are averages over several
         measurements.

	K	T_1T	Δ	χ_{sexp}	χ_{sPauli}
	%	sec K		10^{-6} cgs	10^{-6} cgs
^7Li	+ 0.0249	44 $\pm$ 2	1.57	2.02 $\pm$ 0.1	1.17
^{23}Na	0.1085	5.1 $\pm$ 0.3	1.60	1.04 $\pm$ 0.1	0.65

The experimental value of χ_s was found in both cases greater than
the Pauli susceptibility (table 1). This can be explained when
interactions between conduction electrons are taken into account.
The Korringa constant is also enhanced in these metals as can be
seen in table 1 ($\Delta = K^2 T_1T/\zeta$).

Experimental measurements of the Overhauser effect [13]
allows to show that the relaxation originates purely from the
contact interaction for ^{23}Na, while only 6 % of the measured
rate is of a different origin for Li and can be attributed to the
orbital relaxation. Consequently the value of Δ is directly
associated with the properties of the s electron band, and can
be shown to originate from the electron electron interactions.
In order to understand that point, one must consider the complex
susceptibility $\chi(q,\omega)$ of the electron gas, which is its response
to a time and space dependent magnetic field with Fourier
components $h(q,\omega)$. The Knight shift within this generalized
theory is proportional to $\chi(0,0)$ (i.e. the uniform static
susceptibility), while the relaxation rate involves the imaginary
part of $\chi(q, \omega)$:

$$(1/T_1) \propto kT \sum_q Im\chi(q, \omega_n)/\omega_n \qquad (14)$$

where the summation has to be taken over all q values. When taking
into account the interaction between electrons, in a molecular
field approximation, it is found that the static susceptibility

is enhanced:

$$\chi(0,0) = [\chi_0(0,0)] / (1 - \alpha) \tag{15}$$

where $\chi_0(0,0)$ is the susceptibility for non interacting electrons
and α is a parameter which measures the strength of the interaction
between electrons. The enhancement of Im $\chi(q, \omega)$ and then of $1/T_1$,
is a function of the wave vector which, within the same theory,
decreases from the value $(1 - \alpha)^{-2}$ for $q = 0$ and tends towards
unity for high q values. Consequently when summing equation 14
over all q values it can be seen that the enhancement of $1/T_1$ is
always smaller than that of K^2, yielding Korringa constants
greater than unity. These simple arguments show qualitatively
how such effects can explain why the data for Δ in alkali metals
are always found greater than unity. When more quantitative
agreement is required, the available theories of the enhancement
of the relaxation rate do show that the value of this enhancement
is not only dependent on the strength of electron-electron
interactions but also of their range [14]. Moreover there is no
complete theoretical treatment of the effect of the lattice
potential on the interactions between electrons. As there is
very few other experimental information on these interactions,
it is   impossible at the present time to perform more detailed
analysis. In the case of Li and Na, for which experimental values
for χ_s are available, the results for H_{hf}^s were found to be in
good agreement with those obtained from band calculations. In
the other alkali metals, as χ_s could not be measured directly, the
NMR shift together with similar evaluations of H_{hf}^s can be considered
as a good determination of χ_s and then of the strength of
electron-electron interactions.

From the complexity of this analysis, it can be guessed that
detailed interpretations of experimental data for systems with
a more complicated band structure such as noble metals, and a
fortiori polyvalent metals, are far from reaching within the
present knowledge of the electronic properties of metals.

2. Transition Metals

Most transition metals have very pronounced magnetic properties,
as whether they present low temperature ferromagnetic or antiferromagnetic phases, or they have nearly temperature independent
susceptibilities, much greater than those measured in simple
metals. As quite different physical situations are encountered in
the rare earth series and transition series, only the paramagnetic
state of d transition metals will be considered here, as these
metals have been the most thoroughly studied by NMR[8, 15]. In such
metals a two band model can be applied, which means that a free
electron like s band and a narrow d electron band overlap at the

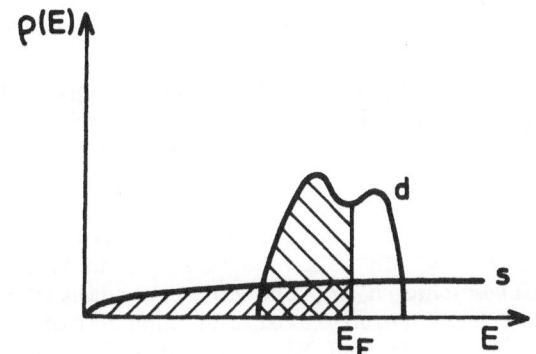

Fig. 3. Two band model for transition metals. The parabolic s
        band overlaps with a narrow d band.

Fermi level (Fig. 3). The high measured susceptibility is then
explained by the correspondingly high d density of states at
the Fermi level, but also because such a narrow d band may yield
correspondingly high orbital susceptibilities χ_{orb}, much greater
than the one expected for the s band. The total measured
susceptibility can be splitted into four parts:

$$\chi = \chi_s + \chi_d + \chi_{orb} + \chi_{dia} \qquad\qquad (16)$$

where χ_{dia} is the diamagnetic susceptibility of the ion cores. As
has been seen in section II, the first three terms in the
susceptibility should give rise to Knight shifts of corresponding
origins

$$K = K_s + K_d + K_{orb} \qquad\qquad (17)$$

where $K_i = \mu_B^{-1} H_{hf}^{(i)} \chi_i$ (i = s, d, orb)

Without more information than K and χ, it is impossible to know
the relative importance of the three terms in equation 17, for
a given metal. An interesting feature is that $H_{hf}^{(d)}$ is experimentally found negative as

- the total Knight shift in Pt and Pd metals, which are paramagnetic, is negative;

- similarly the internal fields as measured by the NMR frequencies of ^{57}Fe, Co and Ni in the pure ferromagnetic metals are found to diminish when an external field is applied, which means that $H_{hf}^{(d)} < 0$.

This negative sign of H_{hf}^d is confirmed by theoretical calculations performed on 3 d ions, which allow to understand that though $|H_{hf}^{(d)}|$ might be much smaller than $H_{hf}^{(s)}$, the correspondingly high d density of states at the Fermi level can then yield situations for which K_d dominates as in Pt and Pd, resulting in a total negative Knight shift. Though the order of magnitude of the various contributions to K might be deduced using theoretical estimates of the hyperfine fields, the best analysis could only be done in cases where important temperature dependences of K and χ could be detected. These are assumed to originate mainly from the temperature variation of K_d and χ_d. This is expected, as the d band is very narrow, and steep variations of $\rho_d(E)$ in an energy range $\sim k_BT$ around the Fermi level may occur in some cases. In Pt and Pd for instance, a direct plot of K versus χ allows to deduce both $H_{hf}^{(d)}$ and χ_d [16, 17] (Fig. 4), while K_{orb} and K_s can then be deduced, using theoretical estimates for $H_{hf}^{(s)}$ and $H_{hf}^{(orb)}$.

The cases of Pt and Pd metals are rather simple as K_d and χ_d greatly dominate, while in less paramagnetic metals such as V, K_{orb} is found to be the most important contribution. For Pt and Pd the values of χ_d obtained by such analysis are found to exceed those expected from specific heat measurements of the density of states at the Fermi level, which points out here again the existence of electron-electron interaction effects. As a consequence the partition of the relaxation rate $(T_1T)^{-1}$ into three terms

$$(T_1T)^{-1} = (T_1T)_s^{-1} + (T_1T)_d^{-1} + (T_1T)_{orb}^{-1} \qquad (18)$$

is even harder to perform as different Korringa constants apply for the s and d band, in which moreover electron-electron interaction exchange enhancements might be important. Current estimates show that though the d terms highly dominate the experimental values for K and χ in platinum, the three contributions to $(T_1T)^{-1}$ are of the same order of magnitude [18], as can be seen in table 2.

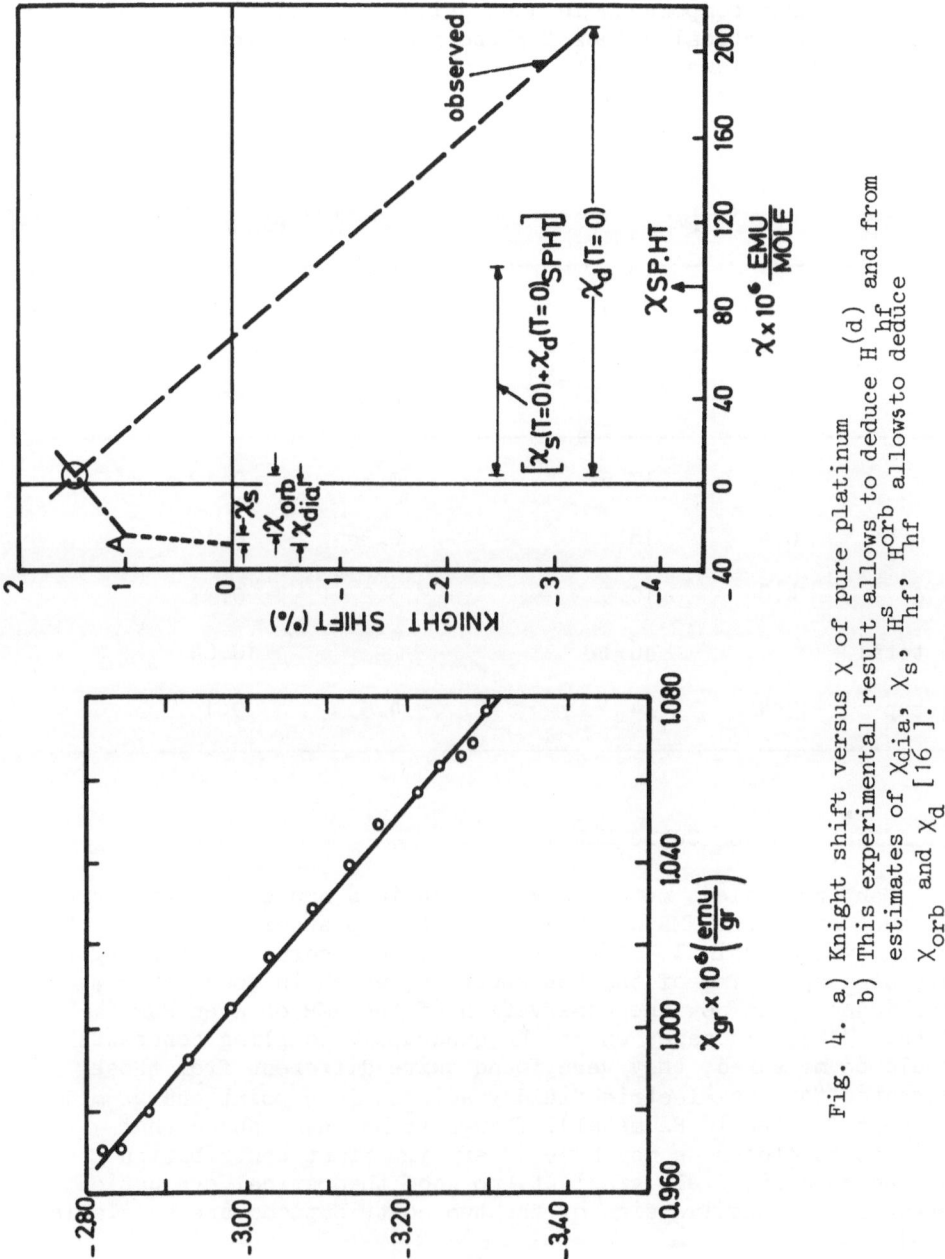

Fig. 4. a) Knight shift versus χ of pure platinum
       b) This experimental result allows to deduce $H_{hf}^{(d)}$ and from
          estimates of χ_{dia}, χ_s, H_{hf}^s, H_{hf}^{orb} allows to deduce
          χ_{orb} and χ_d [16].

Table 2. ^{195}Pt and ^{105}Pd shifts and relaxation rates in the pure
metals. The partition of the experimental shift into
three components is done using the K(χ) diagram and
theoretical values for the s and orb hyperfine fields.
The relaxation rates are obtained then through the
theory of sec. II.

	^{195}Pt		^{105}Pd	
	K	$(T_1 T)^{-1}$	K	$(T_1 T)^{-1}$
	%	$(\sec K)^{-1}$	%	$(\sec K)^{-1}$
s	1	17.5	0.36	0.12
orb	0.4	18.1	0.36	0.15
d	− 4.8	10.5	− 5.2	0.37
total		46.1		0.64
exp	− 3.4	34	− 4.5	9.1

3. Effects in non Cubic Metals

Many polyvalent metals crystallize in a non cubic structure.
In such cases the NMR spectra might show up specific effects. For
nuclei with a spin I > 1/2, the hyperfine quadrupole coupling then
yields a splitting of the NMR spectrum, which in some cases has
forbidden up to now the observation of the NMR or even NQR (^{67}Zn,
^{75}As, ^{185}Re). In cases where the quadrupole coupling constants
could be measured, they were found quite different from those
expected for the electric field gradients in a point charge model
(except for ^{9}Be in Be metal). Though it has been shown that
conduction electrons may have a very important contribution
to the electric field gradient, no good theoretical evaluation,
even of the relative sign of the two contributions are available
[19].

For nuclei with spin I = 1/2, the non cubic structure mainly
yields an anisotropy of the Knight shift due to the dipolar

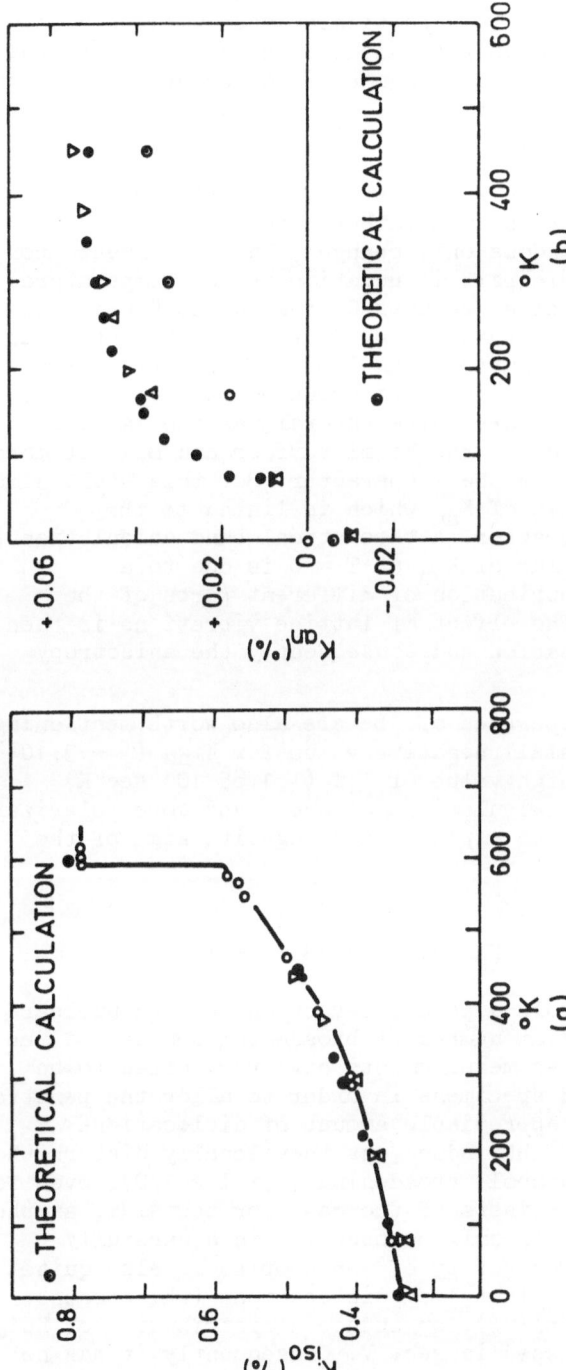

Fig. 5. Experimental temperature dependences of (a) the isotropic
Knight shift and (b) the anisotropic Knight shift of
113Cd in pure Cd metal. The experimental data are compared
with a band calculation of Kasowski and Falikov [21].
Among the various experimental results the ones given by
Δ correspond to single crystal measurements by Sharma and
Williams [20].

hyperfine interaction. This is for instance detected on the NMR of [113]Cd in pure Cd metal. This particular case will be considered here as it has been thoroughly studied, even on single crystals. As Cd crystallizes in an axially symmetric structure, the Knight shift tensor can be characterized by two parameters, the isotropic (K_{iso}) and anisotropic (K_{an}) contributions to the Knight shift. The isotropic part is found to increase regularly from T = 0 K to the melting point, where an abrupt increase ($\sim 30 \%$) is detected upon melting (this is a rather rare situation as in most metals the Knight shift does only change   a few percent upon melting). The anisotropic part is negative at low temperatures, decreases in absolute value, reaches 0 at about 20 K and becomes positive at higher temperatures (Fig. 5) [20]. As Cd metal has a very complicated Fermi surface, an important p character of the wave functions at the Fermi level is expected. An increase of K_{iso} with temperature appears quite natural, as the lattice vibrations should smooth out the Fermi surface and make it more isotropic, increasing then the s character. But this would simultaneously lead a decrease of K_{an} which is linked to the importance of the p character. A theoretical band calculation [21] shows that the small value of K_{an} at T = 0 is due to a compensation of the contribution of different parts of the Fermi surface to K_{an}. The effect of lattice vibrations is then to destroy this compensation and consequently the anisotropy increases.

Finally the NMR properties of [9]Be are also worth mentioning here, as an extremely small negative value for K_{iso} ($\sim - 3.10^{-5}$) and a correspondingly high value of T_1T ($\sim 1.85 \ 10^4$ sec K) are measured. Detailed calculations of the s and core polarization contributions even fail to explain this negative sign of the Knight shift.

SPIN-SPIN INTERACTIONS AND NMR LINESHAPES IN METALS

Apart the usual dipole-dipole interaction between nuclear spins, there exist a great number of broadening sources of the NMR spectra in metals. As measurements have very often to be carried out on powdered specimens in order to allow the penetration of the r.f. field, the appreciable amount of dislocations present in the sample might induce, as they locally disturb the symmetry, a severe quadrupole broadening (for I > 1/2), even for cubic metals. This is for instance the case for tantalum, as the NMR spectra of [181]Ta could only be observed in a carefully annealed foil sample. The purity of the samples is also quite important as impurities might lock dislocations (which cannot then be removed by annealing), or induce a distribution of Knight shifts as will be discussed in sec. V. Consequently it has not always been possible to measure the broadening of the NMR spectra solely due to spin-spin interactions. Nevertheless, it has been understood very early [2, 22] that the conduction electrons were responsible for indirect spin-spin couplings between the nuclear spins, as the NMR line was found in some cases much broader than expected from the dipole-dipole coupling. The theory of these interactions will be considered first, and their effect on the NMR linewidth will be discussed then on some specific examples. The various methods which have been used in order to measure these couplings will be discussed thereafter.

1. The Indirect Spin-Spin Couplings

The origin of the indirect interactions can be understood from simple arguments, as a nuclear spin I_i creates through the hyper-fine coupling $H_{hyf(i)}$ a spatially varying electronic polarization, which in turn interacts with another nuclear moment I_j. Such an effect resumes in an indirect interaction between I_i and I_j. The interaction energy can be calculated by second order perturbation theory:

$$H_{ij} = \sum_{\substack{E_{k\sigma} < E_F \\ E_{k'\sigma'} > E_F}} \frac{<k\sigma|H_{hyf(i)}|k'\sigma'><k'\sigma'|H_{hyf(j)}|k\sigma>}{E_{k\sigma} - E_{k'\sigma'}} + c.c. \quad (19)$$

where $k\sigma$ is the electronic wave function of a state of wave vector k, spin σ and energy $E_{k\sigma}$. The primed parameters apply for non occupied states. It can be seen that though all electrons in the band might contribute to H_{ij}, the most important contribution will come from states such as $E_{k\sigma} \sim E_{k'\sigma'}$, which means states near the Fermi level. This coupling which is bilinear with respect to the nuclear spin operators has a symmetry which depends upon the nature of the hyperfine couplings $H_{hyf(i)}$ and $H_{hyf(j)}$ taken on sites i and j in equation 19.

- A scalar interaction H_{ij}^{exch} is obtained if a scalar interaction (equation 1) is taken on both sites. For a free electron band and a spherical Fermi surface it can be written:

$$H_{ij}^{exch} = (m\Omega^2/8\pi^3)(2k_F)^4\gamma_n^2(H_{hf}^s)^2F(2k_FR_{ij})\, I_i \cdot I_j \quad (20)$$

$$= \hbar\, A_{ij}\, I_i \cdot I_j$$

where m is the electron mass, Ω the atomic volume; the spatial dependence upon the relative positions of the nuclear spins (given by R_{ij}) is isotropic and given by the Ruderman Kittel function

$$F(x) = (x \cos x - \sin x)/x^4 \tag{21}$$

which has an asymptotic $\cos(2k_F R_{ij})/R_{ij}^3$ dependence.

- The only other term which is found to be important is obtained by taking a contact interaction on one nuclear site and the dipole-dipole interaction on the other site. It can be seen that the resulting coupling has the same symmetry as the dipole-dipole interaction in the case of a spherical Fermi surface, hence its name of pseudodipolar coupling [2].

$$H_{ij}^{PD} = \hbar\, B_{ij}^{PD}\, [\mathcal{I}_i \cdot \mathcal{I}_j - 3R_{ij}^{-2} (\mathcal{I}_i \cdot R_{ij})(\mathcal{I}_j \cdot R_{ij})] \tag{22}$$

where B_{ij}^{PD} has a long distance R_{ij}^{-3} oscillatory dependence, and can be expected to be generally smaller than A_{ij}. When the strength of the exchange interaction is greater than the direct dipole-dipole coupling, it may completely determine the shape of the observed spectra, as found in metals like lead, platinum or thallium.

2. Contribution of the Indirect Interactions to the NMR Lineshape in Pure Metals

It is known that complete calculations of the NMR lineshapes cannot be achieved for solids. Nevertheless the first few moments of the line give an indication about the lineshape and width. For a pure dipolar interaction and a concentrated spin system, the lineshape is found to be nearer to Gaussian than a Lorentzian. The peak to peak width of the derivative of the absorption line is then given by:

$$\delta\nu \simeq 2 \langle \Delta\nu^2 \rangle^{1/2} \tag{23}$$

where $\langle \Delta\nu^2 \rangle$ is the second moment of the line. Two different situations for the contributions of the indirect spin-spin interactions to the NMR shape and width do occur.

a) In the case where only one spin species exists in the pure metal, the second moment $\langle \Delta\nu^2 \rangle$ of the line is only due to the dipolar interaction. The scalar interaction contributes to the fourth moment $\langle \Delta\nu^4 \rangle$ and consequently the existence of an exchange interaction results in the decrease of the ratio $\langle \Delta\nu^2 \rangle^2/\langle \Delta\nu^4 \rangle$ with respect to the case of the pure dipolar coupling. This corresponds to a change of lineshape towards Lorentzian and results in an apparent narrowing of the line. A situation of extreme narrowing is reached if the exchange interaction is greater than the dipolar one. A good example of such a case is found for platinum metal as the [195]Pt NMR width is found about five times smaller than the one expected from equation 23.

b) If various isotopes with different gyromagnetic ratios coexist
   in the material, the exchange interactions between unlike
   spins do then contribute to the linewidth. This was observed
   for example for silver and thallium metals which have two
   isotopes with nuclear spin $I = 1/2$. In the case of thallium,
   as the two isotopes ^{203}Tl and ^{205}Tl have gyromagnetic ratios
   which differ only slightly, a situation for which the
   difference of their Zeeman frequencies is rather smaller than
   the indirect Ruderman Kittel interaction can be achieved by
   lowering the exeternal field. A progressive transition from
   an exchange broadening of the two resonance lines, to a
   narrowing of the unique line in low field could be observed [23].

3. Measurements of the Indirect Interactions

   It is hopeless with the few experimental informations which
can be obtained to get values for A_{ij} and B_{ij}^{PD} for all shells
of neighbours. It is usually assumed that only their values
for a few neighbours (at most three shells) j of a nucleous i
can be considered as unknown parameters. At large distance
either they are neglected, or they can be roughly taken into
account through an extrapolation from the near neighbour values
using the theoretical spatial dependence.

a) A first information about these quantities can be found in
   second moment measurements. Those are only meaningful when
   the line is Gaussian like, as otherwise the contribution of
   the wings of the line to the second moment is rather important.
   Such measurements are then mainly performed in cases like
   2.b and immediately indicate then the relative importance
   of the pure dipole-dipole interaction and the indirect
   interactions. More detailed analysis could be performed using
   isotopically enriched samples or through single crystal
   measurements, which allow then to separate the scalar from
   the dipolar interaction through the angular dependence of the
   second moment [24]. Even more information could be obtained
   in niobium single crystals by nuclear acoustic resonance
   measurements. In such a case $(I = 7/2)$, two NAR lines
   corresponding to transitions $\Delta m = 1$ and $\Delta m = 2$ could be
   observed with different contributions of the spin-spin
   interactions to the second moments [25].

b) In cases like 2.a alternative types of measurements involving
   slightly more complicated NMR concepts have been used. The
   exchange coupling does not contribute to the second moment
   of the line when the r.f. level of irradiation H_1 is small.
   But it comes into the local field in the rotating frame and
   then appears in the expressions for the shape (Provotorov
   theory) of the dispersion signal obtained with a strong r.f.
   field. Such a method has been applied for Rb and Cs [26], but

becomes nearly impossible for technical reasons when the spin-spin couplings become greater than a fraction of a gauss.

When the exchange interaction is of the same order of magnitude as the total dipolar interaction, another method has been used, which takes advantage of the fact that like nuclear spins may be rendered unlike by introducing a small amount of impurities in the metal. An impurity induces a long range oscillatory spatial component of the Knight shift, the physical origin of which will be considered in section V. Let us consider   the model case of two nuclear species i and j whose Knight shifts are $K(i)$ and $K(j)$ and which are only interacting through the exchange term $A_{ij} I_i \cdot I_j$, writing

$$\delta = \gamma_n [K(i) - K(j)]$$

and

$$\gamma'_n = \gamma_n \{1 + [K(i) + K(j)]/2\}$$

allows to write the total Hamiltonian:

$$H = \gamma'_n \hbar \; H_0 (I_{iz} + I_{jz}) + \hbar \; \delta H_0 (I_{iz} - I_{jz}) + \hbar \; A_{ij} \vec{I}_i \cdot \vec{I}_j \quad (24)$$

This is a well known situation for the multiplet structure of a resonance line in molecules. For $A_{ij} \gg \delta H_0$ the exchange inter-action narrows the structure and only one line at the frequency $\gamma'_n H_0$ is observed, while if $A_{ij} \ll \delta H_0$ the exchange term does no more commute with the total Zeeman Hamiltonian and then yields a splitting of each line into two lines distant of A_{ij}. Increasing then δH_0 by chosing a specific impurity and the highest available external field H_0 allows to induce the splitting of the lines, which is of course not observable in an alloy as it is completely smeared out by the Knight shift variation from site to site. Nevertheless, taking advantage of the spin echo technique which allows to eliminate this magnetic inhomogeneous broadening, allows to restitute the exchange frequencies in the echo envelope of the spectrum. This method has been successfully applied in Pt, Pb, Sn and Cd [27]. Discussing theoretically the values of the indirect coupling parameters is rather a hard task, as it can be seen from equation 19 that the whole band structure would be necessary. Though an order of magnitude agreement is usually obtained with the simplified Ruderman-Kittel result, complete calculations have only been attempted in a few cases and a rather good agreement has been found between the experimental measurements and band structure calculations for the exchange interaction in Rb, Cs and Pb [28]. It is then clear that the experimental para-meters available from NMR in metals are qualitatively understood in terms of simplified theories. Nevertheless in order to get more refined explanations of the measured parameters for T_1, K, the exchange interaction or the electric field gradients, a very detailed description of the electronic properties of individual metals is required. Though the present status of band structure calculations does give a satisfactory explanation of the experimental data in some specific cases, it has been shown that great details about some specific points such as electron-electron interactions, excited states, etc... are still required in order to get an accurate derivation of the NMR parameters from first principles.

NMR IN DILUTE ALLOYS

Any imperfection in a metal induces spatially extended perturbations of the host metal properties which are usually rather difficult to study by macroscopic experimental techniques. For instance magnetic susceptibility measurements give an average of the susceptibility on all sites in the sample, while NMR as well as other resonant hyperfine techniques (such as Mossbäuer effect) might allow to measure χ on different nuclear sites. Though practically experiments could be attempted in many alloy systems, reliable interpretations of the data can only be achieved on carefully choosen cases for which simplifying assumptions can be made. For instance, as, from equation 6, the Knight shift involves (at least) two parameters H_{hf}^s and χ_s, both might be modified on a local scale. Separate cases for which only one of these quantities is affected by the presence of dilute impurities will be considered here. By "dilute" it is usually meant that the distance between impurities is large enough to consider them as practically isolated in the pure matrix. This allows then, from the theoretical knowledge of the perturbation produced by a single impurity on the host metal, to derive the real alloy properties by superposition of the effects associated with individual impurities.

1. Magnetic Impurities in a Simple Metal Host

Some 3 d transition metal elements, or rare earth elements, bear local moments when diluted in a simple sp metal host such as Ag, Au, Cu or Al. This means that the total spin S of the 3 d or 4 f electrons of the impurity has a paramagnetic free spin behaviour, the impurity magnetization following then a Brillouin function. It is usually assumed that the localised electronic spin of the impurity is coupled with the conduction electron spin $\vec{s}$ through an exchange interaction

$$H^{ex} = - J \ \vec{S}.\vec{s} \ \ \delta(r) \tag{25}$$

The physical origin of this interaction, as well as the
theory of the behaviour of the conduction electron and localised
spin system involve a number of physical concepts which are
outside the scope of the present lectures. An introductory review
article about the physical problems which arise in these alloys
might be found in[29], while a summary of the work performed
with NMR and other hyperfine techniques on such systems is given
in [30].

a) Perturbation theory

The simple approach in which the exchange coupling of
equation 25 is considered as a perturbation with respect to the
free electron states will be briefly considered here as it
allows to introduce the physical effects which are observed
experimentally. Within such an approximation both nuclear spins
(through the contact interaction) and the localised electron spins
are coupled in the same manner with the conduction electrons. The
contact hyperfine Hamiltonian is very often written:

$$H^{ex} = A_s \ \vec{I}.\vec{s} \ \delta(r) \tag{26}$$

with $A_s = H^{(s)}_{hf} \ \gamma_e \gamma_n \ \hbar^2/\mu_B$

which means that for the nuclear spins the metal can be considered
as a free electron gas, the effect of the periodic lattice
potential being merely introduced in the hyperfine coupling
constant A_s. It can then be seen immediately that the EPR shift
Δg and spin lattice relaxation time τ will be deduced from the NMR
Knight shift and T_1 of equation 6 and 7 by interchanging A_s and
γ_n with J and γ_e, which yields

$$\Delta g = J\rho(E_F) \tag{27}$$

$$\tau^{-1} = (\pi/\hbar) \ k_B T \ [J\rho(E_F)]^2 \tag{28}$$

Similarly the result of equation 7 for the indirect interactions
between nuclear spins can be straightforwardly transposed for
the interaction between the impurity electronic moment and a
nuclear spin located at $\vec{r}$ with respect to the impurity. Apart
the classical dipole-dipole coupling between these two magnetic
moments, an indirect exchange coupling:

$$H_{IS} = A_s(r) \ \vec{I}.\vec{S}. = 4\pi E_F A J\rho^2(E_F) \ [(\cos 2k_F r)/(2k_F r)^3]\vec{I}.\vec{S}. \tag{29}$$

is induced by the conduction electrons. In equation 29 only the
large distance asymptotic behaviour has been kept. The influence

of this coupling on the nuclear spin resonance will be quite
different from the interaction between two unlike spins in a
metal, as J is a rather strong coupling (J $\sim$ 1 eV) which corresponds
then to relaxation rates $\tau T \sim 10^{-10}$ sec K from equation 28.
Consequently the electron spin relaxation time is much more
shorter than the nuclear spin Larmor period. The nulcear spin
will then sense:

- An average static local field:

$$H_{eff} = A_s(r) < S_Z > / \hbar\gamma_n \tag{30}$$

which originates in the fact that $\uparrow$ and $\downarrow$ conduction electrons
are scattered differently by the impurity, yielding a spatially
oscillating polarization of the conduction electrons which
adds to the Pauli polarization. The extra local field H_{eff} sensed
by the host nuclei is proportional to the impurity magnetization

$$M = g \mu_B < S_Z > = g \mu_B B_S (g\mu_B H/kT) S \tag{31}$$

where B_S is the Brillouin function for spin S. Consequently
the nuclei in the alloy will experience different local fields
which depend upon their position with respect to the impurities.
The excess shift might be either positive or negative and
decreases as r^{-3}.

- The fluctuating part of the local field, which will contribute
to a spin lattice relaxation process. The relaxation rate for
a nuclear spin at distance r is given by:

$$1/ T_1(r) = 2 \hbar^{-2} A_s^2(r) (k_B T/\hbar\omega_e) <S_Z> \tau/(1 + \omega_e^2\tau^2) \tag{32}$$

where τ is the electron spin relaxation time of equation 28.
Let us point out here that if the NMR of the impurity nucleus
itself is considered, the only difference with the host nuclei
resides in the fact that the coupling between the 3 d electron
spin and the impurity nuclear spin is a core polarization hyperfine coupling which is written:

$$H = A_d \vec{I}.\vec{S} \tag{33}$$

and consequently the shift and T_1 of the impurity NMR are
identical to equation 30 and 31 in which $A_s(r)$ has to be replaced
by A_d.

b) Experiments on dilute magnetic alloys

The simple approach given above does not describe the real physical situation as transition impurities in metals do not generally behave as free spins, with a Curie paramagnetism as is the case of Mn in Cu. Some impurities have a Pauli like susceptibility, nearly temperature independent as Mn in Al, and intermediate behaviours for which χ follows roughly a Curie Weiss law are quite common. A typical example is that of Fe in Cu for which the magnetization is $M \propto H/(T+29)$. In such cases the perturbation approach is not correct at low temperatures, and even the Hamiltonian of equation 25 for the interaction between the impurity and the host electrons might be a bad starting point. It has often been asked whether the constant χ at low temperature, which is one manifestation of the so called Kondo effect, was linked with a net antiferromagnetic polarization of the conduction electrons which would appear at low enough temperatures [31]. It will be shown that the NMR results on the three systems mentioned above allows to answer negatively to this question.

Cu Mn is the typical example for which a perturbation theory should hold. In order to know the order of magnitude of the NMR parameters of the impurity and host nuclei, numerical values have to be introduced in the various equations given above. The Mn ionic moment ($S \sim 5/2$) has a d hyperfine field estimated to be about $- 100$ kG/μ_B. For $T = 1$ K, $H = 20$ kG the impurity magnetization is roughly saturated, yielding $<S_Z> \sim 5/2$. Then the expected local field on the Mn nuclei is, from equation 30 (in which $A_s(r)$ is replaced by $A_d = H^d_{hf}(\hbar^2\gamma_e\gamma_n)\mu_B^{-1}$, as large as $H_{eff} = - 500$ kG which is much greater than the external field. As $J\rho(E_F) \sim - 0.25$ is obtained through various experimental results, equation 28 yields $\tau T \sim 3.7 \ 10^{-11}$ sec K. The relaxation time T_1 for the Mn nuclear spin from equation 31 would then roughly be $T_1 \sim 5 \ 10^{-9} \ T^{-1}$ for $g\mu_B H/kT << 1$. This shows that the impurity NMR is quite impossible to be observed except at high temperatures (~ 1000 K) as T_1 reaches then the μ sec range [32].

As the indirect coupling $A_s(r)$ of the host nuclei with the impurity spin is much smaller than A_d, the effects on the host nuclei are less pronounced. For the nearest neighbours of Mn, $A_d/A_s \sim 20$, which should allow to observe the NMR of these nuclei even at low temperatures. Both $H_{eff}(r)$ and $T_1^{-1}(r)$ decrease very rapidly for nuclei at larger distances from the impurity. The host NMR spectrum is then composed of a number of discrete satellite lines, while nuclei far apart just contribute to a broadening of a central line. Owing to the particular $\cos 2k_F r/r^3$ spatial dependence of $H_{eff}(r)$ this main line is unshifted with

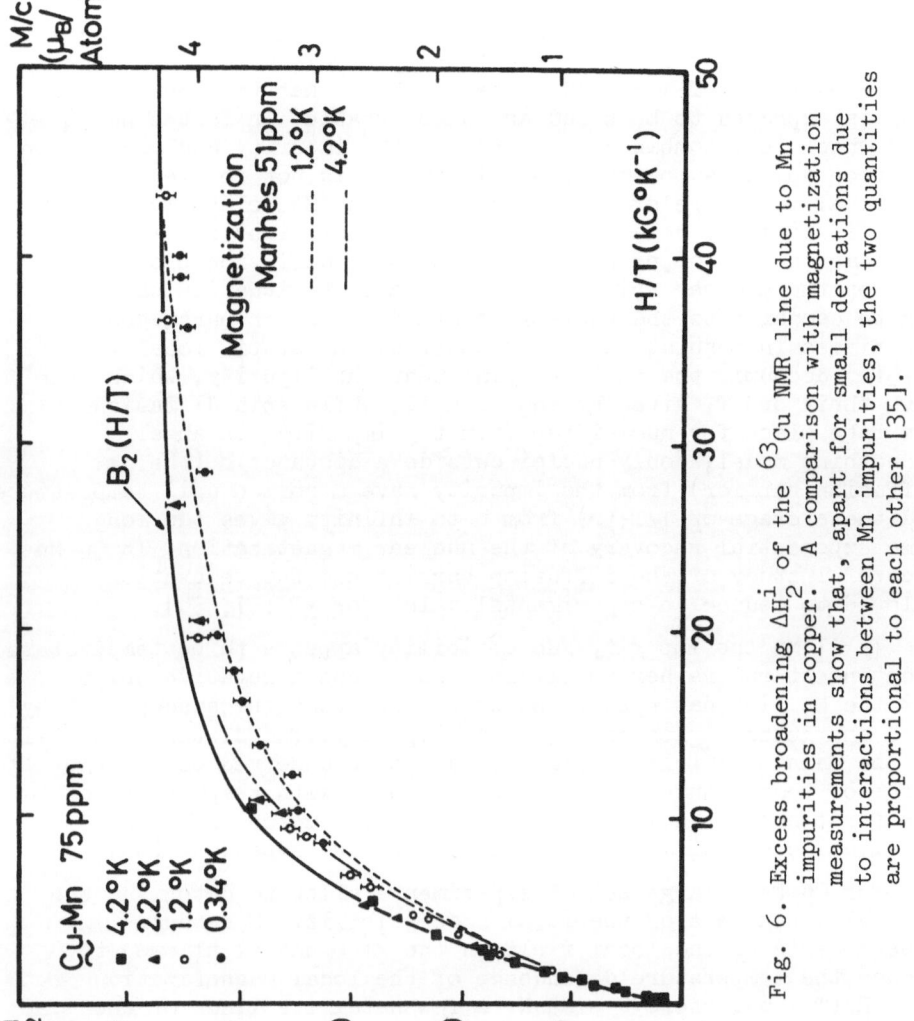

Fig. 6. Excess broadening ΔH_2^i of the ^{63}Cu NMR line due to Mn impurities in copper. A comparison with magnetization measurements show that, apart for small deviations due to interactions between Mn impurities, the two quantities are proportional to each other [35].

respect to the pure copper resonance and has a width proportional
to the concentration of impurities [32, 34]. The broadening of
the main line is also found proportional to $<S_z>$ as can be
seen in figure 6 [35]. Some satellite lines could be resolved
at high enough T ($\sim$ 77 K) but are lost at lower temperature
as they are severely broadened then [36].

A further interest of the host NMR is to allow an indirect
measurement of the impurity electron spin relaxation rate. The
estimated value $\tau T \sim 3.7 \ 10^{-11}$ sec K shows that the impurity
EPR is expected to be broad and unobservable. In fact a narrow
line which is a combined resonance of the impurity and conduction
electron spins is observed, but its width is not related with
the local moment relaxation time. The host NMR relaxation time
$T_1(r)$ is directly related to τ. As the relaxation time is not
the same on all copper nuclear sites, spin diffusion processes
between the nuclear spins will take place. A situation which
can be compared to the nuclear relaxation through paramagnetic
impurities in insulators is encountered. The static local fields
$H_{eff}(r)$ decouple the nuclear spins near the impurity, which then
have their own T_1 given by equation 32, while spin diffusion
can take place for nuclei far from the impurity. In an all
or nothing model, only nuclei outside a distance b (the
diffusion barrier) from the impurity have a common spin temperature
and the average of $1/T_1(r)$ from b to infinity gives the long
time exponential recovery of the nuclear magnetization. In Cu Mn
a careful study of the diffusion barrier and the relaxation
allowed to deduce an experimental value for τT [37, 38].

In Al Mn the impurity susceptibility appears to be temperature
independent and rather small. In such a case a negative shift
for the Mn line has been measured showing that the susceptibility
of d character is localised on the impurity. Two distinct satellites
of the main Al^{27} line (Fig. 7) have been observed on a large
range of temperatures. The slight decrease with temperature of
the first nearest neighbour satellite shift demonstrates that
the local susceptibility is also temperature dependent [39].

For Cu Fe a large set of experiments allow to determine the
spatial dependence of the local susceptibility. Mossbäuer
measurements of the local field on the ^{57}Fe nuclei allowed to
deduce the temperature dependence of the local magnetization
$\chi H \sim H/(T + \theta)$, where θ = 29 K. A few satellite lines of the
Cu main line could be observed and their shift relative to the
main line ΔK also follows the same law $\Delta K \propto (T + 29)^{-1}$ as can
be seen in figure 8 [40] .

The macroscopic magnetization as well as the main ^{63}Cu linewidth do have the same temperature dependence at high T but

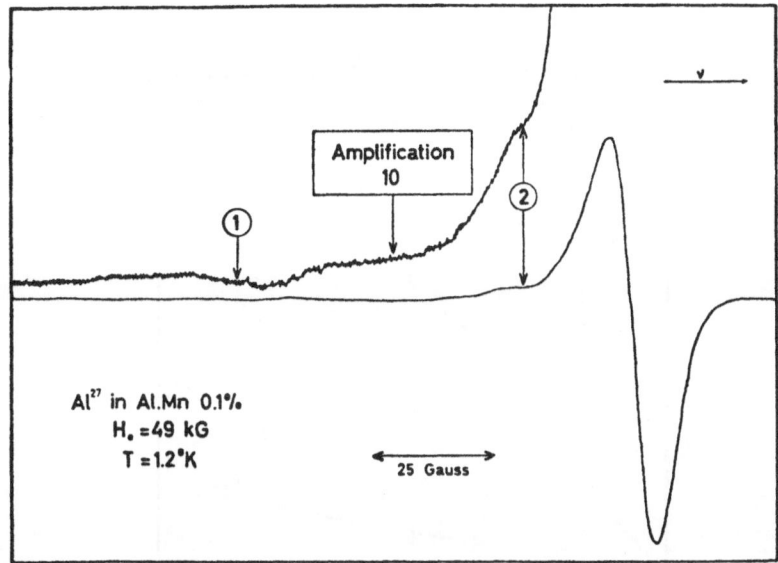

Fig. 7. NMR spectrum of ^{27}Al in an A̲l̲ Mn sample, showing two
        satellite lines. The outer satellite corresponds to the
        1st nearest neighbours of the impurity [39].

deviate below T $\sim$ 20 K (Fig. 9). This excess magnetization and
broadening are explained by the fact that some Fe impurities
are strongly coupled to each other either by a magnetic interaction
or because they form a chemical cluster. These Fe impurities
behave nearly like paramagnetic ions and are then easily polarized
at low T giving an excess magnetization and an excess linewidth
[35, 41]. These experimental results show that, even though the
perturbation theories and equation 30 do not strictly apply, the
spin polarization of the host electron spins is strictly proportional to the impurity magnetization at any temperature. It can
also be shown that the exact spatial dependence of $H_{eff}(r)$ is
related with the symmetry of the interaction potential between
localised and conduction electrons, and with the physical
situation encountered in the given alloy. Comparing the broadening
of the central line and the shift of a few neighbouring shells
of the impurity allowed in some cases to compare $H_{eff}(r)$ with
theoretical shapes expected from simple models [33].

    It has been seen here through these examples that the
interest of NMR measurements is to allow to understand how the
magnetization is distributed in such samples, showing in this
particular case that the magnetization is mainly localised on the

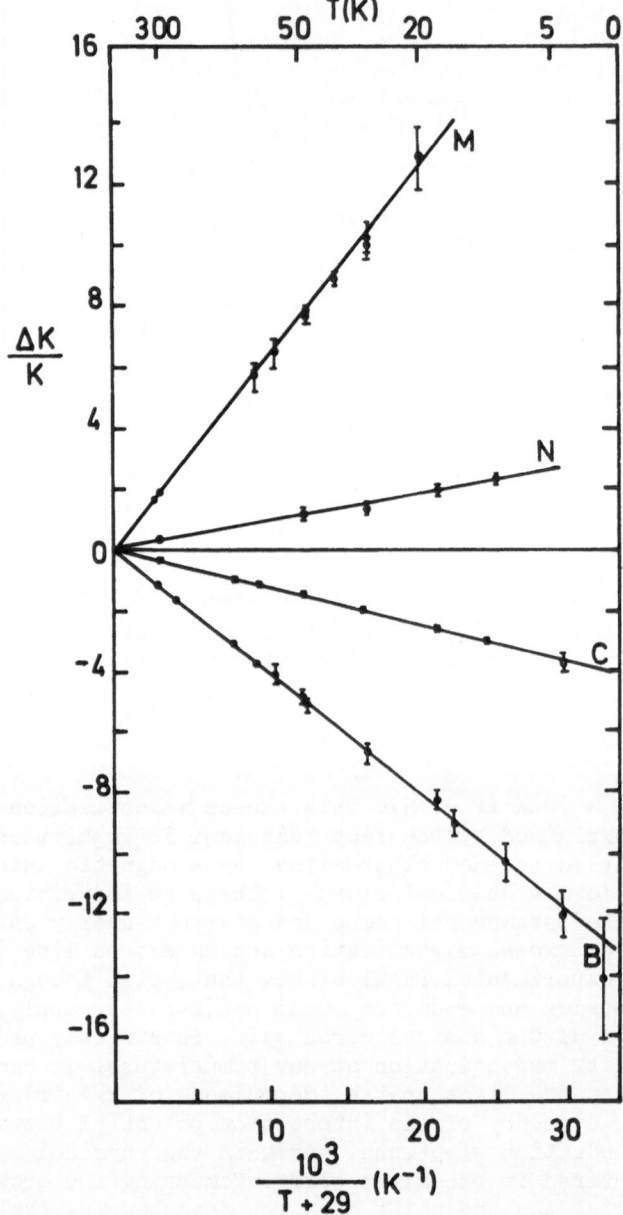

Fig. 8. Relative shifts ΔK/K of four satellites of the ^{63}Cu
resonance in Cu Fe alloys, versus <S$_z$>which is proportional
to (T + 29)$^{-1}$ [40].

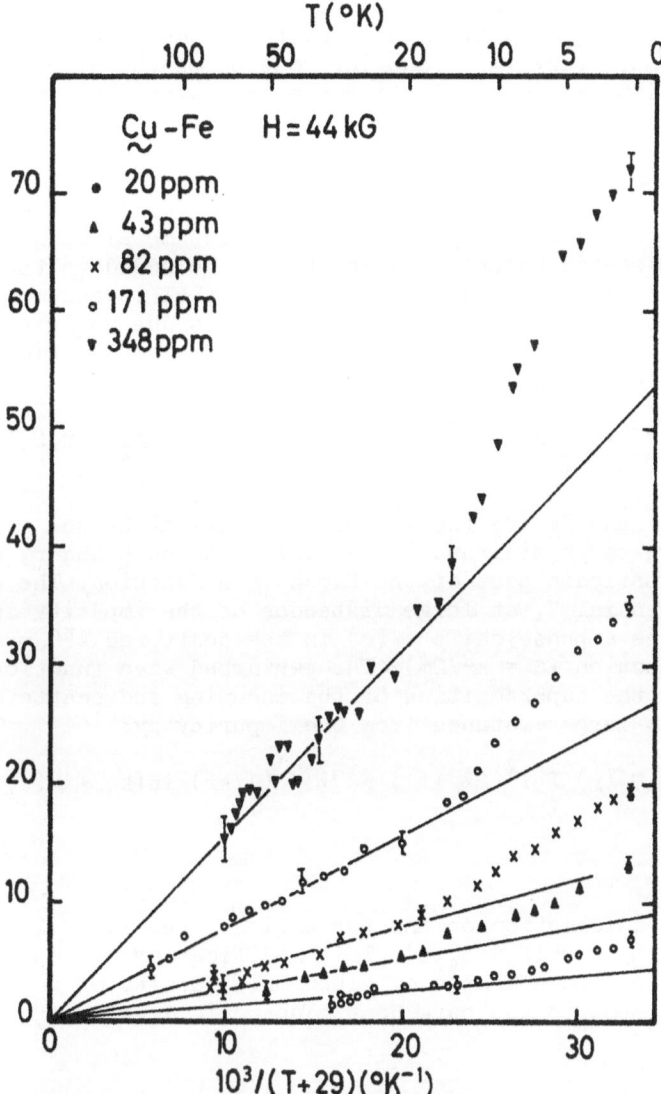

Fig. 9. Broadening of the central line versus $(T + 29)^{-1}$. An excess broadening with respect to the impurity magnetization appears at low temperatures. It corresponds to the spin polarization associated with groups of Fe impurities [35].

impurity sites. Such experiments also allow to separate the
properties of isolated impurities from the contribution of
clusters or groups of impurities to the macroscopic magnetization.
Furthermore very important information on the dynamic properties
of the impurity magnetization can be reached by such a technique,
and are hardly available from other types of experiments.

2. Non Magnetic Impurities in Metals

A non magnetic impurity in a sp host metal mainly produces
a local static potential V which must be added to the periodic
potential of the lattice. If the potential is assumed to be
purely spherical a partial wave analysis due to Friedel [42]
can be performed. A free electron Bloch wave function can be
decomposed in spherical harmonics

$$\psi_{\underset{\sim}{k}}^0(\underset{\sim}{r}) = U_{\underset{\sim}{k}}(\underset{\sim}{r})e^{i\underset{\sim}{k}\cdot\underset{\sim}{r}} = U_{\underset{\sim}{k}}(\underset{\sim}{r}) \sum_{\ell} i^{\ell}(2\ell + 1)j_{\ell}(kr)P_{\ell}(\cos\theta) \quad (34)$$

where j_{ℓ} and P_{ℓ} are the ℓth order spherical Bessel function and
Legendre polynomial and θ the angle between $\underset{\sim}{k}$ and $\underset{\sim}{r}$, and $U_{\underset{\sim}{k}}(\underset{\sim}{r})$
is the periodic part of the Bloch wave function. The effect of
the potential V, at large distances of the impurity is to
introduce a phase shift $\delta_{\ell}(\varepsilon)$ in the scattered ℓ^{th} order partial
wave function ($\varepsilon = k^2/2m$). The perturbed wave function obtained
then by the superposition of the incoming and scattered waves is
given at large distance from the impurity by:

$$\psi_{\underset{\sim}{k}}(\underset{\sim}{r}) = U_{\underset{\sim}{k}}(\underset{\sim}{r}) \sum_{\ell} i^{\ell}(2\ell + 1) e^{i\eta_{\ell}(\varepsilon)}(1/kr)\sin[kr + \eta_{\ell}(\varepsilon) - \ell\pi/2]P_{\ell}(\cos\theta)$$

$$(35)$$

All the information on the perturbing potential V is included
in the phase shifts $\eta_{\ell}(\varepsilon)$. This modification of the wave functions
corresponds to a change of electronic density of wave vector $\underset{\sim}{k}$
with respect to the pure host electronic density $\rho_k^0 = |\psi_k^0(0)|^2$

$$\Delta\rho_k(r) = |\psi_k(r)|^2 - |\psi_k^0(0)|^2 = \rho_k^0 \sum_{\ell}(-1)^{\ell} (2\ell + 1)\sin\eta_{\ell}(\varepsilon)$$

$$\{\sin[2kr + \eta_{\ell}(\varepsilon)]\}/(kr)^2 \quad (36)$$

where angular integration over the direction of k has been taken.
It must be noticed that in order to maintain the charge neutrality
the rearrangement of the charge density in the host has to screen

the excess charge Z of the impurity with respect to the host. This yields to the Friedel sum rule:

$$Z = (2/\pi) \sum_{\ell} (2\ell + 1) \eta_\ell \tag{37}$$

which gives a relation which must be fulfilled by the values of the phase shifts at the Fermi level $\eta_\ell \equiv \eta_\ell (E_F)$.
Those also enter in the expression of the impurity residual resistivity:

$$\Delta\rho_i = (4\pi\hbar c/Ze^2 k_F) \sum_{\ell=1}^{\infty} \sin^2(\eta_{\ell-1} - \eta_\ell) \tag{38}$$

a) Host NMR shift

As for dilute impurities the electronic density of states and thus the Pauli susceptibility of the alloy is not modified with respect to the pure metal, the main change of the host Knight shift occurs through the spatial variation of the s hyperfine field, which is proportional to $<|\psi_k(R_i)|^2>_F$, for the nucleus at site R_i from the impurity.
The excess Knight shift is then given by:

$$\Delta K(R_i)/K = \Delta H^s_{hf}(R_i)/H^s_{hf} = \Delta\rho_{kF}(R_i)/\rho^0_{kF} \tag{39}$$

which points out that $\Delta K(R_i)$ has an oscillatory R_i^{-2} spatial dependence. Experimentally the host NMR is broadened and shifted in the alloys and though satellite lines should be observed, this has not usually been the case, as their relative shift is much smaller than in magnetic alloys. A moment analysis of the spectrum has been performed from equation 36 and 39 for a random distribution of impurities. The obtained shift for the center of gravity of the spectrum is proportional to c, while the square root of the second moment which gives an estimate of the line broadening is found to increase as $c^{1/2}$. As can be seen from experimental data on ^{107}Ag NMR in silver based alloys (Fig. 10), such dependences are found to be valid up to a few atomic per cent [43]. From the present theory the NMR parameters should depend only upon the values of the phase shifts at the Fermi level. The Friedel sum rule (equation 37) indicate that the phase shifts should increase with Z, which explains the experimentally observed increase of $\Delta K/Kc$ with Z. However determinations of this quantity with the use of phase shifts obtained from resistivity data and theoretical expectations are slightly smaller (a factor 2 at most) than the experimental data. Those studied in sec. V, the excess of charge of the impurity

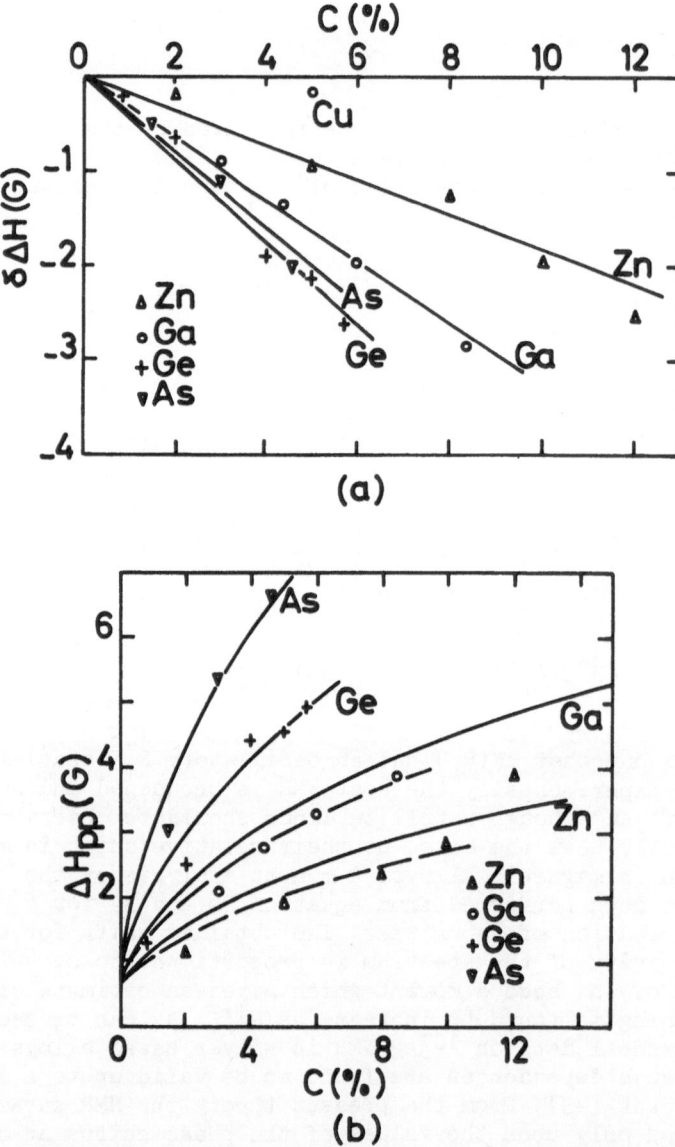

Fig. 10. ^{107}Ag Knight shift relative to pure ^{107}Ag NMR (a) and
linewidths (b) versus concentration for various impurities.
Both effects can be seen to increase with charge
difference Z between solute and host [43].

also yields a similar distribution of $\Delta\rho/\rho$, but usually the relative perturbation of χ_s is much greater than that of H_{hf}^s, which can be therefore neglected.

b) Electric field gradients (EFG)

In a pure cubic metal for which the EFG vanish at each lattice site, the impurities will create through the distribution of charge density a corresponding distribution of the EFG values $q(R_i)$ on the nuclear sites located at R_i from the impurity. It can be shown that the EFG is proportional to the excess charge density at R_i, that is:

$$q(R_i) = (8\pi/3)\ \mu\Delta\rho(R_i) \tag{40}$$

where μ is a factor (= 1 for free electrons), which takes into account the band structure and is determined by the wave function $U_k(r)$, while $\Delta\rho(R_i)$ is the total charge density at site R_i in the case of a free electron gas, that is:

$$\Delta\rho(R_i) = \int_0^{k_F} \Delta\rho_k(R_i)d^3k$$

$$= -\sum_\ell (-1)^\ell \frac{2\ell + 1}{2\pi^2 R_i^3} \sin \eta_\ell \cos (2k_F R_i + \eta_\ell) \tag{41}$$

The total electric field gradient is then given by:

$$q(R_i) = (8\pi/3)\mu\ \alpha \cos(2k_F Ri + \phi)\ /r^3 \tag{42}$$

where α and ϕ are mainly related to the phase shifts. Two types of measurements giving partial information on $q(R_i)$ have been mainly undertaken on Cu and Al based alloys (a host nuclear spin I>1/2 is of course required).

Wipe out numbers

The electric field gradients are usually found so important that only the $-1/2 + 1/2$ transition contribute to the observed ^{63}Cu resonance, as soon as a few hundreds of ppm of impurities are introduced in copper metal. For the near neighbours of the impurity the $\pm$ 1/2 transition is severely broadened [$\Delta H\alpha q^2(R_i)$] and some near neighbour shells of the impurity do not contribute to the resonance. In an all or nothing model which

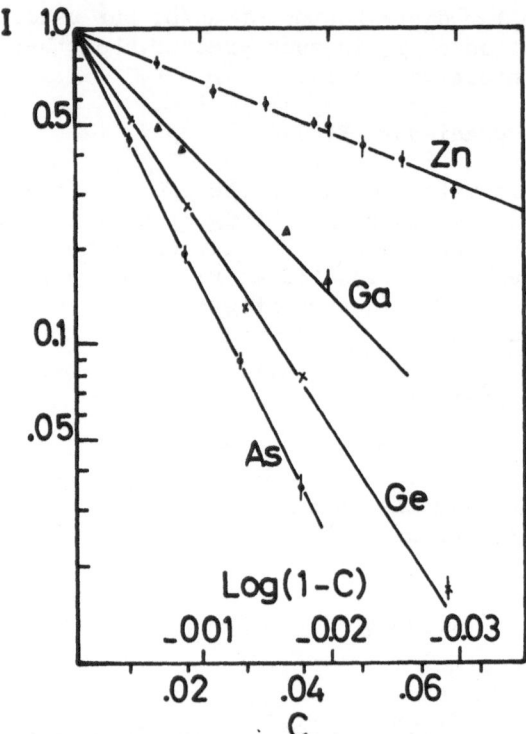

Fig. 11. Logarithm of the ^{63}Cu NMR intensity I versus Log (1 - c)
         for various impurities. The slopes give the wipe out
         numbers which are found to increase with charge
         difference between solute and host [44].

should be quite valid as ΔH has a very abrupt spatial dependence
(αR_i^{-6}), nuclei within a critical distance R_c of an impurity are
wiped out of the resonance. A nucleous has a probability $(1 - c)^n$
to be at a distance greater than R_c from any impurity, where n
is the number of lattice sites within R_c. Consequently the intensity
of the observed line is proportional to $(1 - c)^n$. Experimental
plots of log I versus (1 - c) allow then to deduce the so called
wipe out number n as is illustrated in figure 11. Similar analysis
can be done for the first order effect when the electric field
gradients are smaller, which is the case in some Al based alloys,
but experiments on foil samples are then required in order to avoid

quadrupole wipe out due to lattice imperfections [45].

Direct measurements of $q(R_i)$ on near neighbour shells of the impurity

As the quadrupole frequencies are expected to be small (a few MHz at most), a direct observation of the NQR of various shells of neighbours around an impurity is hopeless because of the lack of sensitivity at these frequencies. A field cycling technique due to Redfield [46] has been rather used in Cu based and Al based alloys. The nuclear spins are polarized in a high external field H_0. The field is then decreased adiabatically in a time shorter than T_1, allowing to cool the dipole-dipole reservoir. If then an audio-frequency field at the NQR frequency of one shell of neighbours is applied, allowing to heat the corresponding nuclei, spin diffusion processes will heat the whole dipole-dipole reservoir. The field is then raised back adiabatically to the value H_0 where the intensity of the measured signal give the final spin temperature of the system. The maximum signal is obtained for no irradiation in low field and decreases of the measured intensity are observed for frequencies $v w_Q(R_i)$, as can be seen on figure 12 [47].

By comparing both types of measurements which deal with values of $q(R)$ at large and small R respectively for the wipe out and field cycling experiments, it can be seen that the spatial dependence of $\Delta\rho(R)$ given by equation 42 does not extend up to the near neighbour shells of the impurity. The measured values of $q(R_i)$ are usually greater than those deduced from equation 42 assuming phase shift values as given by the Friedel sum rule and residual resistivity measurements. This is not extremely surprising as such discrepancies did also occur for the Knight shift measurements in Ag based alloys. It must be pointed out that while the Knight shift data only involve the charge density at the Fermi level and then the phase shifts η_ℓ (equations 36 and 39), the quadrupole data involve the total charge density and moreover details of the band structure through the coefficient μ of equation 40. The difficulties encountered is such refined quantitative analysis of the data are of the same order as those already met for the interpretations of the Knight shift in pure metals.

In the particular case of Al Mn, in which the Mn is a slightly magnetic impurity both the spin polarization effects [33, 39] and the charge density perturbation could be studied in great detail [48, 49].

In such a case only the phase shift η_2 is important at the Fermi level, and the important impurity susceptibility corresponds to the existence of a peaked d density of states at the Fermi level.

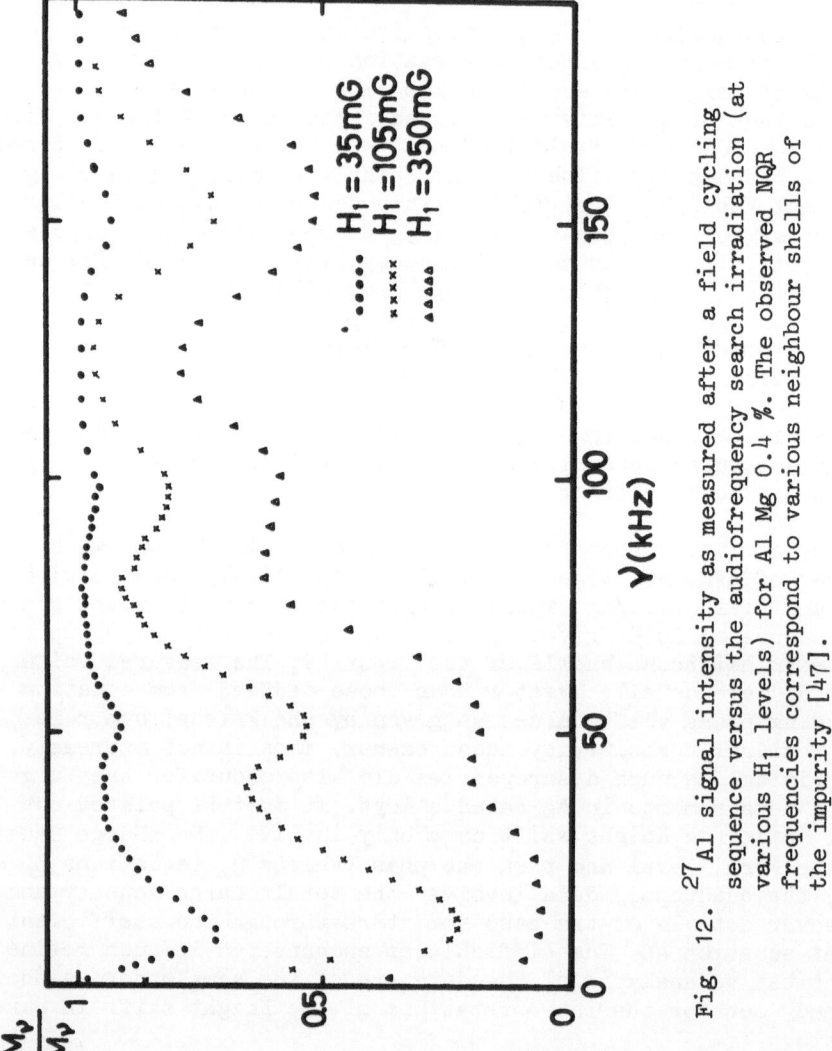

Fig. 12. ^{27}Al signal intensity as measured after a field cycling sequence versus the audiofrequency search irradiation (at various H_1 levels) for Al Mg 0.4 %. The observed NQR frequencies correspond to various neighbour shells of the impurity [47].

The distribution of local fields $H_{eff}(R_i)$ is then due to a spin polarization of the host as discussed in sec. V.   The experimental studies of the spatial dependences of $H_{eff}(R_i)$ and $q(R_i)$ even allow to deduce information on the energy dependence of the phase shift $\eta_2(\varepsilon)$.

CONCENTRATED ALLOYS

In the previous section it has been shown on some dilute alloy systems that local changes of the susceptibility or of the electronic density may be induced by impurities in a metal. In concentrated alloys other type of local effects might be expected, such as for instance a change of magnetic properties of one of the atomic constituents with its immediate environment. Moreover changes of band structure associated for instance with a variation of the total number of free electrons, can modify completely the electronic properties of a binary alloy system throughout the concentration range. In most cases a good knowlegde of the electronic structure of the pure metal constituents and of the properties of the dilute alloys is quite valuable for inferring some properties of the concentrated alloys. NMR experiments might help to understand the relative importance of the local and band effects.

Though experimental data has been taken in a lot of concentrated binary or ternary alloy systems, a succesful interpretation could be provided only in a few cases for which complementary experimental data from other techniques are available, or for which theoretical considerations allow important simplifications. It is then out of question to do a general analysis of alloy systems as each case has its own particularities. Some examples of binary alloy systems showing various effects which might occur will be rather considered here. More details on the analysis which have been performed in concentrated systems can be found in a recent review article [50].

1. Alkali Metals and Noble Metal Alloys

For isoelectronic alloys, the ionic potentials of the two constituents should not be much different; consequently local effects are not expected to dominate specially for simple metals. The simplest case which can be encountered is that of liquid binary alkali alloys which have spherical Fermi surfaces and for which the main modification with composition can be expected to be a volume (or electron density) change. Knight shift data have been taken on a wide range of concentrations in such systems and it is found that in all cases K decreases with the introduction of a heavier constituent. Calculations for $<|\psi(0)|^2>$ being considered show that the s hyperfine field is nearly independent of alloy composition for a given constituent, which allows to conclude that the observed shift changes are mainly related to changes of χ_s. The analysis of the data then shows that the exchange enhancement of the spin susceptibility is a unique function of the electron density in these systems [51].

In Ag Au solid solutions similar analyses are not simple, as the Knight shift data of the pure metals are not well understood, as reliable calculations of $<|\psi(0)|^2>$ are not available. Narath used the specific heat data to determine $\rho(E_F)$ and the measured enhancements of the Korringa value for Ag in order to infer values for χ_s. Such an analysis would indicate that oppositely to the case of alkali alloys the Ag Knight shift variation is due to a change of H_{hf}^s, associated with a change of the s part of the Fermi surface with composition [52].

Owing to the small influence of local effects in such alloys, the complexity of the interpretations of the data is analogous here to the one encountered in pure metals, and the experiments might indeed allow to understand the electronic properties of the pure metals. For alloys with a charge difference Z between the constituents, both local and band structure changes occur simultaneously and there is no satisfactory analysis of the Knight shift data even in simple alloys such as Ag Cd.

2. Transition Metal Alloys

In a simple model which is often used in order to describe the properties of 3 d (or 4 d) metals, it is considered that the shape of the 3 d (or 4 d) band is quite independent of the given metal, the only change from a metal to another being a progressive filling of the d band with increasing atomic number. Within this so called rigid band model, alloying two near neighbour elements of the Mendeleev table, should allow to fill progressively the d band. As the NMR parameters in such transition metals are mainly associated with d electron properties (sec. II), NMR experiments might then permit to check the applicability of the rigid band model (which corresponds to assuming a uniform susceptibility in the alloys). It is for instance found that this model applies quite well for Pt Au alloys, as the negative Knight shift of Pt decreases in absolute value from − 3.4 % in pure Pt and vanishes at 70 % Au. Such a reduction of the d susceptibility is quite compatible with a complete filling of the d band [53].

Another experiment supporting this model can be found in experimental measurements of the ^{51}V relaxation rates in V Cr alloys [54]. The d hyperfine coupling as well as the average orbital degeneration are not expected to change very much with composition. Consequently the d and orbital terms, which dominate the measured relaxation rates, are proportional to $\rho_d^2(E_F)$.

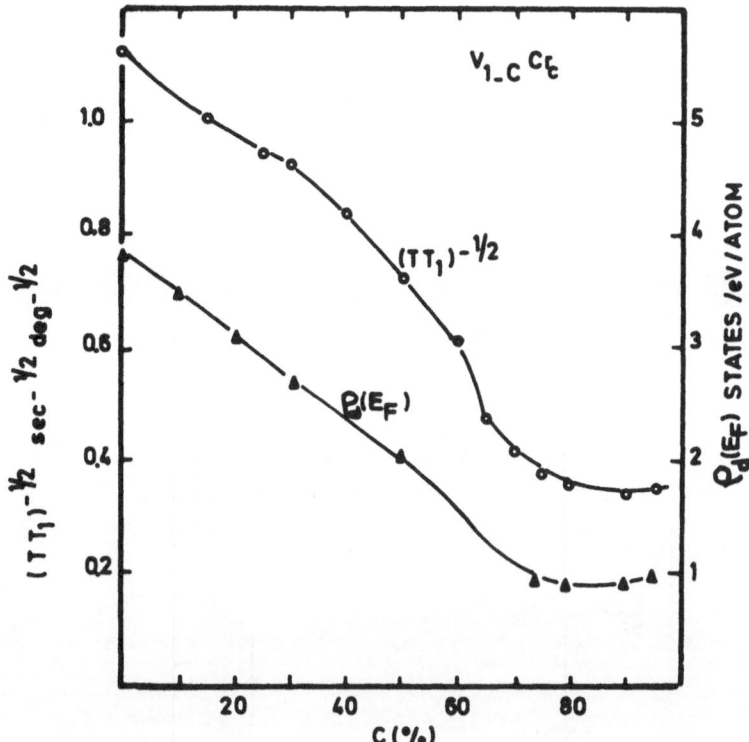

Fig. 13. Variation of the specific heat and $(T_1 T)^{-1/2}$ of the ^{51}V
NMR in the whole concentration range for VCr alloys. It
can be seen that $(T_1 T)^{-1/2}$ is proportional to the d
density of states, which would agree with the
assumption of a rigid band model [54].

The data for $(T_1 T)^{-1/2}$ and the electronic specific heat are found
to have parallel variations in the whole composition range, which
supports seriously the rigid band model (Fig. 13). It must be
noticed that the Knight shift of V is not found proportional to
$\rho_d(E_F)$, which is indeed expected as K is dominated by K_{orb} which is
not proportional to $\rho_d(E_F)$ as was pointed out in sec. II.

The validity of this rigid band model is far from being general
as in some cases the magnetic properties of one of the constituents
might be dominated by local effects. This is for instance the case
for V Fe alloys, as paramagnetic moments on Fe atoms begin to
appear on some Fe sites at concentrations of about 22 %. Such
influence of local environment on the appearance of magnetic moments
on transition metal impurities are known to dominate the properties

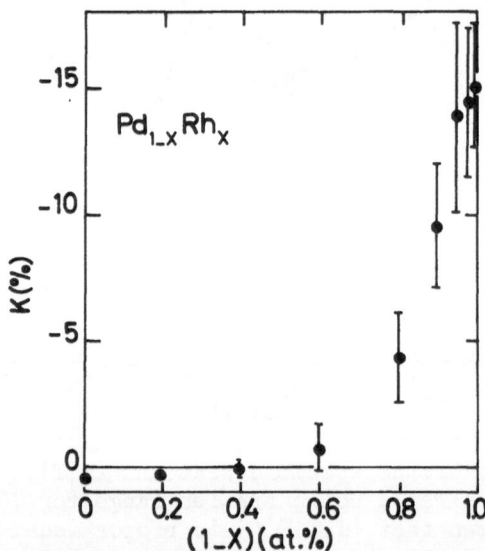

Fig. 14. ^{103}Rh Knight shift in PdRh alloys; the Rh shifts become
           huge and negative in the Pd rich end showing that the Rh
           susceptibility is high. Comparison with the measured
           χ indicates nevertheless that a major part of this
           susceptibility is distributed over the Pd lattice sites
           [55].

of alloys with noble metals. An example has already been given for
Cu Fe alloys in Sec. V. Similarly, groups of 3 cobalt atoms or
of 8 Ni atoms are found to bear magnetic moments in Cu while
isolated impurities are non magnetic.

In the case of Pd based alloys, as this metal has a highly
enhanced susceptibility, very striking deviations from this
rigid band behaviour are observed. In Pd Rh alloys the measured
bulk susceptibility per Rh impurity is about 20 times greater
than the pure atomic susceptibility. Correspondingly the Rh
Knight shift is negative but only 3 times greater than the pure
Pd shift at the Pd rich end of the diagram [55](Fig. 14).

This allows to understand that the measured macroscopic
susceptibility is spatially extended on many Pd sites around Rh.
A similar behaviour was found on Pt Ir alloys as the Knight shift
data for Pt also indicates that part of the impurity susceptibility
is extended on the neighbouring platinum sites [56].

SUMMARY

In these lectures the main concepts which allow to correlate
the NMR data with the electronic properties of metals and alloys
have been outlined. It has been shown that though the Knight shift,
T_1 and indirect couplings are well known in most pure metals, the
main difficulties encountered in attempting to deduce from the
data the interesting information on the electronic properties
are mainly the absence of reliable theoretical calculations of the
hyperfine fields or of the spin susceptibilities. Though NMR brings
some valuable information on both the static and dynamic parts
of the spin susceptibilities, other independent experimental
information on this important quantity, as well as on the electron-
electron interaction effects are required in order to enhance the
present understanding of the electronic properties of metals.
Most efforts have been rather devoted recently to the study of
dilute alloys as quite important information on the local properties
of these systems can be obtained. Though in the present lectures
a detailed description of the origin of the magnetism of d
impurities in s metals could not be given, the important
contributions of NMR experiments to the understanding of the sd
interaction and of the problem of localisation of the magnetic
electrons have been pointed out. It has also been shown that in
any case an analysis of the NMR results did require important
simultaneous data from other experimental techniques. The progress
which has been done in the field of dilute alloys these last few
years is for a great part linked to the great variety of
experimental techniques which have been used in order to study
identical systems.

Though much NMR work on concentrated alloys is available, it has been shown that, at present, a qualitative understanding, at best, can be reached in some cases. The experimental techniques and theoretical concepts which have been improved in order to understand the dilute alloy cases can be expected to allow significant progresses on more concentrated alloys in the future. It should be noticed that very important problems, such as ferromagnetism, superconductivity, magnetic properties of small particles, have also been at the origin of a great number of NMR studies, but have not been mentioned here.

It must be finally pointed out that many present experimental studies are turned towards oxides or compounds which present peculiar physical properties such as a transition from metal to insulator with pressure or (and) temperature, or which behave like uni-or two dimensional metals.

REFERENCES

1. W.D. KNIGHT, Phys.Rev. 76, 1259 (1949)

2. N. BLOEMBERGEN and T.J. ROWLAND, Acta Metall. 1, 753 (1953)

3. A. ABRAGAM, Principles of Nuclear Magnetism, Clarendon, Oxford (1961)

4. C.P. SLICHTER, Principles of Magnetic Resonance, Harper and Row, New York (1963)

5. J. WINTER, Magnetic Resonance in Metals., Clarendon, Oxford (1971

6. W.D. KNIGHT, Solid State Physics, ed. F. Seitz and D. Turnbull, 2, Academic, New York (1956)

7. L.E. DRAIN, Met.Rev. 11, 195 (1967)

8. A. NARATH, Hyperfine interactions, ed. A.J. Freeman and R.B. Frankel, 287, Academic, New York (1967)

9. R.G. BARNES, Magnetic Resonance, ed. C.K. Cougan, N.S. Ham, S.N. Stuart, J.R. Pilbrow and G.V.H. Wilson, 63, Plenum Press, New York (1970)

10. J. KORRINGA, Physica 16, 601 (1950)

11. D. AILON and C.P. SLICHTER, Phys.Rev. 137, 2995 (1966)

12. R.T. SCHUMACHER and C.P. SLICHTER, Phys.Rev.101, 58 (1957)

13. R. HECHT and A.G. REDFIELD, Phys.Rev. 132, 972 (1963)

14. For more details on the theories for the electron-electron enhancement of $\chi(q, \omega)$ see R.W. SHAW and W.W. WARREN, Phys.Rev. B3, 1562 (1971) and references herein

15. V. JACCARINO, Theory of Magnetism in Transition Metals, ed. W. Marshall, Academic Press (1967)

16. A.M. CLOGSTON, V. JACCARINO and Y. YAFET, Phys.Rev. 134, A650 (1964)

17. J.A. SEITCHIK, A.C. GOSSARD and V. JACCARINO, Phys.Rev. 136, A1119 (1964)

18. Y. YAFET and V. JACCARINO, Phys.Rev. 133A, 1630 (1964);
    A. NARATH, A.T. FROMHOLD and E.D. JONES, Phys.Rev. 144, 428 (1966)

19. More details on the non cubic metals can be found in [9]

20. S.N. SHARMA and D.L. WILLIAMS, Phys.Rev.Lett. 25A, 738 (1967)

21. R.V. KASOWSKI and L.M. FALICOV, Phys.Rev.Lett. 22, 1001 (1969)

22. M.A. RUDERMAN and C. KITTEL, Phys.Rev. 96, 99 (1954)

23. Yu.S. KARIMOV and I.F. SCHEGOLEV, Soviet Phys.JETP 14, 772 (1962)

24. S.N. SHARMA, D.L. WILLIAMS and H.E. SCHONE, Phys.Rev. 188, 662 (1969)

25. J. BUTTET, private communication

26. J. POITRENAUD, JPCS 28, 161 (1967)

27. H. ALLOUL and C. FROIDEVAUX, Phys.Rev. 163, 324 (1967)
    H. ALLOUL and R. DELTOUR, Phys.Rev. 183, 414 (1969)

28. S.D. MAHANTI and T.P. DAS, Phys.Rev. 170, 426 (1968)
    L. TERLIKKIS, S.D. MAHANTI and T.P.DAS, Phys.Rev.Lett. 21, 1796 (1968)

29. H. ALLOUL and P. BERNIER, Ann.Physique 8, 169 (1973-74)

30. A. NARATH, Crit.Rev. in Solid State Sciences 3, 1 (1972)

31. D.C. GOLIBERSUCH and A.J. HEEGER, Phys.Rev. 182, 584 (1969)

32. R.E. WALSTEDT and W.W. WARREN, Phys.Rev.Lett. 31, 365 (1973)

33. H. ALLOUL, J.Phys.F(Metals), to be published (1974)

34. R.E. WALSTEDT and L.R. WALKER, Phys.Rev. B9, 4857 (1974)

35. H. ALLOUL, J. DARVILLE and P. BERNIER, J.Phys.F(Metals), to be published (1974)

36. N. KARNEZOS and J.A. GARDNER, Phys.Rev. B9, 3106 (1974)

37. P. BERNIER and H. ALLOUL, J.Phys.F(Metals) 3, 869 (1973)

38. H. ALLOUL and P. BERNIER, Ibid. 4, 870 (1974)

39. H. ALLOUL, P. BERNIER, H. LAUNOIS and J.P. POUGET, J.Phys.Soc. Japan 30, 101 (1971)

40. J.B. BOYCE and C.P. SLICHTER, Phys.Rev.Lett. 32, 61 (1974)

41. J.L. THOLENCE and R. TOURNIER, Phys.Rev.Lett. 25; 867 (1970)

42. J. FRIEDEL, Phil.Mag. 43, 153 (1952)

43. T.J. ROWLAND, Phys.Rev. 125, 459 (1962)

44. T.J. ROWLAND, Phys.Rev. 119, 900 (1960)

45. G. GRÜNER, C. HARGITAI, Phys. Rev. Lett. 26, 772 (1970)

46. A.G. REDFIELD, Phys.Rev. 130, 589, (1963)

47. M. MINIER, Phys.Rev. 182, 437 (1969)

48. G. GRÜNER, Solid State.Commun. 10, 1039 (1972)

49. C. BERTHIER and M. MINIER, J.Phys.F(Metals) 3, 1169 (1973)

50. W.W. WARREN, Proceedings of Twin Symposia on Charge Transfer in Alloys and Electronic Structure of Alloys, Philadelphia (1973)

51. J.P. PERDEW and J.W. WILKINS, Phys.Rev. B7, 2461 (1973)

52. A. NARATH, Phys.Rev. 163, 232 (1967)

53. C. FROIDEVAUX, F. GAUTIER and I. WEISMAN, Proceedings of the International Conf. on Magnetism, Nottingham, 390 (1964)

54. J. BUTTERWORTH, Proc.Phys.Soc. 83, 71 (1964)

55. A. NARATH and H.T. WEAVER, Phys.Rev. B3, 616 (1971)

56. C. FROIDEVAUX, H. LAUNOIS and F. GAUTIER, Solid State Commun. 6, 261 (1968)

Chapter III

METHODS: FOURIER TRANSFORM, MULTIPLE PULSE, DOUBLE RESONANCE

THE FOURIER TRANSFORM IN NMR

I. WHY AND HOW

P. Van Hecke[*]

Katholieke Universiteit Leuven

Leuven, België

INTRODUCTION

Up to very recently, the Fourier Transform (FT) technique
was a typical tool for high resolution spectroscopists in complex
liquid systems.

Although Fouriertransform and its use in NMR as a sensitivity
enhancement tool were known since longer time (Ernst, 1966) and
although an algorithm had been developed for a fast computation
of this transform (Cooley, 1965), one had to wait another few
years for a real breakthrough among high resolution spectroscopists.
This came with the use of (on-line) minicomputers, time-averagers
and further instrumental developments. At the same time, ^{13}C
spectroscopy, which was benefitting from the same technical
evolution (superconducting magnets with high stability and
homogeneity, computer software) required a maximum of sensitivity,
along with maximum resolution for its complex spectra. This was
another trigger for a fast development of the FT spectroscopy.
This sensitivity enhancement is however only a particular aspect
of the FT technique. Various and mainly recent developments
(multiple-pulse line narrowing, cross-polarization rare nuclei
detection) are typically pulse experiments so that if we want
spectral information we will have to perform Fourier transformation.

---

[*] "Aangesteld Navorser" of the "Nationaal Fonds voor Wetenschappelijk Onderzoek"

In general, in every application where we need spectral information, for whatever reason, starting from time domain information (free induction decay), we will perform a Fourier transform.

It seems then quite appropriate to develop this subject here as it is becoming a common technique for most NMR spectroscopists in their study of chemical and biological systems, in the liquid as in the solid phase.

DEFINITIONS

If we define

$$F(\omega) = \int_{-\infty}^{\infty} f(t)e^{-j\omega t}dt \tag{1.1}$$

then $F(\omega)$ is known as the Fourier transform or Fourier integral of $f(t)$:

$$F(\omega) = F[f(t)]$$

Similarly, F^{-1} is the inverse operation of obtaining $f(t)$ when $F(\omega)$ is given, that is

$$f(t) = F^{-1}[F(\omega)] = 1/2\pi \int_{-\infty}^{\infty} F(\omega)e^{j\omega t}d\omega \tag{1.2}$$

$f(t)$ is called the inverse Fourier transform of $F(\omega)$. $F(\omega)$ and $f(t)$ are called Fourier transform pairs.

In other words, we assume that any given function has two equivalent representations: one in the time domain, $f(t)$, and the other in frequency domain, $F(\omega)$. Then eq. (1.1) transforms $f(t)$ in the frequency domain and eq. (1.2) transforms $F(\omega)$ in the time domain. In other words eq. (1.1) analyzes (decomposes) the time function into a frequency spectrum and eq. (1.2) synthesizes the frequency spectrum to regain the time function.

Let us, as an introductory example, consider the Fourier transform (spectrum) of a cosine wave function:

$$F(\omega) = F[\cos\omega_0 t] = \frac{1}{2}F[e^{j\omega_0 t}] + \frac{1}{2} F [e^{-j\omega_0 t}] \tag{1.3}$$

Straightforward calculation then yields:

$$F(\omega) = \pi\delta(\omega - \omega_0) + \pi\delta(\omega + \omega_0) \tag{1.4}$$

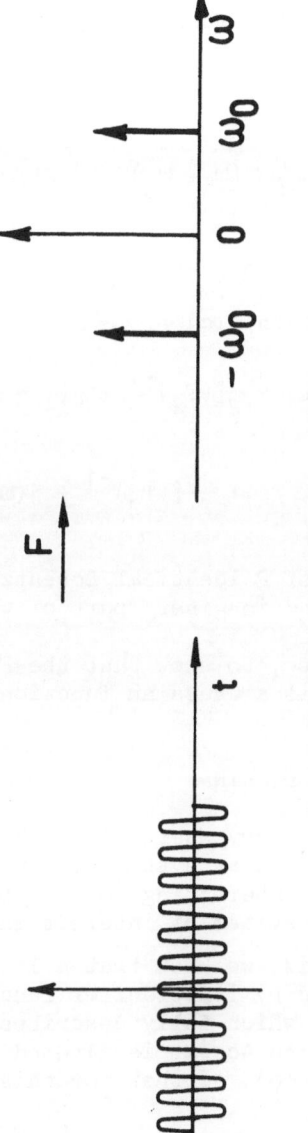

Fig. 1. The Fourier transform spectrum of the function $\cos \omega_0 t$.

Hence, the frequency spectrum consists of 2 infinitely short pulses (δ-function) at $\omega = \omega_o$ and $\omega = -\omega_o$ (Fig. 1).

A more realistic wave function in magnetic resonance, is the exponentially damped cosine function $\exp(-t/T_2) \cdot \cos\omega_o t$. The Fouriertransform of this function is easily calculated using a generalization of previous result, known as the modulation theorem:

if $F[f(t)] = F(\omega)$

then $F[f(t)\cos\omega_o t] = \frac{1}{2}F(\omega - \omega_o) + \frac{1}{2}F(\omega + \omega_o)$ (1.5)

Here, $f(t) = \exp(-t/T_2)$, the Fourier transform of which is the well-known Lorentz function $F(\omega) = [(T_2)^{-1} + j\omega]^{-1}$. Using the modulation theorem, we then find

$$F[\exp(-t/T_2) \cdot \cos \omega_o t] = \frac{1}{2}[(T_2)^{-1} + j(\omega - \omega_o)]^{-1}$$

$$+ \frac{1}{2}[(T_2)^{-1} + j(\omega + \omega_o)]^{-1}$$

This spectrum consists of 2 identical Lorentz lines, at $\omega = \omega_o$ and $\omega = -\omega_o$, the real and imaginary part of which are easily calculated.
It is left as an exercise, to show that the Fourier transform of a Gaussian function is a Gaussian function.

TIME DOMAIN - FREQUENCY DOMAIN

Fourier analysis is very useful in either frequency - or time domain analysis of linear systems (networks, filters, and so on). It is therefore interesting to make here the analogon with a spin system, the system of interest in magnetic resonance.

From system analysis, we know that a linear system is completely characterized by its transfer function or frequency response function $T(\omega)$, which fully describes how the amplitude and phase of an excitation $A\cos\omega t$ is altered by the device into an output signal $B\cos(\omega t + \phi)$, so that for this particular case,

$$T(\omega) = \frac{B}{A} e^{i\phi} \qquad (2.1)$$

In general if a signal $V_1(t)$, with Fourier transform $S_1(\omega)$, is applied to the input of a linear system with transfer function $T(\omega)$, then the Fourier transform $S_2(\omega)$ of the output signal $V_2(t)$ can be written as:

$$S_2(\omega) = T(\omega)S_1(\omega) \tag{2.2}$$

$V_2(t)$ is then synthesized from $S_2(\omega)$ by inverse Fourier transformation. Still from system analysis, it is known that the frequency response function $T(\omega)$ is the Fourier transform of the unit pulse response $I(t)$, the transient response – in time – to an infinitely sharp pulse $\delta(t)$. The output signal $V_2(t)$ can then immediately be calculated by convolution of $V_1(t)$ with the pulse response function $I(t)$

$$V_2(t) = \int_{-\infty}^{\infty} V_1(\tau) \, I(t - \tau)d\tau \tag{2.3}$$

For a spin system, the frequency response function is called the spectrum. Indeed, it gives us the information on what frequencies are absorbed (or irradiated) when subjected to continuous excitation. The unit impulse response is called the "free induction decay" which, very obviously, gives the time development of the behaviour of the spin system after a pulse excitation.

In 1957, Lowe and Norberg rigorously proved that in the high temperature approximation (a normally fulfilled condition), the frequency spectrum obtained by the continuous wave irradiation method – using a vanishingly small radiofrequency field – and the free induction decay after a pulse of high intensity (H_1 much larger than the local field), are Fourier transforms of each other, whatsoever the system, gas, liquid or solid.

PULSED FOURIER TRANSFORM OR CONTINUOUS WAVE

We want to briefly discuss here the specific and particular use of Fourier transform as sensitivity enhancement tool in (complex) liquid systems.

As both approaches, continuous and pulsed irradiation were proved to be able to reveal the same information with regard to the spin system, spectroscopists didn't have any reason to get excited about one or the other method, except for the specific use of pulse methods for relaxation measurements, where the latter have been popular since very long time, as they allow a straight-forward determination of various kinds of relaxation times. In 1966 however, Ernst and Anderson pointed out the potentialities of the pulse method (with Fourier transform of the response) to enhance the sensitivity of the magnetic resonance experiment, or conversely, to greatly reduce the acquisition time needed to reach a given signal-to-noise ratio.

A very crude approach of the problem shows us immediately the advantages of the pulse Fourier transform method. In CW (continuous wave) experiments, the irradiated frequency is swept slowly through the range of interest, while recording the frequency (field) response. The main inefficiencies of this method lie in the fact that the response to one frequency at a time is analyzed (time problem) and hence only a very small fraction of the spinsystem is excited (sensitivity problem).

A much more efficient approach would be the use of a multi-channel spectrometer, using as many transmitters and associated receivers as needed to cover the whole spectrum under investigation with a resolution of the order of the width of the individual lines. If the total spectral width is F(Hz), and the natural linewidth is Δ(Hz), the required number of channels would be

$$N = F/\Delta \qquad (3.1)$$

Given the same total time for the single-scan experiment and the multichannel experiment, it is obvious that we will be able to spend a N much longer time on each channel in the multichannel experiment, or a saving in time of 1/N. Since sensitivity increases as the square root of the observation time[*], the sensitivity gain G will be

$$G = N^{1/2} = (F/\Delta)^{1/2} \qquad (3.2)$$

Example: a multichannel experiment on ^{1}H at 100 MHz, covering a spectrum of 1 kHz (10 ppm), with a line-width of 0.1 Hz, would give a gain of 100 in sensitivity over a conventional slow passage experiment. The saving in time, to get the same S/N ratio, would be 10,000 (1/N).

The method is particularly rewarding for complex spectra - like most high resolution spectra are.

---

[*] For white and 1/f noise, it is shown rigorously (Ernst, 1965), that the r.m.s. value of the noise voltage increases with $n^{1/2}$, where n is the number of scans. The signal amplitude, increases like n, so that S/N increases like $n^{1/2}$, or $t^{1/2}$ as the number of scans is proportional to the time.

This multichannel spectrometer is quite unrealistic. However, a strong radiofrequency pulse can excite at once the entire range of precession frequencies, characteristic for the spin system under study, replacing in a very convenient way the whole string of transmitters of the multichannel spectrometer. Using Fourier transform, one easily calculates that the spectrum $F(\omega)$ of a rectangular pulse of width τ, and amplitude 1, is given by (see figure 2a):

$$\tau \, \sin \frac{1}{2} \, \omega\tau \, / \, \frac{1}{2} \, \omega\tau \qquad\qquad\qquad (3.3)$$

A pulse of width τ, has a spectral range extending to $1/\tau$, on the ν-scale. Making use of the modulation theorem, eq. (1.5), and of eq. (3.3), the spectrum of a pulse of frequency ν_0, and width τ, is

$$\frac{1}{2} \, \tau \{ [\, \sin \tfrac{1}{2}(\omega-\omega_0)\tau]/[\tfrac{1}{2}(\omega-\omega_0)\tau]+[\sin \tfrac{1}{2}(\omega+\omega_0)\tau]/[\tfrac{1}{2}(\omega+\omega_0)\tau] \} \quad (3.4)$$

The pulse covers a spectrum extending up to $1/\tau$ around ν_0 (Fig. 2b). In order to obtain an excitation spectrum of constant amplitude throughout the NMR frequency range F of interest, it is necessary to stay well inside the $1/\tau$ range:

$$|F| \ll 1/\tau \qquad\qquad\qquad\qquad (3.5)$$

The time response of the spin system to this broadband excitation, can be detected in a receiver, with the appropriate bandwith, as an interferogram of superposition of all frequencies irradiated by the spin system. This interferogram being in the time domain, we will recover the previous frequency spectrum of the multichannel experiment, by Fourier transformation of this interferogram, as these two signals form a Fourier transform pair (Lowe, 1957). This pulse Fourier-transform experiment is the analogon of the multichannel experiment, so that the sensitivity enhancement is given by the same crude approximation (eq. 3.2). Ernst (1966) made a more accurate calculation of the S/N ratio obtainable by the pulse Fourier transform method and that by the single scan (slow-passage) experiment, performed over a same total time T_t, assuming white noise and matched filters to maximize the S/N. The n responses after each pulse are added coherently ($nT = T_t$) before taking the Fourier transform. This ratio is found to be:

$$(S/N)_p/(S/N)_s = 0.799 \, (F/\Delta)^{1/2} \, G(T/T_1)$$

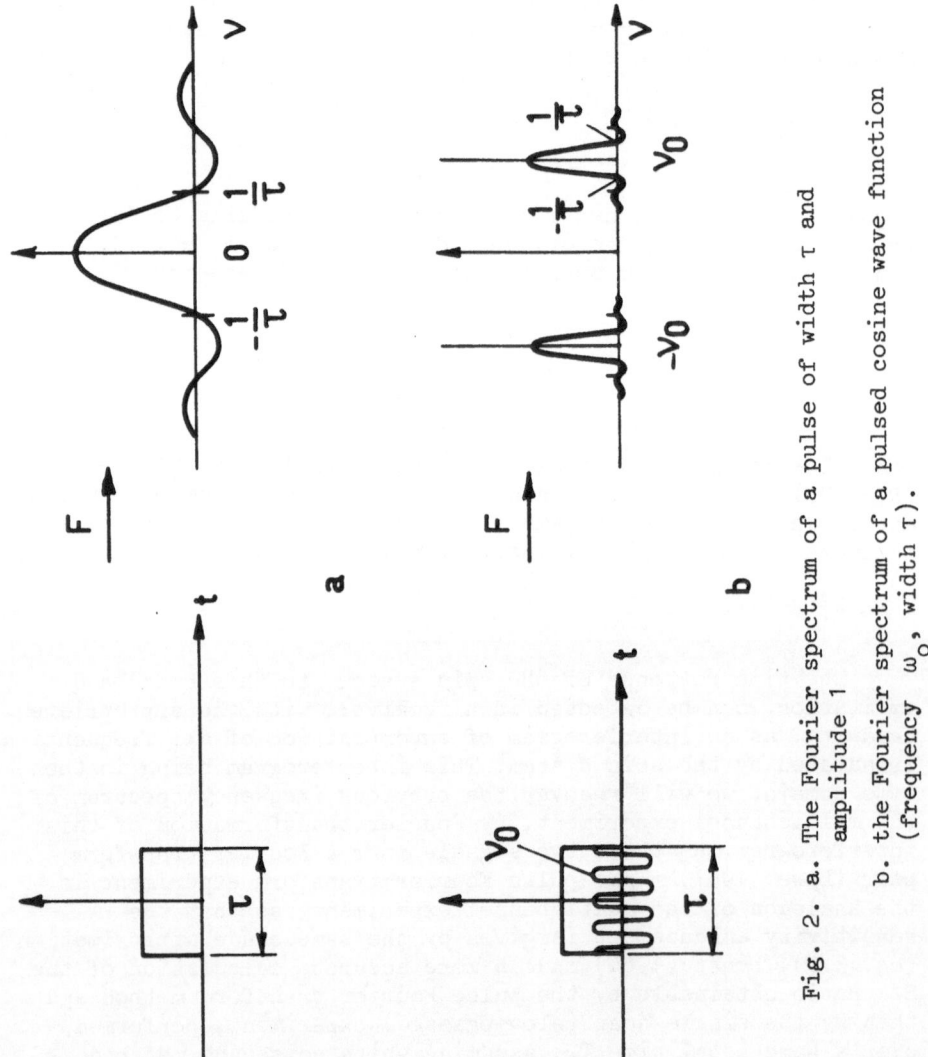

Fig. 2. a. The Fourier spectrum of a pulse of width τ and amplitude 1

b. the Fourier spectrum of a pulsed cosine wave function (frequency ω_0, width τ).

which is essentially the previous result (eq. 3.2). $G(T/T_1)$ is a slow varying function of T/T_1, of the order of 1. The gain is equal to the square root of the total available time to the time actually spent within a single line of width Δ, using the ordinary sweep technique.

Thesetheoretically tremendous savings are reduced to some more realistic numbers, if one recalls that true slow passage experiments are seldom used, and that usually a intermediate or even fast sweep is used, improving the sensitivity, at the expense of distortion and broadening. Enhancements in signal-to-noise ratio on this basis have turned to be about one order of magnitude (i.e. two order of magnitudes in time saving). Moreover, one has to make the following important restriction. The optimum repetition rate of a repetitive experiment, with respect to the S/N ratio is of the order of T_1. In the case of spectral sweep, this rate can only be approached if the ratio T_1/T_2 is high (this is particularly important for a complex spectrum), in other words, if the condition $T_1/T_2 \gtrsim F/\Delta$ is fulfilled. When this is not the case (i.e. $T_2 \approx T_1$), then the sweep rate will cause unacceptable broadening. It is in the latter situation that the pulse experiment is particularly rewarding in time (and gain) saving. A situation however which only occurs in liquids.

In conclusion, as the free induction decay after a pulse, and the absorption spectrum in CW irradiation, form a Fourier transform pair, they both contain the same information and are equivalent. However, if we keep in mind the rate at which the signal comes out of the noise, then, the FID (followed by a FT) will give us a tremendous enhancement, either in time (100) or in quality (10) in the case of complex high resolution spectra (Δ small, N big).

So far for this particular application. The use of FT is, as already mentioned earlier, of much wider application and this can, in a most general way, be formulated in the following terms. The necessity for us to take the Fourier transform of the pulse response is due to our inability to handle in our mind a complex time-domain interferogram, which synthesizes all frequencies irradiated by the spin system. If we want to see the whole information at once, we have to make the spectral analysis, equivalent to Fourier transforming into a frequency "spectrum".

DISCRETE FOURIER TRANSFORM (DFT)

Until now we considered continuous interferograms and their continuous Fourier transforms. The experimental situation however,

will limit the observation time of the FID and, equally, only a
finite or discrete number of samples will be taken from this FID.
The discrete Fourier transform, over a finite time interval will
introduce limitations, as we will discuss now, on the resolution
and on the upper limit of the transformed frequency spectrum.

The discretely sampled free induction decay f(t), over
a finite time domain T, can be written as

$$f^{*}(t) = \sum_{n=0}^{N-1} f(t)\, \delta(t - n\Delta t)$$

for N sampling intervals Δt.
The symbol $*$ stands for "discrete", and δ is the delta function.
Performing the Fourier transform and making use of the properties
of the delta function, yields

$$F[f^{*}(t)] = \frac{1}{N} \sum_{n=0}^{N-1} e^{-i\omega n \Delta t}\, f^{*}(n\Delta t) \tag{4.1}$$

A discrete number of samples in the time domain will give
a discrete number of points in the frequency spectrum. It is
convenient to write:

$$\omega = k\Omega \tag{4.2}$$

where Ω is associated to the lowest frequency one is able to
resolve, due to the limiting of the sampling to a total time

$$N\Delta t = T \tag{4.3}$$

Hence,

$$\Omega = 2\pi/N\Delta t = 2\pi/T \tag{4.4}$$

We can now write for the Discrete Fourier Transform (DFT)

$$F^{*}(k\Omega) = \frac{1}{N} \sum_{n=0}^{N-1} e^{-ikn\Omega\Delta t} f(n\Delta t) \tag{4.5}$$

where k = 0, ... N-1. The number of discrete frequencies in the
spectrum being equal to the number of samples in the time domain.
$1/N$ is the normalization constant. One can further prove that the
inverse discrete Fourier transform exists:

$$f^*(n\Delta t) = \sum_{k=0}^{N-1} e^{ikn\Omega\Delta t} F(k\Omega)$$

The use of the finite discrete Fourier transform, introduces some limitations that we discuss in the next two sections.

WINDOWING AND RESOLUTION

Sampling a signal over a finite time domain, from 0 to T, is looking at the signal through a window of length T, which results in multiplying our interferogram by a pulse function of unit amplitude and length T. All knowledge about the wavevorm before and after a time T is lost, which means that the frequency spectrum will be discontinuous around $1/T$.

If our interferogram consists of a unique cosine function, than we know (see section 3) that the Fourier spectrum of a windowed (pulsed) cosine of frequency ν_o and length T, is given by

$$T[\sin \pi(\nu-\nu_o)T]/ [\pi(\nu-\nu_o)T] \qquad\qquad (5.1)$$

The infinitely sharp peak (δ-function) at ν_o (and $-\nu_o$), associated to the cosine wavefunction extending to infinity   has been broadened to a width $1/T$ by limiting its observation to a time T (Fig. 2b).
$1/T$ is the ultimate resolution of the frequency spectrum and is the smallest frequency step in the discrete Fourier spectrum (eq. 4.4). The function (5.1) displays oscillations around ν_o, at frequencies $1/T$ apart. However due to the discrete sampling in the time domain, the discrete points in the frequency spectrum are at $1/T$ intervals, so that the side-oscillations due to windowing will not appear (the function is zero for these points).

The effect of finite sampling (windowing) was illustrated on a cosine wave function. NMR free induction decay signals however are decaying functions of time, due to the interaction of the spins with the static magnetic field or interactions intrinsic to the spin system. Their spectral lines are accordingly broadened (Gaussian, Lorentzian,. ...). In this case, one has to take care not to introduce additional broadening by truncating too soon the free induction decay. The acquisition time (T) should be long enough to satisfy $T > T_2^*$, in which case broadening effects due to windowing, will be negligible.
However, the larger the ratio T/T_2^* and consequently the better the resolution $(1/T)$, the poorer the sensitivity, since the S/N degrades

towards the end of the signal. This is a conflicting situation
between resolution and sensitivity.

SAMPLING RATE AND FOLDING

     The spacing Δt between two samples in the time domain is
defining the sampling rate (number of samples / unit time), $F=1/\Delta t$.
Sampling theory learns that the sampling rate F must be at least
twice the highest frequency present in the waveform, for the
waveform to be defined completely (Nyquist theorem). If the
NMR spectrum under investigation contains frequencies higher than
F/2, the sampling theorem is not fulfilled and the frequencies
above F/2 will be converted in the frequency range F/2, 0.
This is illustrated in the next example using cosine functions
(Fig. 3a). The waveforms (time-domain) are sampled at a
frequency F = 10 kHz (Δt = 100 µs); the highest frequency should
not be higher than 5 kHz for a faithful representation. A cosine
wavefunction, at a frequency of 8 kHz (3 kHz above the max.
frequency F/2, set by the Nyquist criterion) is seen to be
aliased (folded) to a frequency of 2 kHz (3 kHz below F/2).
The frequency of 5 kHz, or generally $1/2 \Delta t$, is called the
foldover frequency, because any frequency above it is folded
down by a same amount, below it.

     Aliasing can be recognized by inspection of the frequency
spectrum for different sampling rates F; lines moving over
the spectrum are the result of folding; the sampling rate has
to be increased accordingly.

     To illustrate these concepts of sampling rate (F),
resolution (1/T) and number of points (N), let us take 2
examples:
- ^{13}C liquid spectrum of progesterone at 23 MHz (Breitmaier, 1971).
  The spectrum covers about 200 ppm, i.e. 5 kHz. The sampling
  rate F will have to be at least 10 kHz. Having at our disposal
  4096 "channels" to store the data (memory of averager or
  computer), the total sampling time is:

     $$4096 \times \mathcal{V}(10 \text{kHz}) \overset{\sim}{=} 0.4 \text{ s}$$

which is of the order of T_2^* (no visible effects of truncation),
and the resolution is $(0.4 \text{ s})^{-1}$ = 2,5 Hz. If the resolution
were not sufficient, then we would have to increase the observation time, this is, the number of data points, and thus the
capacity of the memory as the smallest interval of 100 µs, is
set by the Nyquist criterion.

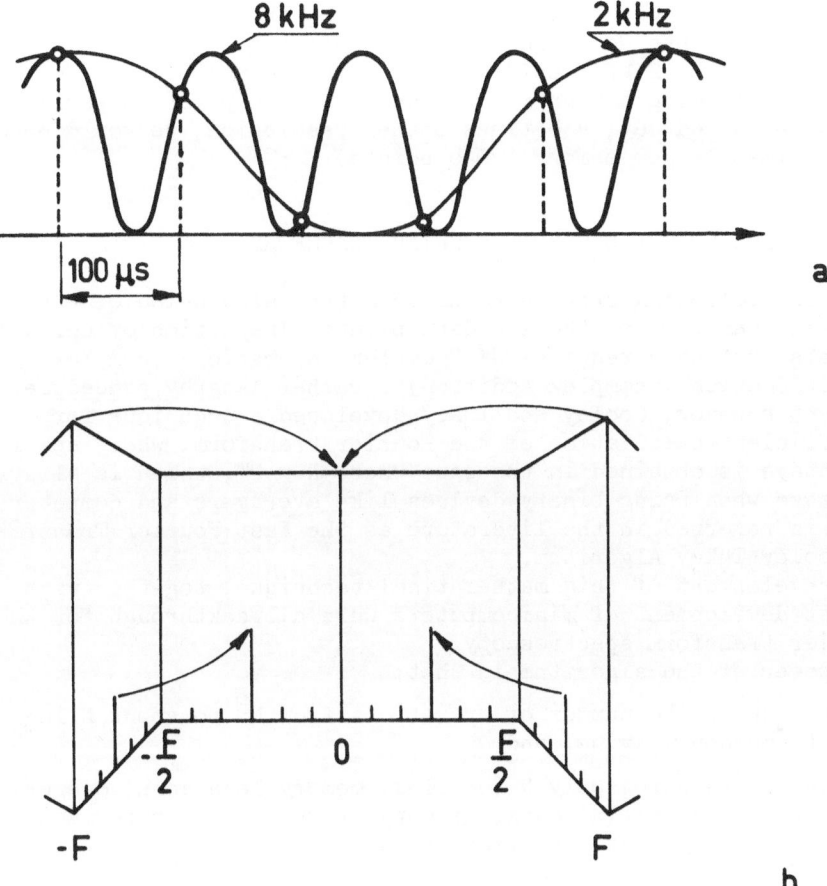

Fig. 3. a. Illustration of folding. A cosine wave function at a
        frequency 8 kHz, sampled at a rate of 10 kHz, is
        seen to be folded down to a frequency of 2 kHz about
        the foldover frequency (5 kHz)

     b. A pictorial representation of the folding of a
        frequency $F/2 + f_i$, to a frequency $F/2 - f_i$.

- ^{31}P solid spectrum of P_4S_3 at 100 MHz (Gibby, 1972). The spectrum
shows several well-resolved peaks due to large chemical shift
interactions, extending over about 700 ppm. At 100 MHz this
represents a spectrum of 70 kHz, requiring a sampling rate of
140 kHz (7 μs/point). This can usually not be accomplished by
standard averagers or computers and requires a fast digitizer.
With 256 points as taken by the authors, this gives a
resolution of 2 kHz, sufficient for this experiment. If we
were performing a high resolution (multiple pulse) experiment,
and we wanted say, a 8 times better resolution, we would need
a 8 times larger memory (2048 points).

CALCULATION OF THE DISCRETE FOURIER TRANSFORM

     The following step is to perform the calculation of the
Fourier transform of these N data points. Inspection of eq.(4.5)
reveals that this requires N^2 "complex" operations (complex
multiplication + complex addition) a rather lengthy procedure.
In 1965 however, Cooley and Tukey developed a technique for
an efficient calculation of the Fourier transform, where special
advantage is obtained in the case where $N = 2^m$, which is always
the case when using binary devices like averagers and computers.
This is referred in the literature as the Fast Fourier Transform
or Cooley-Tukey Algorithm.
The development of this mathematical technique, together with
a fast development of minicomputers were a breakthrough for the
Fourier transform spectroscopy.
The power of the algorithm is that:

1) it reduces the number of operations from N^2 to about $N \log_2 N$
   a tremendous saving, and

2) that it requires only N (complex) memory locations, overwriting
   however the initial data, (a copy of which has to be made) to
   store intermediate and final results.

     For the case of N complex points in the time domain (the
data of the resp. output of 2 quadrature phase detectors form
a pair of numbers which may be regarded as a complex number),
Fourier transformation will deliver N real and N imaginary points
of the spectrum, using N complex locations.
For the case of N real points in the time domain (single phase
detector, the most encountered situation up to now), it is
possible to perform the Fourier transform, yielding N/2 real
and N/2 imaginary points, using N real locations (Bergland, 1969).

     The algorithm is now a standard item in the software of
most computer systems (it can also be performed hardware) and
there are as many versions as users.

REFERENCES

G.D. BERGLAND, IEEE, Trans.Audio and Electroacoustics AU-17, 138 (1964)

E. BREITMAIER, G. JUNG and W. VOELTER, Angew.Chem.Internat.Edit. 10, 673 (1971)

J.W. COOLEY and J.W. TUKEY, Math.Comp. 19, 297 (1965)

R.R. ERNST, Rev.Sci.Instrum. 36, 1689 (1965)

R.R. ERNST and W.A. ANDERSON, Rev.Sci.Instr. 37, 93 (1966)

M.G. GIBBY, A. PINES, W.K. RHIM and J.S. WAUGH, J.Chem.Phys. 56, 991 (1972)

I.J. LOWE and R.E. NORBERG, Phys.Rev. 107, 46 (1957)

THE FOURIER TRANSFORM IN NMR

II. SIGNAL PROCESSING AND INSTRUMENTAL REQUIREMENTS

P. Van Hecke[*]

Katholieke Universiteit Leuven

Leuven, België

FILTERS AND DETECTORS

We defined the Fourier transform (FT), we know the real
advantages of using it, we discussed the discrete Fourier
transform (DFT) over a finite time domain and some of its draw-
backs and finally we referred to an algorithm for a fast
calculation of this discrete Fourier transform (FFT) (cfr.
previous lecture, referred as I). Now we completed the
description of the main steps in performing Fourier Transform
Spectroscopy, it is useful to discuss:

1) how to optimize the quality of our signals (in the time and
the frequency domain), either by instrumental improvements
or mathematical manipulations;

2) what corrections we have to make on the signals for a faithful
representations of the input.

This discussion will mainly involve instrumental considerations,
so that we will start again right at the source of our signals:
the receiving system which amplifies the signals induced in the
radiofrequency coil by the precessing magnetization. This
magnetization is often represented as a complex function which
it is convenient to think of as a superposition of magnetizations
precessing at different frequencies (associated to different

---

[*] "Aangesteld Navorser" of the "Nationaal Fonds voor Wetenschappelijk Onderzoek"

spin groups) each displaced by a small amount $\Delta\omega_i$ from the spectrometer frequency ω, and damped according to a real envelope function $f_i(t)$:

$$M = M_x + iM_y = \sum_i f_i(t) \exp[j(\omega + \Delta\omega_i)t - \phi_i] \qquad (1.1)$$

A receiver coil along the x-axis in the laboratory frame will detect only the real part of the magnetization M:

$$M_x = \sum_i f_i(t) \cos[(\omega + \Delta\omega_i)t - \phi_i] \qquad (1.2)$$

This waveform is now linearly amplified. A detector will decode the information carried at the frequencies $\omega + \Delta\omega_i$, to give the real envelopes $f_i(t)$, in the following way: if the input signal (1.2) is mixed with a cosinusoidal signal at the spectrometer frequency, the output of the detector will be, for the ith spin group (isochromat):

$$f_i(t) \{\cos[(2\omega + \Delta\omega_i)t - \phi_i] + \cos[\Delta\omega_i t - \phi_i]\}$$

This is accomplished in a phase detector. After filtering at the frequencies ω, 2ω using a low pass filter, like a RC filter, the signal is reduced to:

$$f_i(t) \cos[\Delta\omega_i t - \phi_i] \qquad (1.3)$$

which is the envelope of the original induction signal modulated by the frequency-offset particular to each of the irradiated signals (or spin groups).

From this, we immediately see that a frequency $\Delta\omega_i$ above the spectrometer frequency, cannot be distinguished from a frequency $-\Delta\omega_i$, below the spectrometer frequency. And this is a real problem if we take the Fourier transform. Indeed, using the modulation theorem (equation I.1.5), it is clear that the spectrum will consist of two lines F_i at frequencies $\Delta\omega_i$ and $-\Delta\omega_i$ around the spectrometer frequency ω(or $\Delta\omega = 0$). Hence we will not be able to distinguish in our transformed spectrum, the lines "below" and "above" the spectrometer frequency, as each one has a symmetrical ghost line. In order to avoid this, one chooses (shifts) the spectrometer frequency ω well outside the range of all frequencies $\Delta\omega_i$ in the spectrum (either below or above it), so that we know that all $\Delta\omega_i$ will have the same sign. The positive part of the Fourier transformed spectrum will then be a true representation of the experimental spectrum under investigation.

There are some disadvantages to this off set method:

1. The Nyquist theorem learns that the sampling rate should be at least twice the largest value of $\Delta\omega_i/2\pi$. This rate will be twice as high when the spectrometer is off set on one side of all $\Delta\omega_i$, than in the case where the spectrometer was adjusted in the middle of the $\Delta\omega_i$ spectrum. With the same number of points we would then be able to sample over twice a longer time, improving the resolution by a factor 2. Any extra-offset should also be minimized.

2. The same problem arises for the intensity of the 90° pulse, which in any circumstance should be adjusted in such a way that

$$\gamma H_1 \gg |\Delta\omega|_{max} \qquad (1.5)$$

so that the 90° condition is fulfilled for all isochromats of the spectrum. By offsetting the spectrometer frequency, we will need double the value of H_1, i.e. 4 times the power. This will usually not be a problem in liquids. In solids however, mainly in multiple pulse techniques where short pulses are used, this might be a more stringent requirement.

3. Although we solved, by offsetting the spectrometer frequency, the sign ambiguity for the signals, this is not true for the noise. Indeed, all noise frequencies, say below $\Delta\omega = 0$ (i.e. ω) will be imaged above $\Delta\omega = 0$. Having this in mind, we can reduce the noise power with a factor 2, by putting a rf crystal filter before detection (Allerhand, 1973) which will pass all frequencies above $\Delta\omega = 0$ (ω) and attenuate all frequencies below, improving the r.m.s. noise voltage and consequently the S/N ratio by a factor $\sqrt{2}$.

Another (and better) way to solve the ambiguity in the sign of $\Delta\omega_i$ is the use of 2 phase detectors whose reference signals are in quadrature, $\cos\omega t$ and $-\sin\omega t$. This can be realized in a convenient way (see fig. 1a) by using commercially available quadrature hybrids which deliver, usually over a broad frequency band, 2 outputs which are in phase quadrature (within about 1°). The output signals of those 2 detectors:

$$f_i(t) \cos [\Delta\omega_i t - \phi_i]$$

and

$$f_i(t) \sin [\Delta\omega_i t - \phi_i]$$

can then be treated as the real and imaginary part of the complex signal, which contains all the information induced in the coil by the complex magnetization.

These two inputs form an appropriate input for a complex Fast Fourier Transform, which will deliver a complex spectrum: the real part of which is the absorption signal, the imaginary part, the dispersion signal. This spectrum, centered around $\Delta\omega = 0$, will be free of any image or "ghost" lines, as each line has been defined (in amplitude and phase) by a complex number (i.e. a pair of real numbers). This is readily seen from the complex modulation theorem:

$$F\ [f_i(t)e^{j\Delta\omega_i t}\ ] = F_i(\Delta\omega - \Delta\omega_i)$$

As at the same time, and for the same reason, the noise will not be imaged about $\Delta\omega = 0$, we will have the same reduction in noise power (which means the same improvement in S/N ratio by a factor $\sqrt{2}$) as in the case of a single-detector making use of a rf filter before the detector.

Quadrature phase detectors, which are broadband devices or at least working at some fixed intermediate frequency, are more convenient to use than rf crystal filters (in front of the detector), as these have to be replaced for every change in frequency, and require an accurate adjustement and tracking of the spectrometer within its bandpass curve, for maximum results. Although advantages of a quadrature phase detector above a single detector are obvious, few people reported its use (Ellett, 1971; Jeener, 1964; Redfield, 1971; Stejskal, 1974). Care has to be taken for a good phase and amplitude balance of the two detecting devices. The deviation of a few degree from perfect quadrature found on commercial components and the few percents deviation on gain, have to be trimmed for in order to avoid weak "ghost" lines in the Fourier spectrum. Several methods can be used: introduction of stray impedances, computer corrections or use of a 2 pulse cycle and routing of the data (Stejskal, 1974). With a quadrature detector accurate to $1°$ in phase and 1% in amplitude, no reflections are visible in ^{13}C spectra of average sensitivity (Stejskal, 1974).

The detected signal(s) is (are) now routed to the digitizer, where they will be sampled at a rate F, where F satisfies the Nyquist criterium ($F > 2|\Delta\omega_i|_{max}/2\pi$). As frequencies above $F/2$ are folded down below $F/2$, the noise frequencies above $F/2$ are converted down and cannot be separated from the signal anymore. Therefore it is important to put a low-pass filter in front of the digitizer, in order to suppress all frequencies above $F/2$. Of course, the filter will have to be matched to the spectral width F, in order not to distort the incoming signal by over-filtration (introducing frequency dependent intensities and

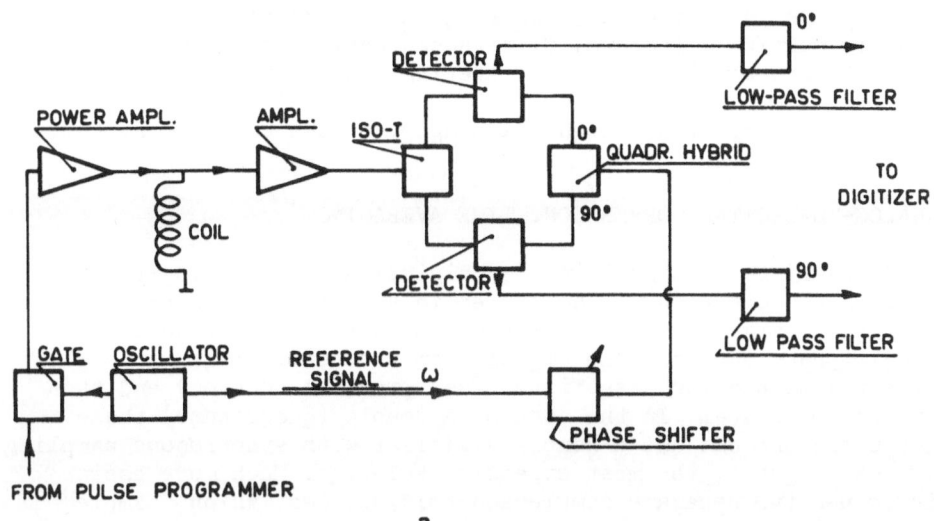

a

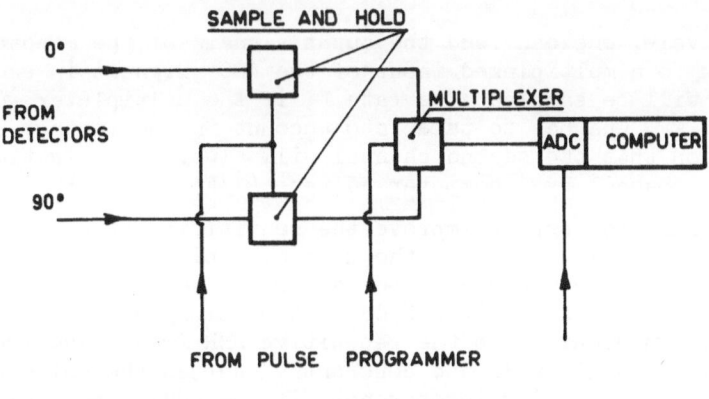

b

Fig. 1.a. The quadrature phase detector schema

b. Sequential digitization of the two quadrature phase
signals.

shifts). A perfect filter would pass all frequencies from 0 to
F/2 with no alteration or time delay, and reject all higher
frequencies. Commonly used RC filters have a roll-off such that
the voltage is halved when the frequency is doubled (6 dB/octave);
tunable multipole (octave) filters can achieve a steeper roll-
off (12 dB/octave). Optimum filtering can always be done after
sampling (see furhter). This filter in front of the digitizer
is usually part of the signal averager, or the computer system.

ANALOG-TO-DIGITAL CONVERSION. TIME AVERAGING

     The filtered signal is then digitized by an Analog-to-
Digital Converter (ADC) and stored in some memory locations of
either the signal averager or the computer. In the case of a
single phase detector the signal from the detector is routed
directly to the low-pass filter (see previous section) and then
to the ADC system. In the case of a double (quadrature) phase
detector, one can use a double digitizer with synchronous sampling
of both signals, the most expensive solution. Much more easier
is to use two separate sample-and-hold devices, which "sample"
the signals synchronously at a rate F; the data are then
sequentially fed to the ADC by a multiplexer (Fig. 1b). (Note
that if one samples the free precession signal at a rate F,
the multiplexer will have to enter data in the computer at a
rate 2F).
Alternatively, one can feed the ouput signals of the 2 phase
detectors to a multiplexed input of the ADC. Signals in each
detector will be sampled at a rate F, if the multiplexer works
at a rate 2F. One has to take into account in the further
computation that the second channel will always lag, in time,
the first channel by 1 computer address (1/F).

     In order to further improve the sensitivity of the free
precession signals, one uses the method of time averaging: the
experiment is repeated many times and responses are algebraically
added together after digitization in the memory of a signal
averager, synchronized to the repetitive NMR free induction
decays. The signals will add coherently whereas the noise will
add randomly due to its incoherency. If the power spectral
density of the random noise has the form

$$W_n(\omega) = W_o\ \omega^{-|\lambda|}$$

where $0 \leq \lambda \leq 1$,
(two limiting cases are, white random noise, $\lambda = 0$, and 1/f
noise, $\lambda = 1$), then the S/N ratio is proportional to the square
root of the number of scans (Ernst, 1966). This also holds for

linear combinations of such noise power spectral density functions, a behaviour found in practical situations: white thermal noise in presence of drifts and instabilities. That low frequency noise (1/f) is reduced, is a particular merit of this method.

It should be emphasized that time averaging is complimentary to filtering. Indeed, filtering is based on the specific spectral properties of signal and noise, whereas time averaging utilizes the differences in coherence properties of a repetitive signal and of noise. Filtering operates in the frequency domain, time averaging operates in the time domain.

The number of bits of the ADC (which sets the amplitude resolution) and the number of bits of each memory location (or "word") will determine the ultimate S/N one is able to reach. One wants a dynamic range of the ADC as large as possible, to be able to detect signals buried in the noise, whilst accepting, at the same time, very large signals. The smallest digitization quantum of the ADC will depend on how accurate we have to digitize the noise. Both sources of ADC errors: the systematic deviation of the average caused by the quantization process and the random variations caused by the random noise of the signal (and the ADC) (Ernst, 1971) will be made negligible by signal averaging if the condition

$$v_{rms} \geq 2d$$

is fulfilled, where v_{rms} = the rms noise voltage

d = the digitization quantum.

In other words, if a sufficient amount of noise voltage is present in (or added to) the signal, the characteristic step transfer function of the ADC is smoothed out to a linear function after sufficient time averaging of the input signal. If the ADC is filled with a signal of amplitude v_{max}, a dynamical range of $2v_{max}/v_{rms}$ will be required, to recover at the same time a signal buried in the noise, A typical signal with a S/N of 100 requires a dynamical range of at least 8 bits. The 10 bits ADC on most systems will be more than adequate.

The word length of the memory will be larger than the range of the converter by the number of accumulations (scans) to be added in the signal averager or computer. In most small computers the memory will overflow, and double precision storage will be needed. To avoid the problem of overflow, some averaging systems (hardware or software) make use of an algorithm for weighted averaging (Trimble, 1968). The average stored in the memory can be, at any time, written as

$$\text{new value} = \text{old value} + \frac{\text{new sample} - \text{old value}}{\text{number of samples}}$$

In binary devices where division by an arbitrary integer is lengthy, the number of samples, N, is approximated by the closest power of 2. Division by a power of 2 is much easier to perform (K shifts to the right) than division by an arbitrary integer N. This is called the "stable averaging"algorithm. It is shown that for a given number of repetitions, the S/N ratio obtained using this algorithm is nearly equal to that of the exact average (negligible loss). This stable average has the main advantage of giving at any time a full scale display, with the noise being gradually reduced. The computer word length is now limiting the obtainable signal to noise ratio.

OPTIMIZATION OF THE STORED SIGNAL

The signal now stored in core, some necessary corrections and further improvements of the signal can be performed in the averager or the computer, either hardware or using software. The first correction to be made, once the FID stored in the memory, is to restore the zero DC baseline. Indeed a constant signal in the time domain would give rise to a peak at zero frequency in the frequency domain.

Sensitivity and Resolution Enhancement

If we want to maximize the S/N ratio of the signal stored in the memory (sensitivity enhancement), we will simulate the effect of a filter matched to the shape of the line, which is easily accomplished by a computer.

In the case of an even line shape function, the weighting function of the matched filter is identical with the line shape function, or, the frequency response function of the matched filter is the Fourier transform of the line shape function, which is proportional to the envelope of the FID of a single line. The use of a matched filter in Fourier transform spectroscopy means to multiply the FID signal point by point with the envelope function of the FID of a single line. In other words, the points of the FID are weighted accordingly to the local S/N ratio. This holds for white random noise only.

In the case of Lorentzian lines, for instance, with relaxation time T_2^*, the envelope of the decay goes like $\exp(-t/T_2^*)$.

The S/N ratio, being proportional to the signal height, ·
deteriorates toward the end of the trace. The best S/N ratio,
without introducing appreciable distorsion of the line, is
reached by weighting each point of the FID with its own S/N
ratio, i.e. multiplying the FID by

$$e^{-t/T_2^*}$$

which is rigorously equivalent to the use of a matched filter. This
operation is trivial for a computer.

Most computer software provide a general routine of exponential multiplication where the coefficient in the exponent is
entered as parameter. Negative coefficients are used for sensitivity enhancement.

Positive coefficients (weighting function increasing with
time) can also be entered in order to enhance resolution at
the expense of signal-to-noise ratio.

Resolution enhancement reduces the linewidth in the spectrum
by mathematical means. The optimum frequency response
function for the resolution enhancement filter is the
inverse of the Fourier transform of the line shape function.
For a Lorentzian shape this results in a point by point
division, of the free induction decay signal by $\exp(-t/T_2^*)$,
the free induction decay of a single line.

In the extreme case, an exponentially decaying FID could be
weighted by with an exponential function that increases
with the same time constant as the FID, giving a free precession
signal of constant amplitude which after Fourier transformation
would be a delta function.
It is clear that, the S/N ratio increasing with time, sensitivity
is sacrified and a delta function is out of question!

Apodization

A correction (software routine) called apodization is
used to attenuate the effects of truncation of the FID after
a time T (in cases, where for sensitivity reason T is shorter
than T_2^*). Windowing gives rise to small side lobes, 1/T apart
around each frequency in the spectrum. These are normally not
seen as the discrete frequency points in the frequency spectrum
are at multiples of 1/T where those lobes are zero. This is
however not always true. The ultimate resolution 1/T is limited
by the time of truncation T, which is not taken much longer than
the FID as one would just store noise in the remaining locations
and degrade the sensitivity of the transformed signal. One can
artificaly increase the resolution without degrading the S/N
ratio, by accumulating the FID over a time T, and fill up the
next locations with zero's up to a total time T'. The resolution

now will be 1/T' instead of 1/T. The FID, however, has been windowed for T, giving rise to side oscillations going through zero at intervals 1/T. The points in the frequency spectrum now being at intervals 1/T', the oscillations around each peak will appear in the spectrum.

These side lobes, can be attenuated by apodization. This consists of multiplying the FID by a function which around T, goes to zero much slower than does the discontinuity created by windowing. Such a function could be, for instance, the exponential weighting function for sensitivity enhancement, discussed in the previous section. A frequently used apodizing routine in software, is the multiplication of the FID signal with a trapezoidal function. For a triangular function, the extreme case, the Fourier transform goes like

$$(\frac{\sin\omega T}{\omega T})^2$$

a function where the side lobes are much more attenuated.

PHASE CORRECTIONS ON THE SPECTRUM

The time-domain function has now been optimized. After performing the Fourier transform, using the FFT algorithm, both the real and imaginary part of the transformed spectrum will normally be a mixture of absorption and dispersion modes. This because it is not possible to adjust in advance, the phase of the spectrometer detector to obtain a pure absorption and dispersion mode.
The mixing of real and imaginary components originates not only in the frequency independent phase shifts due to the adjustment of the phase detector but also in frequency dependent shifts introduced by the various filters and by the delay between the exciting pulse and the initial data accumulation time (due to finite pulse widths and recovery time of the receiver). That any delay in the time domain introduces a phase shift proportional to the frequency, is a straightforward consequence of the shift theorem in Fourier transform theory:

$$\text{if } F[f(t)] = F(\omega)$$
$$\text{then } F[f(t-t_d)] = e^{-j\omega t_d} F(\omega) \tag{4.1}$$

In other words, the complex Fourier transform is rotated over an angle ωt_d in the complex plane. The magnitude of the Fourier components in the spectrum remain unchanged, only the phase of each component is shifted linearly with frequency, by an

amount ωt_d, for a given delay t_d.
For a cosine wave function of frequency ω_o we would have from
eq. (I.1.4) and (4.1)

$$F[\cos\omega_o(t-t_d)] = e^{-i\omega t_d} \pi[\delta(\omega-\omega_o) + \delta(\omega+\omega_o)]$$

$$= \pi\{\delta(\omega-\omega_o)e^{-i\omega_o t_d} + \delta(\omega+\omega_o)e^{i\omega_o t_d}\}$$

The component at $\omega = \omega_o$ is rotated over an angle $\omega_o t_d$, the one at
$\omega = \omega_o$ over $-\omega_o t_d$.

A routine is included in the computer software to correct
for phase distortion, so that the real and imaginary part of
the transformed FID signal can be rotated until a pure
absorption (and dispersion) signal is obtained. If $F_r(\omega_k)$ and
$F_i(\omega_k)$ are the real and imaginary parts of the Fourier transform
of the interferogram, then the phase correction (rotation in
the complex phase) involves the calculation of

$$F_a(\omega_k) = F_r(\omega_k) \cos\phi_k + F_i(\omega_k) \sin\phi_k$$

$$F_d(\omega_k) = F_r(\omega_k) \sin\phi_k + F_i(\omega_k) \cos\phi_k$$

where F_a and F_d are the (pure) absorption and dispersion
signals. The phase correction angle ϕ_k is usually well approximated
by a frequency dependent function up to first order:

$$\phi_k = A + B(k/N) \qquad k = 0, \ldots N-1$$

The correct angle ϕ_k, i.e. the correct values for A and B are
determined by iterative interaction between computer display
and operator, while adjusting a "phase knob" on the computer
or by entering values of A and B in the software, until a pure
absorption spectrum is obtained. This can also be done by
computer maximization of the ratio $A_u(\phi)/A_l(\phi)$ where $A_u(\phi)$ is
the area above, A_l the area below the baseline of the
spectrum $F_a(\omega_k)$. As explained earlier, B takes into account
the frequency dependent phase shifts, and represents the
difference in degrees of rotation between the lowest and highest
frequency components ω_k.

It is interesting to note that if the spectrum was containing
"ghost" lines (lines of higher frequencies folded down around F/2,
due to a too slow sampling rate F) they would show up as being
out of phase as these frequencies will have been shifted over an

amount proportional to their real value.

Magnitude Spectrum

    The problem of phase correction can be avoided if we
calculate the magnitude spectrum $F_m(\omega)$:

$$F_m(\omega) = [F_r^2(\omega) + F_i^2(\omega)]^{1/2}$$

$$= [F_a^2(\omega) + F_d^2(\omega)]^{1/2}$$

This spectrum is that of the amplitude of the magnetization, losing
all phase information. It is the spectrum obtained by analog devices,
like by a "spectrum analyzer."

    However, this method has some serious drawbacks. First of all
the line shapes have long tails (wings), due to the presence of
the dispersion mode (which is zero at the centre of the line).
Secondly, when two lines overlap, the dispersive components tend
to cancel in the overlap region (the dispersion mode is not
additive like the absorption mode), so that frequencies and
intensities of both lines are distorted.

REFERENCES

A. ALLERHAND, R.F. CHILDERS and E. OLDFIELD, J.Mag.Reson. 11,
272 (1973)

J.D. ELLETT Jr., M.G. GIBBY, U. HAEBERLEN, L.M. HUBER, M. MEHRING,
A. PINES and J.S. WAUGH, in "Advances in Magnetic Resonance"
(J.S. WAUGH, ed.) 5, 117, Academic Press, New York (1971)

R.R. ERNST, in "Advances in Magnetic Resonance" (J.S. WAUGH, ed.)
2, 1, Academic Press, New York (1966)

R.R. ERNST, J.Magn.Reson. 4, 280 (1971)

J. JEENER, H. EISENDRATH and R. VAN STEENWINKEL, Phys.Rev. 133,
A478 (1964)

A.G. REDFIELD and R.K. GUPTA, in "Advances in Magnetic Resonance"
(J.S. WAUGH, ed.) 5, 81, Academic Press, New York (1971)

E.O. STEJSKAL and J. SCHAEFER, J.Magn.Reson. 14, 160 (1974)

C.R. TRIMBLE, Hewlett-Packard Journal, April 1968 (see also Hew-
lett Packard Journal, November 1969 and June 1970)

LINE NARROWING BY MULTIPLE PULSE TECHNIQUES

I. OBJECTIVES AND PRINCIPLES

U. Haeberlen

Max Planck Institut

Heidelberg, Deutschland

The claim is that multiple pulse techniques are a method by which it is possible to selectively remove the effects of dipolar interactions, more specifically: homonuclear secular dipolar interactions

$$\hbar H_D = \hbar \sum_{i<k} b_{ik} (\vec{I}^i \cdot \vec{I}^k - 3 I_z^i I_z^k)$$

of the spins from nmr spectra of solids, while retaining the effects of other interactions, in particular chemical shifts including their anisotropic parts:

$$\hbar H_C = \hbar \omega_o \sum_i \sigma_{zz}^{(i)} I_z^i$$

The major objectives of the experimenter are these:

(1) Demonstration that the method works at all.
(2) To make the experiment work better and better, that means: to strive to improve the spectral resolution, to understand what affects the resolution.
(3) Actual measurement of chemical shift tensors.

There are certainly more objectives but we shall confine ourselves to these.

The first objective is a settled one: The method works. The second is not settled. The resolution we can obtain routinely is not good enough for routine work of chemical and physical interest. The third is obviously the ultimate one. Our success or failure

in measuring chemical shift tensors will decide whether or not
multiple pulse techniques will eventually establish themselves
as a really useful tool for investigating matter.

With this objective in mind let us briefly consider the
problems involded in measuring chemical shift tensors, specifically
of nuclei with spin I = 1/2.

H_C is a single-spin operator and leads in single crystals
to line spectra. The positions of the lines in the spectra allow
a determination of the $\sigma_{zz}^{(i)}$s. If we measure $\sigma_{zz}^{(i)}$ for at least 6
nondegenerate orientations of a crystal we have enough information
to calculate $\underset{\sim}{\sigma}^{(i)}$. We define that it is $\underset{\sim}{\sigma}^{(i)}$ we are interested in.
H_D is a many – spin – Hamiltonian. It leads in general to a single
broad bellshaped resonance line.

It depends upon the relative size of the largest b_{ik}'s and the
differences $\omega_o(\sigma_{zz}^{(i)} - \sigma_{zz}^{(i')})$ whether or not it is possible to
resolve spectral lines and thus to measure shift tensors.

The b_{ik}'s (essentially internuclear distances) are intrinsic
properties of the sample whereas the chemical shifts are proportional
to the applied field B_o. Hence by increasing B_o we can increase
the interesting Hamiltonian (H_C) relative to the disturbing one
(H_D).

In the strongest fields currently available for nmr (6 – 8 T)
relative (isotropic) chemical shifts and shift anisotropies
exceed in typical samples the homonuclear dipolar broadening for
all nuclei except protons and ^{19}F.

For the measurement of chemical shift tensors, multiple pulse
line-narrowing techniques are only needed for protons and ^{19}F
nuclei.
(This does not mean, however, that there are no applications
at all of multiple pulse techniques to other nuclei.)

To understand the principles of multiple pulse line narrowing
techniques recall how standard Fourier Transform NMR works:
The sample is magnetized in a – preferably large – static magnetic
field B_o. Then it is excited with a short intense rf-pulse.
Thereafter, the sample induces a voltage in an rf-coil. This
voltage is amplified, sampled at equal intervals in time, and
phase sensitively detected. We obtain something like in figure 1.
The sampled signal values are Fourier-transformed and give the
spectrum of the sample. To prepare the discussion of multiple
pulse techniques consider the samples we take from the nmr
signal:

The N-th sample is proportional to

$$\langle I_y(Nt_s)\rangle_R = \text{tr } \rho_R(N t_s) I_y$$

provided the phase Φ of the local oscillator is chosen appropriately.

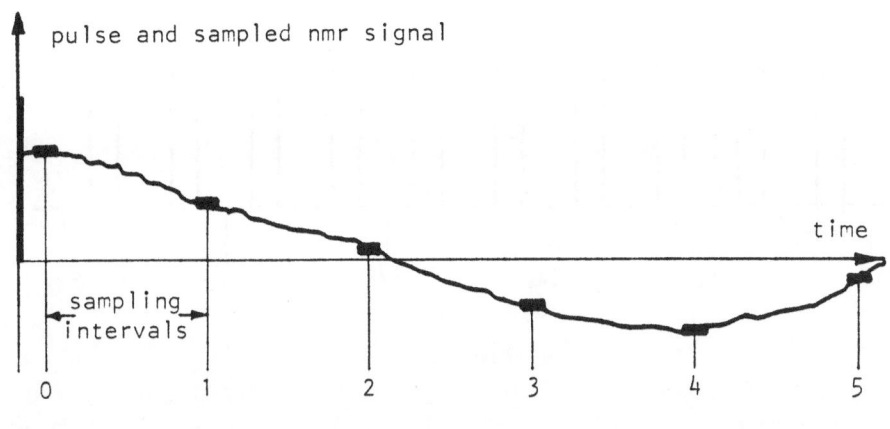

Fig. 1

(See problem 1 for definitions!)

All the information needed to calculate the spectrum is contained in $\rho_R(Nt_s)$. Prior to the pulse

$$\rho_R(t < 0) = \rho_{eq} \propto I_z = \sum_i I_z^{(i)}$$

Immediately after the pulse

$$\rho_R(0_+) \propto e^{i(\pi/2)I_x} I_z e^{-i(\pi/2)I_x} = I_y$$

This gives the zeroth sampling point.

From now on $\rho_R(t)$ evolves under the influence of the rotating frame internal Hamiltonian $\hbar H_{R,int} = \hbar H_D + \hbar H_C$ according to the v. Neumann equation

$$\dot{\rho}_R(t) = - i [H_{R,int}, \rho_R(t)]$$

From now on I shall drop the index R.

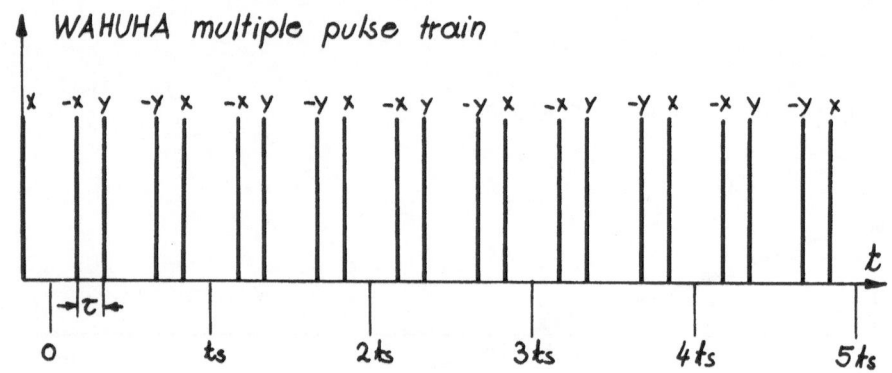

Fig. 2

The formal solution of the v. Neumann equation is

$$\rho(t) = U(t)\, \rho(0)\, U^+(t) \quad \text{with} \quad U(t) = e^{-iH_{int}t}$$

(We suppose H_{int} does not depend on time).
In particular

$$U(Nt_s) = e^{-iH_{int}Nt_s}$$

Alternatively to saying that all the information for the spectrum is contained in $\rho(Nt_s)$ we can also say: it is contained in $U(Nt_s)$. Now consider a multiple pulse experiment. We proceed as before – magnetize the sample, excite it with a pulse and sample the nmr signal at times Nt_s. However, inbetween the sampling points we apply additional pulses, e.g. as shown in figure 2. The sampling points are the same as before. We consider again $\rho(Nt_s)$ or $U(Nt_s)$. Obviously $\rho(0)$ is the same as before.

$$U(t_s) = e^{-iH_{int}\tau} \cdot e^{i(\pi/2)I_x} \cdot e^{-iH_{int}\tau} \cdot e^{-i(\pi/2)I_y} \cdot e^{-iH_{int}2\tau} \cdot e^{i(\pi/2)I_y}$$

$$\cdot e^{-iH_{int}\tau} \cdot e^{-i(\pi/2)I_x} \cdot e^{-iH_{int}\tau}$$

This expression can directly and simply be written down. Nevertheless it looks terrifying and it may be amazing that it is useful at all. To handle it, first shift the factor $e^{-i(\pi/2)I_y}$ to the right until it meets and cancels $e^{i(\pi/2)I_y}$. Whenever, on its way to the right, it passes spin operators I_x and I_z, it transforms them into $-I_z$ and I_x, respectively. Then shift $e^{i(\pi/2)I_x}$ to the right. Each I_y passed is transformed into $-I_z$, each I_z into I_y.

To provide proofs for these rules is the subject of problem 2. After these steps

$$U(t_s) = e^{-iH_{int}^z \tau}\, e^{-iH_{int}^y \tau}\, e^{-iH_{int}^x 2\tau}\, e^{-iH_{int}^y \tau}\, e^{-iH_{int}^z \tau}$$

where $H_{int}^j = \sum_{i<k} b_{ik}(\vec{I}^i \cdot \vec{I}^k - 3I_j^i I_j^k) + \omega_o \sum_i \sigma_{zz}^{(i)} I_j^i$

$U(t)_s$ now looks as if the spin system was not subjected to a time independent internal Hamiltonian plus additional r.f. pulses, but to a time dependent internal Hamiltonian $H_{T,int}(t)$ and no additional pulses. (The index T stand for toggling frame and will become apparent later on). The same holds true for $U(Nt_s)$. The crucial point is that the operators from the pulses cancel completely between successive sampling points. This is a consequence of the so-called cyclic property of the multiple pulse sequence chosen. The apparent time dependent internal Hamiltonian $H_{T,int}(t)$ is periodic modulo t_s.

:: When observed at proper sampling points only, the spin
:: system subject to a multiple pulse sequence behaves as if it was
:: under the influence of a periodically time dependent internal
:: Hamiltonian and as if there were no pulses.

Now suppose τ is made so small (it is an experimental parameter!) that $H_{int}\tau$ is small compared with unity. Under this condition we can forget in a first order approximation about the non-commutativity of th operators H_{int}^x, H_{int}^y, and H_{int}^z and can combine them all in one single exponent:

$$U(t_s) \approx e^{-i(H_{int}^x + H_{int}^y + H_{int}^z) \cdot 2\tau} = e^{-i\bar{H}_{int} \cdot t_s}$$

$$U(Nt_s) \approx e^{-i\bar{H}_{int} \cdot Nt_s}$$

where

$$\bar{H}_{int} = (1/3)(H_{int}^x + H_{int}^y + H_{int}^z)$$ is the average internal

Hamiltonian.

We have thus the result:

::   The sampled signal values in a standard FT experiment express
:: the evolution of the spin system under the (time independent)
:: internal Hamiltonian H_{int}, and carry information about H_{int}.
::   The sampled signal values in a multiple pulse experiment
:: express the evolution of the spin system under the (time
:: independent) average Hamiltonian $\bar{H}_{int}$, and do carry information
:: about $\bar{H}_{int}$.

It remains for us to inspect $\bar{H}_{int}$:

$$\bar{H}_D = 0 \quad (!)$$

$$\bar{H}_C = \omega_o \; \sum_i \sigma_{zz}^{(i)} (1/3)(I_x^i + I_y^i + I_z^i) = \omega_o \; \sum_i \sigma_{zz}^{(i)} (1/\sqrt{3}) I_Z^i$$

where

$$I_Z^i = (1/\sqrt{3})(I_x^i + I_y^i + I_z^i)$$

The multiple pulse spectrum is thus free from dipolar
broadening. The standard and multiple pulse chemical shift
Hamiltonians are very similar. They differ in two minor points
only: the multiple pulse chemical shift Hamiltonian contains
an extra factor (scaling factor) of $1/\sqrt{3}$ and the spin operator
is not along the rotating frame z-axis, but along the rotating
frame (111)-axis.

These are the principles of line narrowing with multiple
pulse techniques. The decisive point is that with our pulses
we are able to modify the effective Hamiltonian. The pulse
sequence I have chosen - the so-called WAHUHA-sequence - is
by no means unique in killing H_D and letting survive H_C. There
are many others which are all equivalent on the level of average
Hamiltonians. This level is - I stress - a first approximation.
The average H-approximation does not tell us how well dipolar
effects are suppressed. We only know that this approximation
should work the better the smaller we make $H_{int} \cdot \tau$. So there is
a need for a fuller description of line narrowing. Such a need
arises also from the following questions. There is a host of
sequences all resulting in $\bar{H}_D = 0$, $\bar{H}_C \neq 0$. Which of them are
more, which are less powerful? Are there guidelines to design
still better sequences? Furthermore: so far we argued with highly
idealized rf-pulses. To what extend do real rf-pulses which we
are able to generate with our transmitters affect our conclusions

and, eventually, our experimental results?

These questions are obviously of great importance for everyone who considers to set up a multiple line narrowing apparatus.

To approach these questions we must study briefly what has become known as

Average Hamiltonian Theory

Consider the rotating frame Hamiltonian of a spin-system:

$$\hbar H = \hbar H_1(t) + \hbar H_{int}$$

where $H_1(t)$ describes the pulses of the multiple pulse train. The motion of the spin-system under the influence of $H_1(t)$ alone is tractable and, in a sense, not interesting. It has nothing to do with the internal properties of the sample. We disregard that motion by way of an interaction – representation – ansatz

$$\rho_R(t) = U_1(t) \, \rho_T(t) \, U_1^+(t)$$

where

$$U_1(t) = T \cdot \exp \left\{ -i \int_0^t H_1(t') dt' \right\} \quad (T = \text{time ordering operator})$$

which means: we view the spin system from a new frame, the so called toggling frame. The toggling frame flips back and forth with respect to the rotating frame in the rythm of the pulses. ρ_T is the toggling frame density matrix. It evolves according to

$$\dot{\rho}_T(t) = -i[H_{T,int}(t), \rho_T] \qquad (*)$$

where

$$H_{T,int}(t) = U_1^+(t) \, H_{int} \, U_1(t) \qquad (**)$$

$H_{T,int}(t)$ is, of course, identical with the apparent time dependent internal Hamiltonian I was speaking of in the elementary introduction. We can now give an interpretation: $H_{T,int}(t)$ is the internal Hamiltonian viewed from the toggling frame. Equation ** gives a general rule how to calculate $H_{T,int}(t)$.

The formal solution of equation * is

$$\rho_T(t) = U_{int}(t) \rho_T(0) \, U_{int}^+(t)$$

with

$$U_{int}(t) = T \exp \{-i \int_0^t H_{T,int}(t') \, dt'\}$$

Hence:

$$\rho_R(t) = U(t) \, \rho_R(0) \, U^+(t)$$

with

$$U(t) = U_1(t) \, U_{int}(t)$$

Now we choose $H_1(t)$ (that means: the pulse train consisting of ideal or real pulses) to be cyclic. By cyclic we mean:

$$H_1(t) \quad \text{is periodic modulo } t_C \ (= t_s)$$

and, in addition

$$U_1(t) \quad \text{is also periodic modulo } t_C.$$

This means

$$U_1(Nt_s) = 1 \quad \text{as} \quad U_1(0) = 1.$$

The consequences of $H_1(t)$ being cyclic are

(1) $H_{T,int}(t)$ becomes periodic modulo t_C

(2) $U(Nt_C) = U_{int}(Nt_C) = T \exp \{-i \int_0^{Nt_C} H_{T,int}(t') dt'\}$

$$= \{T \exp [-i \int_0^{t_C} H_{T,int}(t') dt']\}^N$$

$$= \{U_{int}(t_C)\}^N$$

This result means: in order to describe the state of the spin system at any integer of the cycle time t_C it is sufficient to know its short time evolution over one cycle.

To obtain the result in the form of a single exponential it is convenient to apply the Magnus formula:

$$U_{int}(t_C) = T \exp \left[-i \int_0^{t_C} H_{T,int}(t') dt' \right] \qquad \text{(definition)}$$

$$= \exp [-i\, F\, t_C] \qquad \text{(ansatz)}$$

$$= \exp [-i(\bar{H} + \bar{H}^{(1)} + \bar{H}^{(2)} + \ldots)t_C] \qquad \text{(expansion)}$$

where

$$\bar{H} = (1/t_C) \int_0^{t_C} H_{T,int}(t)\, dt$$

$$\bar{H}^{(1)} = (-i/2t_C) \int_0^{t_C} dt_2 \int_0^{t_2} dt_1\, [H_{T,int}(t_2), H_{T,int}(t_1)]$$

$$\bar{H}^{(2)} = (1/6t_C) \int_0^{t_C} dt_3 \int_0^{t_3} dt_2 \int_0^{t_2} dt_1 \;.$$

$$\{[H_{T,int}(t_3),[H_{T,int}(t_2),H_{T,int}(t_1)]]$$

$$+[H_{T,int}(t_1),[H_{T,int}(t_2), H_{T,int}(t_3)]]\}$$

etc....

We can now make rigourous some of our previous somewhat loose statements:

(1) The time dependent internal Hamiltonian which we generate with our pulse train is nothing else but the internal Hamiltonian viewed from the toggling frame, $H_{T,int}(t)$.

(2) Our previous "average Hamiltonian" $\bar{H}_{int}$ is, rigourously, F. The lowest order approximation of F is the average of $H_{T,int}(t)$ over the cycle.

This average Hamiltonian theory enables us to study the efficiency of multiple pulse sequences (not necessarily pulse sequences) beyond the point of just averages.

Furthermore we have not provided the tools for the study of the effects of all pulse imperfections we can possibly think of. Minimizing - often compensating - their disturbing effects is an active field of research and in my next lecture I shall outline how one proceeds in practice.

LINE NARROWING BY MULTIPLE PULSE TECHNIQUES

II. THE REAL WORLD OF PULSES

U. Haeberlen

Max Planck Institut

Heidelberg, Deutschland

PHASE TRANSIENTS

The pulses which we have underlaid my first lecture have been supposed to be ideal in a number of respects:

(1) pulsewidth $t_w \to 0$,

(2) flipangle $\beta \equiv \gamma B_1 t_w = \pi/2$ (exactly!)

(3) exact quadrature rf-phases,

(4) no phase transients,

(5) no timing errors.

We cannot discuss all possible nor even all important pulse imperfections in one lecture. We must make a selection and start with phase transients.

Phase transients are inevitable companions of rf-pulses generated in tuned passive circuits and in non-linear amplifiers. To understand what phase transients are suppose we route an rf-pulse through a phase sensitive detector (PSD) (Fig. 1). We expect at the output of the PSD a video pulse the amplitude of which varies as cos Φ when the reference phase Φ of the local oscillator (L.O.) is varied. In particular, we expect that there is some phase Φ_o for which we have a zero output.

Well, if you check in this way the pulses of your pulse spectrometer you will most likely find something different (Fig. 2).

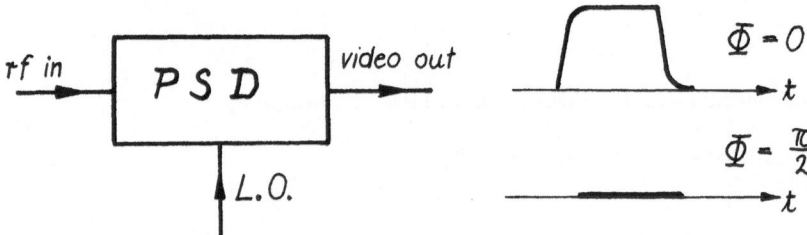

Fig. 1

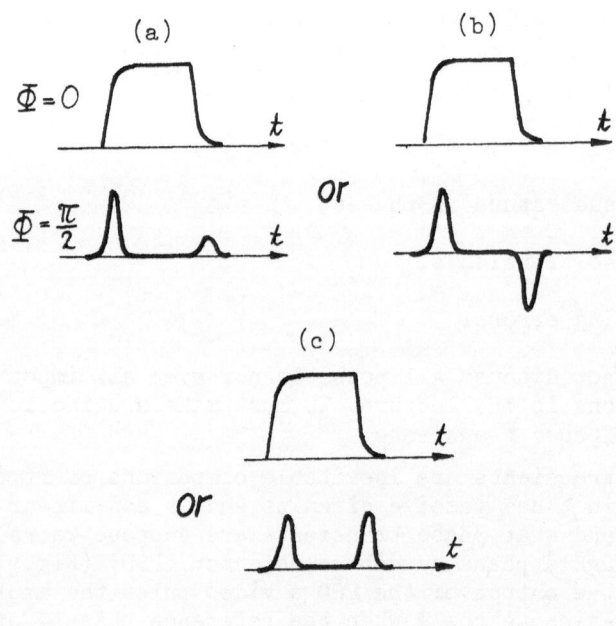

Fig. 2

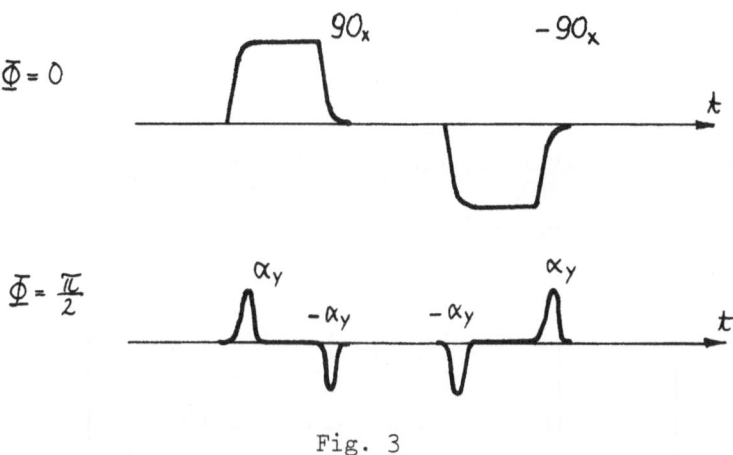

Fig. 3

The "spikes" obtained for $\Phi = \pi/2$ mean that the pulses contain out – of – phase components during the rise and fall of the pulses. These out – of – phase components are called phase – transients.

To see why they are annoying consider a pulse pair $(P_x^{90} - \bar{P}_x^{90} -)$ with phase transients of type (b) (Fig. 3).

The theoretician who does not know about phase transients expects the total rotation resulting from the two pulses to be

$$R_{ideal} = e^{-i(\pi/2)I_x} \, e^{i(\pi/2)I_x} = 1$$

In reality:

$$R_{real} = e^{i\alpha I_y} \underbrace{e^{-i(\pi/2)I_x} \, e^{-i2\alpha I_y} \, e^{i(\pi/2)I_x}}_{e^{2i\alpha I_z} \, = \, \bullet} \, e^{i\alpha I_y} \underset{\sim}{e^{2i\alpha(I_y+I_z)}} \neq 1$$

In a long pulse train these extra rotations accumulate and cause

an oscillation of the nmr signal even when the spectrometer is
set exactly on resonance. In multiple pulse spectra they cause
apparent resonance shifts and false line intensities. Therefore,
phase transients are disturbing.

Phase transients are virtually unavoidable but when the
spectrometer (above all, but not only: the probe) is properly
tuned and if all amplifiers are properly biased (!) the leading
and trailing phase transients are symmetric (type c)(Fig. 2).

If they are symmetric, they are harmless:

$$R_{real}(\text{symmetric ph.tr.}) =$$

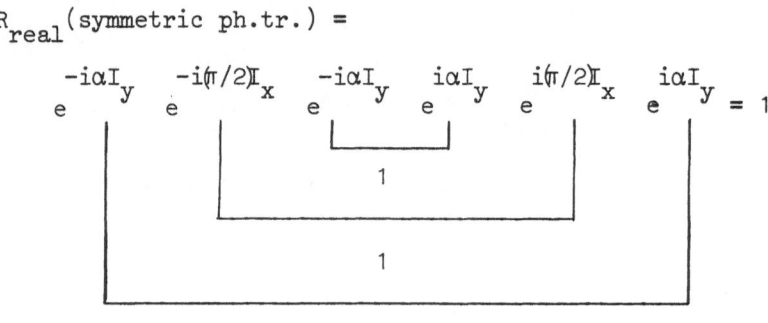

$$e^{-i\alpha I_y} e^{-i(\pi/2)I_x} e^{-i\alpha I_y} e^{i\alpha I_y} e^{i(\pi/2)I_x} e^{i\alpha I_y} = 1$$

Important:

✕   If you want to do multiple pulse experiments make sure
✕ that the phase transients of your pulses are symmetric.

I can only state here but cannot go through the calculation
that phase transients do not affect the suppression of dipolar
interactions by multiple pulse trains up to terms which are of
first order in the phase transient flip angles α. This is
reassuring!

FINITE PULSE WIDTHS, FLIP ANGLE ERRORS

Flip angle errors ε are defined by $\beta = \beta_o + \varepsilon$.
β = actual flip angle, β_o = ideal flip angle. This discussion will
lead us to

- compensation schemes for $t_w \neq 0$, $\varepsilon \neq 0$

- and to criteria by which we can distinguish "better" from
  weaker pulse sequences.

Figure 4 shows a WAHUHA multiple pulse sequence with pulses
of finite width t_w. Also shown is $U_1(t)$ which we need for
calculating

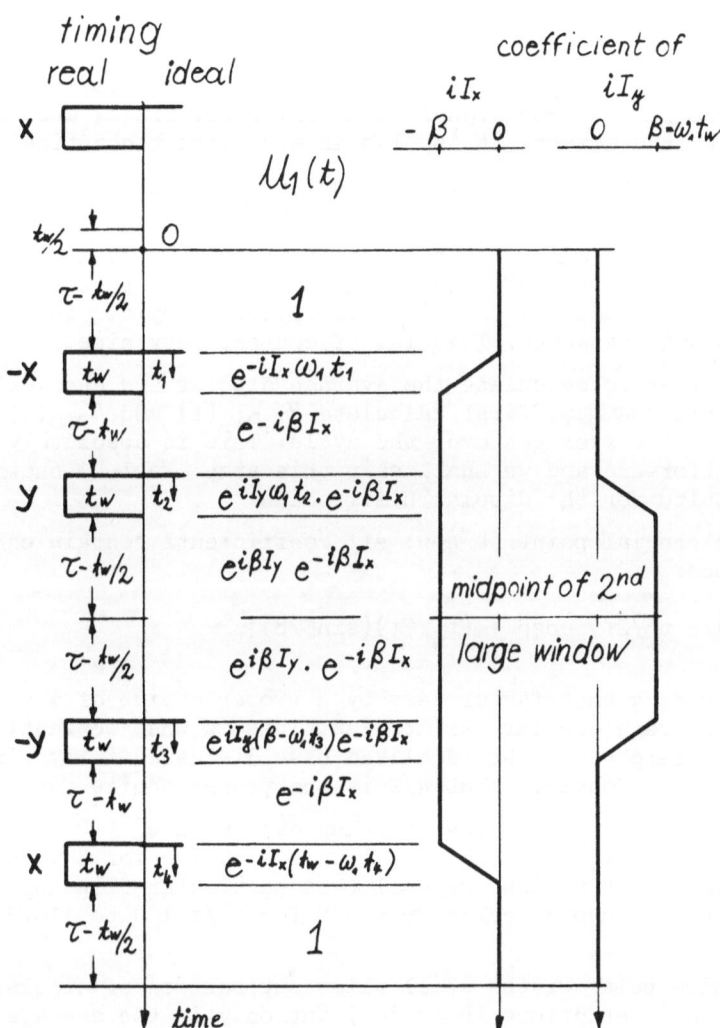

Fig. 4. WAHUHA sequence with pulses of finite width.

$$H_{T,int}(t) = U_1^+ \, H_{int} \, U_1$$

Recall: it is the average of $H_{T,int}(t)$ which governs multiple
        pulse spectra.

We choose the origin of time t in the middle of the window
between the preparation pulse and the first pulse of the multiple
pulse sequence proper. This is a reasonable experimental choice –
we can sample the nmr signal in a convenient place, and, as you
will see in a moment, it is also an excellent theoretical choice.

At the right hand side of figure 4 are indicated the factors
that go with I_x and I_y in these exponentials.

One sees immediately that the cycle remains symmetric. The
consequence is that all first order corrections to the
average Hamiltonian vanish identically, as they did for the
ideal WAHUHA sequence. This is, of course, very nice.

In order to calculate the average dipolar and chemical shift
Hamiltonians we must first calculate $H_{T,Dip}(t)$ and $H_{T,C}(t)$ and
then take the averages over the cycle. This is absolutely
straightforward and we shall skip this step. Table 1 shows
the results for the dipolar Hamiltonian.

The crucial point is now: all coefficients contain one factor
in common:

$$[(1 - t_w/2\tau)\cos\beta + (t_w/2\tau)(\sin\beta/\beta)]$$

By making this factor zero by a proper choice of β we get
a zero average dipolar Hamiltonian. This is what we want!
If t_w is zero as in the idealized case discussed in my first
lecture it is obvious that $\pi/2$ is the proper choice for β.

It is important to realize that even with $t_w > 0$ – up to the
natural limit $t_w = \tau$ – there is always a choice for β which makes
this expression vanish. We call this particular flip-angle β_0.
β_0 increases monotonically from 90° for $t_w/\tau \to 0$ to 116.24° for
$t_w/\tau \to 1$.

※   Finite pulse widths still allow suppression of dipolar
※ spin-spin interactions in solids. Not only is the average
※ dipolar Hamiltonian zero, but also the first order correction
※ term ($\propto 1/T_{2rigid} \cdot \tau/T_{2rigid}$).

The message for those who want to do multiple pulse
experiments is: there is no need to build or to buy (!) giant
power transmitters in an effort to make t_w ever and ever smaller.
Power levels of, say, 1 kW are definitely sufficient.

Table 1.   Average of $U_1^+ (\vec{I}^i \cdot \vec{I}^k - 3\, I_z^i I_z^k)\, U_1$

First column: operators contained in this average
Second column: accompanying coefficients.

operator	coefficient
$I_x^i I_x^k$	$[(1 - t_w/2\tau)\cos\beta + (t_w/2\tau)(\sin\beta/\beta)]\cos\beta$
$I_y^i I_y^k$	$[\ldots\ldots\ldots\ldots\ldots\ldots\ldots]\cos^3\beta$
$I_z^i I_z^k$	$-[\ldots\ldots\ldots\ldots\ldots\ldots\ldots](1 + \cos^2\beta)\cos\beta$
$\frac{1}{2}(I_x^i I_z^k + I_z^i I_x^k)$	$-[\ldots\ldots\ldots\ldots\ldots\ldots\ldots]2\sin 2\beta$
$\frac{1}{2}(I_x^i I_y^k + I_y^i I_x^k)$	$-[\ldots\ldots\ldots\ldots\ldots\ldots\ldots]2\sin^2\beta$
$\frac{1}{2}(I_y^i I_z^k + I_z^i I_y^k)$	$-[\ldots\ldots\ldots\ldots\ldots\ldots\ldots](1 + \cos^2\beta)2\sin\beta$

Table 1 is also valuable for discussing flip angle errors, i.e. deviations ε of β from β_o ($\beta = \beta_o + \varepsilon$). As I mentioned already flip angle errors (common to all pulses) are a consequence of rf-inhomogeneity and of power droop. Therefore, a discussion of flip angle errors is a discussion of the sensitivity of multiple pulse sequences to rf-inhomogeneity and to power droop.

Note that for $t_w/\tau \to 0$ not only the square-bracket-expression in table 1 vanishes for $\beta_o = 90°$, but also four of the extra factors. Around $\varepsilon = 0$, three of the coefficients vary as ε^2, one as ε^4 and only two vary linearly with ε.

For $\varepsilon \ll 1$ - this we may safely assume under normal circumstances - these last two coefficients convey most of the residual dipolar broadening resulting from rf-inhomogeneity. Now note that the spin operators associated with these coefficients both contain I_y linearly. This is the hook for a

Compensation scheme

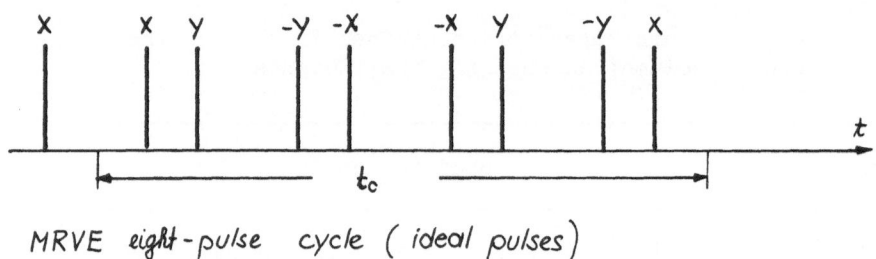

MRVE eight-pulse cycle (ideal pulses)

Fig. 5

Consider again the WAHUHA sequence, but exchange the x and -x pulses. This causes the signs of all the I_x-operators in $U_1(t)$ to be reversed and the consequence is that the signs of the disturbing coefficients are reversed. All others remain unchanged. If we combine a WAHUHA cycle with another one with reversed x and -x pulses we arrive at the Mansfield-Rhim-Vaughan-Elleman eight-pulse cycle. The coefficients of the first four of the spin-operators in a table 1 for the MRVE eight pulse cycle are the same as for the WAHUHA cycle, but they vanish identically for those where I_y appears linearly. That means: those coefficients which for the WAHUHA four pulse cycle were most sensitive to flip angle errors (rf-inhomogeneity, power droop) are now identically zero (Fig. 5).

In addition, there are now two choices for β which make the average dipolar Hamiltonian zero:

(1) β_0 as for the WAHUHA cycle and

(2) $\beta_1 = \pi/2$

They coincide for $t_w/\tau \to 0$.

Figures 6 and 7 show how the remaining coefficients vary with β for various ratios t_w/τ. Recall: these coefficients express the residual dipolar broadening in multiple pulse experiments! Note, in particular, the range of β over which all these coefficients remain below 1 %.

Important conclusion!:

    The MRVE - eight pulse cycle is substantially superior to

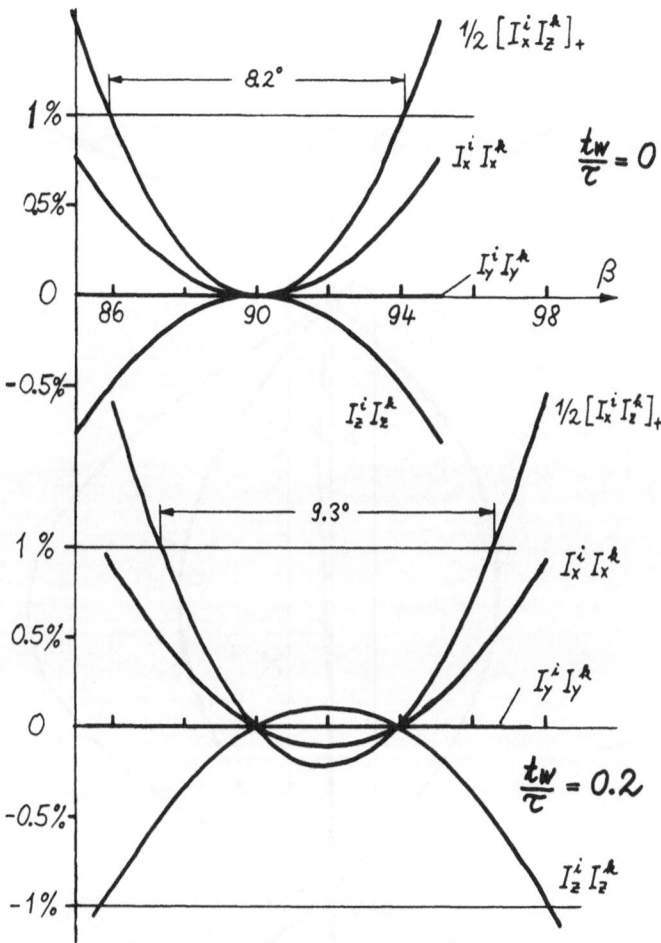

Fig. 6. Coefficients of various spin-operators in the average of $U_1^+(I^i \cdot I^k - 3I_z^i I_z^k)U_1$ for the MRVE eight pulse sequence as a function of the flip angle β. $[I_x^i I_z^k]_+$ is an abbreviation for $(I_x^i I_z^k + I_z^i I_x^k)$.

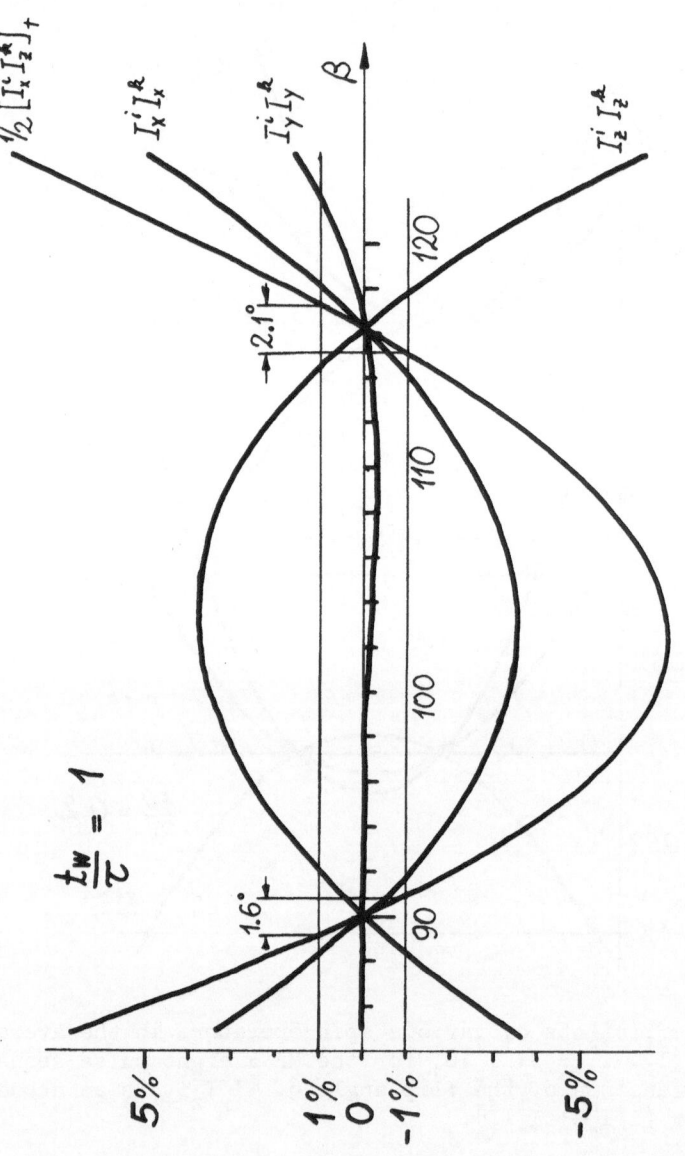

Fig. 7. Same as figure 6 but now $t_w/\tau = 1$.

× the WAHUHA four pulse cycle with respect to the sensitivity of the
× dipolar line broadening against flip angle errors, that means against

×   - rf-inhomogeneity over the sample,

×   - power droop, and

×   - alignment of pulse widths.

    Rf-inhomogeneity leads also directly to line broadening.
The reader will have suspected this already since the rf-field
contributes to the effective field in the rotating frame.

    Formally this broadening arises through the dependence of the
chemical shift scaling factors on the flip angle ε.

    For $t_w \to 0$ this dependence is:

$$S(\beta) \to S(\beta_o + \varepsilon) = S(\pi/2 + \varepsilon) = (1/\sqrt{3})\,(1 + {}^2/_3\varepsilon) \quad \text{WAHUHA}$$

$$= (\sqrt{2}/3)\,(1 + {}^1/_2\varepsilon) \quad \text{MRVE}$$

    The broadening is proportional to $\Delta\omega$, the distance of the
resonance line from the origin. It is comparable for the
WAHUHA and MRVE sequences.

    To get an idea of the importance of this line broadening
mechanism suppose the distribution of the rf-field strength
$B_1(\vec{r})$ over the sample is Gaussian with a root mean square width

$$\sqrt{\overline{\varepsilon^2}} = 0.052 \cong 3°$$

This is not very much!

    Furthermore suppose $\Delta\omega = 2\pi \cdot 2000$ Hz.
Again this is not very much in view of the fact that even proton
spectra spread over a range of $2\pi \cdot 90 \cdot 30 = 2\pi \cdot 2700$ sec^{-1} (90 MHz-
spectrometer, shift-anisotropy range 30 ppm). These numbers lead
to a line width of as much as $2\pi \cdot 80$ Hz!

    From this result we learn:

×   rf-inhomogeneity is a serious problem through its direct
× line broadening effect. The residual dipolar line-broadening
× is much less important, provided we use the compensated cycle.

    As a problem I ask you to think about designing a multiple
pulse sequence useful for solid state line narrowing spectroscopy
where the linear dependence of the chemical shift scaling factor
on the flip angle error ε is compensated.

    The problem may be solved by making use of also 270°-pulses
instead of 90°-pulses only (cf. A.N. Garroway, P. Mansfield, and
D.C. Stalker, Phys. Rev., in print).

We have now learned how to deal with phase transients and with flip angle errors common to all pulses.

We must be aware also of flip angle errors and phase errors different for the four different types of pulses we use in our sequences. After all, we all adjust them individually.

There is also a scheme to compensate (in first order) such errors (see P. Mansfield and U. Haeberlen, Zeitschrift f. Naturf. 28a, 1081, 1973). It involves phase switching of the basic spectrometer frequency and requires a twofold extension of the cycles. We have implemented in Heidelberg both compensation schemes. They work very well in the following sense: our spectrometer has 10 multiple pulse alignment controls:

4 pulse width controls,

4 phase controls,

1 transmitter power control (screen grid voltage). This is effectively a common flip angle control,

1 phase transient control (bias voltage of transistor driver amplifier).

When we use the phase compensated version of the MRVE eight pulse cycle the multiple pulse response of our samples is very insensitive to substantial variations of the first 9 of these 10 controls.

The spectral resolution is also insensitive to the 10th control. This means: the results cannot be made better by more efforts in aligning the spectrometer. The experiment works as well as it possibly can under the given physical conditions.

LINE NARROWING BY MULTIPLE PULSE TECHNIQUES

III. EXPERIMENTAL ASPECTS

U. Haeberlen

Max Planck Institut

Heidelberg, Deutschland

A multiple pulse experiment starts with the selection of
the sample - unless it is to be a mere test experiment. In
that case the sample is usually a "single crystal of CaF_2 doped
with paramagnetic impurities and oriented with its (111)-axis
along the applied field"! The following considerations influence
the decision:

(1) What is possibly interesting?
    For our current work we defined the determination of chemical
    shift tensors of protons in characteristic bonds (hydrogen,
    alephatic, aromatic, olefinic ... bond) as being interesting.

(2) Which suitable compound contains protons in the desired bond?
    The emphasis is on 'suitable'. What makes a compound suitable
    for a proton - multiple - pulse investigation?

    (a) The compound itself and the space group in which it
        crystallizes should be such that the spectrum is not
        overcrowded with lines.
        (Problem: What is the number of nmr-lines in a single
        crystal spectrum of a spin 1/2 nucleus?)

    (b) It is allowed to contain only a very limited selection
        of atoms, otherwise heteronuclear dipolar broadening
        is fatal. C, S, O, Ca are excellent, K and some others
        are tolerable.

    (c) It should be a solid at room temperature.

    This is desk work. If it resulted in a certain compound, the
experimental work starts. It may have to start with synthesizing
the compound. But even if one can buy it one usually must grow

single crystals. This is often very tough. The next step consists
in orienting the single crystals. We do this either on an optical
goniometer exploiting natural growth and cleavage planes or,
if this is not possible, by x-ray diffraction. This also is often
a frustrating task. What I want to convey is: *the really tedious
* part of multiple pulse line narrowing experiments has now largely
* shifted away from work at the spectrometer itself to preparing
* the samples.

Now suppose that we do have an oriented single crystal
prepared for the nmr-experiment. How do we get multiple pulse
spectra?

(1) Shim B_O for acceptable homogeneity.

(2) Check the rf-pulses for symmetric phase transients. Use an
open-ended rf-cable, inserted into the probe itself, as
antenna. If the phase transients are not symmetric, retune
the probe. If the phase transients are still not symmetric,
change bias of saturated transistor amplifier.

(3) Adjust x, -x, y, -y pulse widths for $\beta = 90°$. The transmitter
power must be down from its peak value by at least 10 %.
Pulse width adjustments are done using $[P_j^{90°} - P_j^{90°} -]_n$ pulse
trains, j = x, -x, y, -y. A liquid, single-line sample (e.g.
water) is used. B_O is set exactly on resonance (for field
control we use an external nmr field lock synchronized with
the master clock of the spectrometer). The phase sensitively
detected nmr signal should behave as shown in figure 1.
Usually it does not: the upper and lower traces decay and
show painfully clear that the rf-homogeneity should be
improved.

(4) Adjust antiphases for the x and -x pulses. This is done on
the same sample, with B_O still exactly on resonance, with a
"phase alternated sequence": $[P_x^{90°} - P_x^{-90°} -]_n$. The oscilloscope
trace should look as shown in figure 2. Unlike in step 3, the
nmr signal has now no tendency to decay: the $[P_x^{90°} - P_x^{-90°} -]_n$
pulse train is "compensated" automatically against flip angle
errors and, hence against rf-inhomogeneity.
However, this train is sensitive against phase transients.
If during step 2 the phase transients have not been carefully
symmetrized, a "beat" in both the upper and center traces
will develop which cannot be made to vanish by shifting the
phase of the -x pulse relative to the phase of the x pulse.

(5) Adjust the $\pm$ y pulses in quadrature to the $\pm$ x pulses.
A Carr-Purcell-pulse train with B_O set somewhat off resonance
is used for that purpose. The procedure is explained in the
legend of figure 3. It not only allows a sensitive alignment
of quadrature rf-phases, but also leads to the correct setting
of the reference phase (denoted by α in the figure) of the
phase sensitive detector.

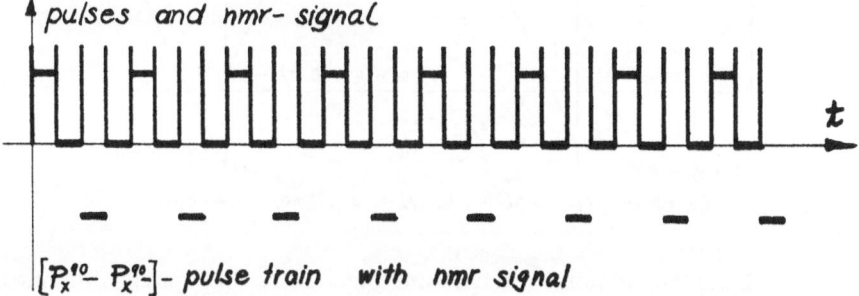

Fig. 1

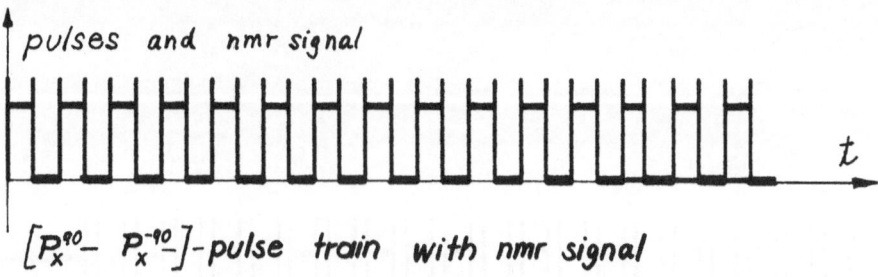

Fig. 2

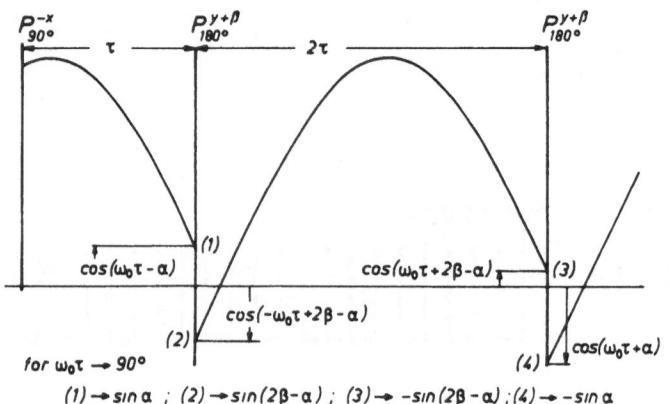

Fig. 3. Scheme to adjust quadrature phases in two pulse channels, e.g., the -x and y channels. β and α are the phase errors of the y pulse channel and of the reference of the phase sensitive detector. The -x channel phase error is zero per definition. Adjusting for (1) = (4) nullifies α , subsequent adjusting for (2) = (3) which then will be equal to (1) and (4) nullifies β. Note the sine-dependence of (1), ..., (4) on α and β when $\omega_0 \tau$ is chosen to be ≈90°.

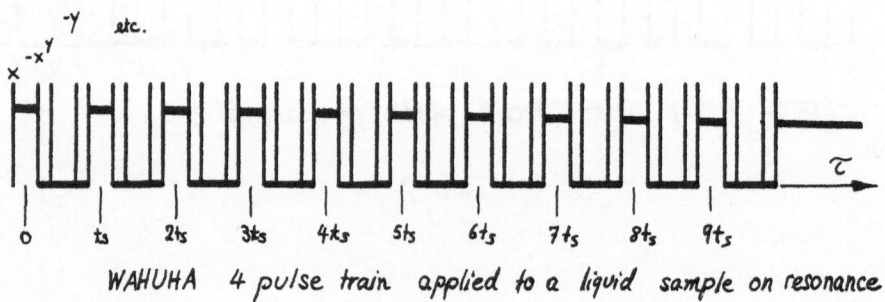

WAHUHA 4 pulse train applied to a liquid sample on resonance

Fig. 4

(6) Repeat step 3 for the y-pulses which temporarily have been made π-pulses.

(7) Apply step (4) on the +y and -y-pulses.

(8) We are now ready to switch on all four types of pulses using the WAHUHA - timing. Typically we choose τ = 4 μsec. We still use the single-line liquid sample.
Figure 4 shows how the oscilloscope traces should look like. The top trace usually has no tendency to decay, but both traces have a tendency to oscillate due to asymmetric phase transients as well as to phase and flip angle errors. We tolerate such oscillations up to a certain frequency (100 Hz, say) and rely on the compensation schemes to prevent the misalignments which are responsible for the oscillations to adversely affect the final results. Now we shift the magnetic field away from resonance both up   and downfield. Beats appear. They should be equal in their initial amplitude. The top trace should oscillate around +1/3, the bottom trace around -1/3. The top trace has the same phase both up and downfield, contrary to the bottom trace (see figure 5).

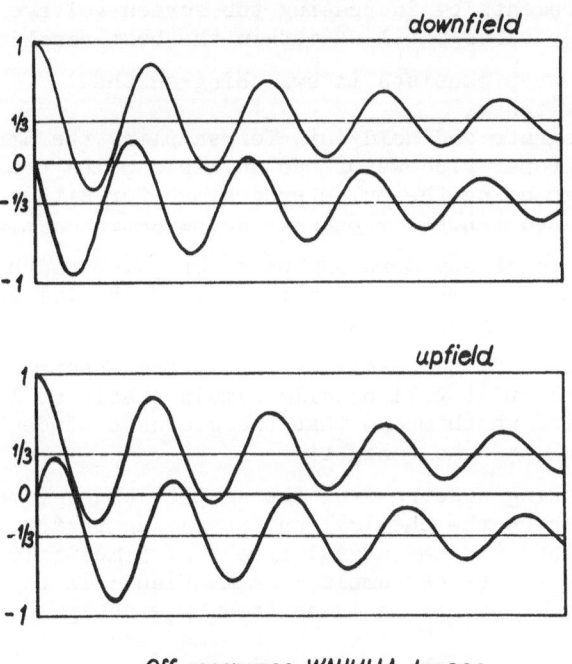

Off-resonance WAHUHA traces

Fig. 5

Problem: confirm this by calculating appropriate rotating
frame expectation values! How will the traces look when
the reference phase of the L.O. is shifted by 90°?
If the sequence is not properly aligned the top- and bottom-
traces are usually grossly different in amplitude and the
alignment procedure has to be started over again.
Problem: How look off – resonance MRVE – traces?
Deduce from the result an advantage of the MRVE sequence
over the WAHUHA sequence!
Hint: consider the signal/noise-ratio. How can the WAHUHA
cycle be modified to obtain the same advantageous property?

(9) Flip angle fine adjustment.
We saw in the previous lecture that for $t_w/\tau \neq 0$ (0.2 is a
typical number) the optimum flip angle is slightly larger
than $\pi/2$. To find the optimum we now (for the first time!)
need a solid sample. For ^{19}F-work CaF_2 is a natural choice.
We don't know about a similarly convenient test compound for
protons. Usually we resort to a powder sample of malonic acid.
At ambient temperature this compound has an acceptable
spin-lattice relaxation time of about 10 seconds. Its spectrum
consists of two lines, the multiple pulse signal correspondingly displays a beat structure. We focus our attention on this
beat structure when we now increase the rf-power in very
small increments by increasing the screen voltage of the
transmitter tubes until we obtain the best resolution.

(10) The final step consists in switching on the
  – phase switcher,
  – the integrate-and-hold unit for sampling the nmr signal
    in the proper windows of the multiple pulse trains,
  – and programming the pulse programmer for either the phase
    compensated WAHUHA or phase – compensated MRVE sequences.

We do not touch any more any phase or pulse width control
after we have programmed and running one of the multiple
pulse line narrowing sequences.

We are now ready to take spectra. If the spectrometer had been
allowed to warm up it will usually remain stable at least over
a 10 hour period which means that the graduate student or postdoc
can work during the whole night!

Between taking spectra from the samples under investigation
we regularly check the chemical shift scaling factor. To this
end we accumulate the nmr signal from e.g. 5 multiple pulse
shots from our reference sample – adamantane – in the computer.
Between each shot the field is shifted by exactly 10 ppm. The
compound-nmr-signal is Fourier-transformed and yields a 5-line
"spectrum" in which each line is shifted relative to the
previous one by exactly 10 ppm. The separation of the lines in Hz

allows a quick, simple and very precise measurement of the
scaling factor.

The last figure shows two of the nicest multiple pulse
spectra we have seen so far.

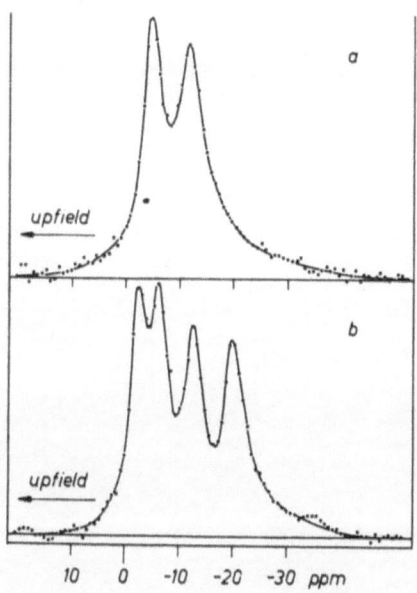

Fig. 6. Multiple pulse spectra of a single crystal of malonic acid.
(b) general orientation of B_O with respect to the crystal
lattice; all four possible lines are resolved.
(a) special orientation of B_O. There is a pairwise
coalescence of lines.

PROBLEMS + REFERENCES ABOUT LINE NARROWING BY MULTIPLE PULSE

TECHNIQUES

U. Haeberlen

Max Planck Institut

Heidelberg, Deutschland

(1) Describe how a phase sensitive detector (PSD) works. A black
box description is sufficient. If necessary, inform yourself!
Show that the PSD output in a pulsed nmr-experiment is
proportional to $<I_x>_R \sin\phi + <I_y>_R \cos\phi$, where ϕ is the
phase of the local oscillator. $<I_j>_R$ is the j-component of the
expectation value in the rotating frame of the total nuclear
spin angular momentum $\vec{I}$. Where definitions are missing, provide
them! Convince yourself that expectation values in the
rotating frame are logically defined by $<A>_R = \text{tr}\rho_R A$ where
$\rho_R = U\rho U^+$ (with $U = e^{-i\omega_0 t}$) is the rotating frame density
matrix. Note the difference between rotating frame and
laboratory frame expectation values. (The latters are given
by $<A>_{LAB} = \text{tr}\rho_R A_R = \text{tr}\rho A$).

(2) Prove the following equation:
$$e^{-i\alpha I_x} e^{iHt} e^{i\alpha I_x} = \exp \{e^{-i\alpha I_x} (iHt) e^{i\alpha I_x}\}$$

Generalize the result!

(3) Calculate the WAHUHA average Hamiltonian for the following
nuclear spin interactions: scalar and pseudodipolar indirect
spin-spin interactions, heteronuclear dipolar couplings,
quadrupolar couplings. By using irreducible spin-operators
establish a general rule. In the first lecture it is
shown how these averages are obtained for homonuclear dipolar
couplings and chemical shifts.

(4) Derive the equation of motion of the spin density matrix in
the toggling frame, $\rho_T(t)$, by starting from $\hbar H = \hbar H_1(t) + \hbar H_{int}$,
the equation of motion of the spin density matrix $\rho_R(t)$ in

the rotating frame, and the ansatz

$$\rho_R(t) = U_1(t) \; \rho_T(t) \; U_1^+(t) \quad \text{with}$$

$$U_1(t) = T \exp \left\{ - i \int_0^t H_1(t') \, dt' \right\}$$

(5) Derive the first non-trivial term of the Baker-Campbell-Hausdorff-formula

$$e^{-iA} \, e^{-iB} = e^{-i(A+B)-(1/2)[A,B]+\ldots}$$

This formula is a simple special case of the Magnus expansion which is of great importance for us.

(6) Calculate the (homonuclear) dipolar and chemical shift average Hamiltonians for the Mansfield-Rhim-Vaughan-Elleman 8 pulse cycle.

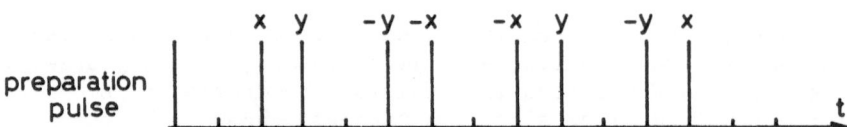

What is the scaling factor?

(7) A cycle is called symmetric if

$$H_{T,int}(t) = H_{T,int}(t_C - t)$$

A lemma states that all odd terms in the Magnus expansion vanish for symmetric cycles.
Prove this lemma for the first odd term $(\bar{H}_{T,int}^{(1)})$
Check whether the WAHUHA cycle is symmetric!
Note: there is a free choice where to start the cycle!
Check whether the MRVE-eight pulse cycle (cf. problem 6) can be defined symmetrically. Consequences?

(8) Derive the lowest order non-vanishing purely dipolar correction term to the average Hamiltonian for the WAHUHA four pulse cycle. Hint: it is $\bar{H}_{Dip}^{(2)}$. Make use of the symmetry properties of the cycle! What conclusions can be drawn from your result with respect to the dependence upon the cycle time of the suppression of dipolar line broadening by this multiple pulse sequence?

(9) Convince yourself that $\bar{H}_{T,int}$ (MRVE) contains no terms linear in I_y even when the flip angle β is not equal to $90°$.

REFERENCES (and suggestions for further reading)

1. U. HAEBERLEN and J.S. WAUGH, Phys.Rev. 175, 453 (1968)
   Coherent Averaging Effects in Magnetic Resonance (contains average
   Hamiltonian theory with applications to many nmr line narrowing
   procedures)

2. P. MANSFIELD, J.Phys. C4, 1444 (1971)
   Symmetrized Pulse Sequences in High Resolution NMR in Solids
   (contains systematic development of compensation schemes)

3. M. MEHRING, Z.Naturf. 27a, 1634 (1972)
   Arbitrary Pulse Width in the Four-Pulse NMR Experiment

4. P. MANSFIELD and U. HAEBERLEN, Z.Naturf. 28a, 1081 (1973)
   Phase Compensation in Multi-Pulse NMR Experiments

5. U. HAEBERLEN, J.D. ELLETT and J.S. WAUGH, J.Chem.Phys. 55, 53 (1971)
   Resonance Offset Effects in Multiple Pulse NMR Experiments

6. W.K. RHIM, D.D. ELLEMAN and R.W. VAUGHAN
   Analysis of Multiple Pulse NMR in Solids
   I : J.Chem.Phys. 59, 3740 (1973)
   II : J.Chem.Phys. 60, 4595 (1974)
   (I contains a thorough investigation of what limits the resolu-
    tion in multiple pulse spectra of CaF_2 under various conditions.

   II deals in an almost exhaustive manner with pulse imperfections.)

7. ^{19}F - powder studies: see, e.g.: M. MEHRING, R.G. GRIFFIN and J.S.
                          WAUGH, J.Chem.Phys. 55, 746 (1971)

   ^{19}F - single crystal study: see, e.g.: R.G. GRIFFIN et.al., J.Chem.
                          Phys. 57, 2147 (1972)

   ^{1}H - powder study: see, e.g.: U. HAEBERLEN and U. KOHLSCHÜTTER,
                          Chem.Phys. 2, 76 (1973)

   ^{1}H - single crystal study: see, e.g.: U. HAEBERLEN, U. KOHLSCHÜT-
                          .TER, F. KEMPF, H.W. SPIESS and H. ZIMMERMANN,
                          Chem.Phys. 3, 248 (1974)

HIGH RESOLUTION DOUBLE RESONANCE DIRECT DETECTION OF RARE

NUCLEI IN SOLIDS. METHOD AND TECHNIQUE

P. Van Hecke[*]

Katholieke Universiteit Leuven

Leuven, België

INTRODUCTION

In rigid solids, the dipolar interaction among spins dominates,by orders of magnitude, the various weaker spin interactions, which give rise to chemical shifts, Knight shifts and spin-spin couplings.
These weak interactions, which originate in the coupling of the nuclear spins with the surrounding electron cloud, are an invaluable source of information for structural determinations and chemical binding.
In the last few years, tremendous efforts have been done to reduce and suppress this dipolar broadening in solids, giving rise to the new and fast extending field of High Resolution NMR in Solids, opening a wide range of application in solid state physics, chemistry and biology.

The various methods used to suppress the dipolar interaction among spins are classified according to which specific part of the dipolar Hamiltonian they are working on. The truncated dipolar Hamiltonian, for a spin system consisting of 2 spin species I and S, can be written as (Van Vleck, 1948):

$$H_d^o = H_{II}^o + H_{SS}^o + H_{IS}^o$$

---

[*] "Aangesteld Navorser" of the "Nationaal Fonds voor Wetenschappelijk Onderzoek"

$$H^{O}_{II} = \gamma^{2}_{I}\hbar^{2} \sum_{i\ j}^{N_{I}} r^{-3}_{ij} P_{2}(\cos\theta_{ij})(\bar{I}_{i}\cdot\bar{I}_{j} - 3I_{iz}I_{jz})$$

$$H^{O}_{SS} = \gamma^{2}_{S}\hbar^{2} \sum_{m\ n}^{N_{S}} r^{-3}_{mn} P_{2}(\cos\theta_{ij}) (\bar{S}_{m}\cdot\bar{S}_{n} - 3S_{mz}S_{nz})$$

$$H^{O}_{IS} = 2\gamma_{I}\gamma_{S}\hbar^{2} \sum_{i=1}^{N_{I}} \sum_{m=1}^{N_{S}} r^{-3}_{im}P_{2} (\cos\theta_{im})I_{iz}S_{nz}$$

In magic angle experiments (for a review, see Andrew, 1971) a time dependence is introduced in the spatial coordinates, by rotating the whole sample at a sufficiently high rate so that

$$<P_{2} [\cos\theta (t)] > = 0$$

for all i,j,m,n, making all dipolar interactions zero, along with all anisotropic interactions transforming like second rank tensors: the tensor character of the chemical shielding is lost, and at the same time, all related structural information.

In multiple pulse experiments (Haeberlen, 1967 and this conference; for a review see also P. Mansfield,1971) a time dependence is introduced in the spin coordinates, by a suitable pulse sequence, which produces a rotation of the spin operators so that

$$<\bar{I}_{i}\cdot\bar{I}_{j} - 3I_{iz}(t)I_{iz}(t) > = 0$$

Dipolar interaction is removed for one spin species and the anisotropic chemical shift is preserved but reduced by a factor $\sqrt{3}$.

PRINCIPLES

A third approach, the most recent one (Pines, 1972, 1973), does not introduce time-dependency, but speculates on a very straightforward way to achieve reduction of the dipolar Hamiltonian, that is, by making r the distance between like

nuclei big enough. Indeed H_d^o decreases with r^{-3}! This situation
arises for all chemically or isotopically diluted spins, $N_S \ll N_I$,
like (the chemically and biologically important) ^{13}C, ^{15}N and
^{29}Si. ^{13}C is 1.1 % abundant, with a 4 times smaller gyromagnetic
ratio than the proton one. ^{15}N is 0.37 % abundant, with a 10
times smaller gyromagnetic ratio. This, at the same time, sets
the class of spin systems which are attainable by the method.

There are two serious problems arising from such situation:

1. The diluted or rare spins S are in most cases, embedded in
   a matrix of concentrated spins I (which is essential for this
   method) so that the S spins will experience a very strong local
   field induced by the surrounding aboundant I spins, resulting
   in a strong broadening of the S line.
   In other words, in the dipolar Hamiltonian of interest for
   the S spins, the term H_{IS}^o will now largely dominate the term
   H_{SS}^o, which had been made small by virtue of its r^{-3} dependence.

2. The price we have to pay for the reduction in dipolar interaction between S-spins, is a loss in the number of S spins and
   consequently in NMR sensitivity.
   This is the same problem as in liquid studies of rare nuclei,
   where one has to use Fourier transform and signal averaging
   techniques in high magnetic fields, to get a maximum in
   sensitivity. In solids where T_1 is pretty long, accumulation
   becomes rather painful and conventional methods useless.

The high resolution method we describe here, makes use of
2 premises: dilution of a spin system S and a matrix of abundant
spins I. In the exceptionally favorable situation where $N_I = 0$, no
broadening by I spins will occur. This is for instance the case
in $CaCO_3$ and CS_2, for which Lauterbur (1958) and Pines (1971)
recorded the ^{13}C spectrum by conventional (high resolution)
techniques.

Luckily enough, both problems of resolution and sensitivity
can be solved, using established methods and techniques.

The dipolar broadening due to the I spins, can be removed
by strong irradiation of the I spins at their resonance frequency
(Sarles, 1958). Like in the case of J-coupling in liquids,
this induces a spin decoupling.

The sensitivity problem is approached by adapting an idea
established by Hartmann and Hahn (1962) for the sensitive
detection of rare nuclei using a double resonance method. The
specific schema used here, is based on the Lurie-Slichter (1964)
experiment, with, however, a direct observation of the S spins.
These ideas along with a full description of the high resolution
method are best discussed using the language of spin thermodynamics.

DESCRIPTION OF THE METHOD

To each of the spinsystems, I and S, we associate an energy reservoir in thermal equilibrium. The relaxation times characterizing equilibrium between these systems are resp. T_{IS}, T_{1I} and T_{1S}.
The inverse temperature of reservoir I is $\beta_I = 1/kT_I$, and of reservoir S, $\beta_S = 1/kT_S$.
Suppose both reservoirs are at equilibrium with the lattice at an inverse temperature $\beta_L = 1/kT_L$, in an external magnetic field H_o. This will give rise to a magnetization

$$M_S = \beta_L C_S H_o \qquad (1) \qquad \text{with} \qquad C_S = [\gamma_S^2 \hbar^2 S(S+1)N_S]/3k \quad (2)$$

for the S-spin system

and

$$M_I = \beta_L C_I H_o \qquad (3) \qquad \text{with} \qquad C_I = [\gamma_I^2 \hbar^2 I(I+1)N_I]/3k \quad (4)$$

for the I-spin system.

The energy associated to each spin reservoir is then given by

$$E_S = -\beta_L C_S H_o^2 \text{ , resp. } E_I = -\beta_I C_S H_o^2$$

In the absence of any r.f. irradiation, the I and S spin systems are disconnected from each other, in other words, no cross-relaxation is taking place between them.

The magnetization of the rare spins S given by eq. (1) can be observed in the free induction decay following a 90° pulse. This can then be repeated after a time of the order T_{1S} and the signals can be accumulated for sensitivity enhancement, in the presence of a strong irradiation of the I-system (decoupling) yielding a high-resolution NMR signal (Fig. 1).
However due to the long T_{1S} and to the dilution of the S-spins this method is in general not feasible. Indeed, if we compare for instance a ^{13}C with a ^{1}H FID, for which $\gamma_I/\gamma_S = 4$ and $N_I/N_S = 100$, the ^{13}C magnetization will be 1600 times smaller, in the same field, and another factor 4, at the same frequency.

Using the polarization of the I-reservoir, we can now enhance this S-signal, in the following way.

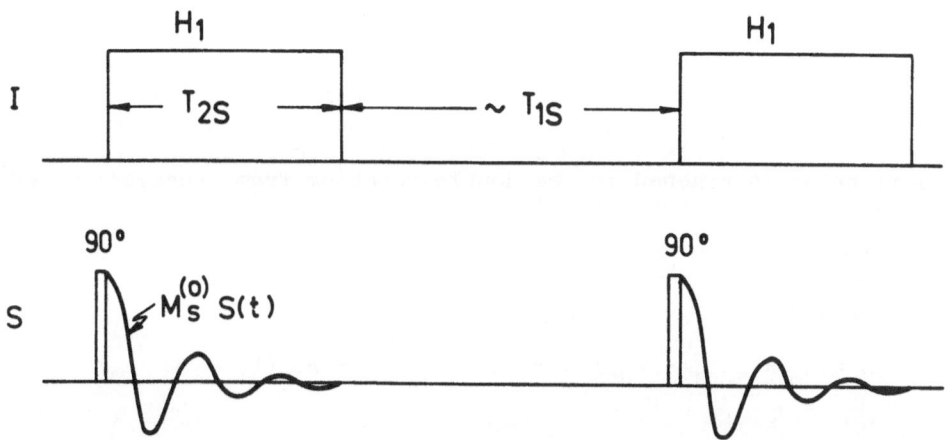

Fig. 1. Conventional high-resolution S free induction decay with I-spin decoupling. Signals are coherently added and the experiment is repeated at a rate $\sim 1/T_{1S}$.

Let us cool the polarized I-reservoir (and "keep it cool"). A convenient way to achieve this is by "spin locking" (Solomon, 1950) M_I along the x-axis in the I-rotating frame, by applying a 90° pulse (at the I-resonance frequency) followed by a long 90°-phase shifted pulse of amplitude H_{1I}, so that M_I is brought parallel to H_{1I} in the rotating frame.
This method conserves M_I, originally along H_0, but locks it along a much smaller field H_1; this is equivalent to cooling the I-spin system, to a rotating frame temperature β_0:

$$\beta_0 = \beta_L (H_0/H_{1I}) \tag{5}$$

In other words, we transferred the Zeeman order in the laboratory frame, into Zeeman order in the rotating frame.

In the next step, we bring the cold I-reservoir in contact with the warm S-reservoir, by closing the thermal link between I and S. The NMR equivalent of this, is to allow cross-polarization between the two spin reservoirs. This is achieved by applying simultaneously two strong r.f. fields H_{1I} and H_{1S} at the I and S

resonances. If the amplitude of these r.f. fields satisfy the
Hartmann-Hahn condition (Hartmann, 1962):

$$\gamma_I H_{1I} = \gamma_S H_{1S} \tag{6}$$

i.e. both magnetizations, M_I and M_S, precess at the same frequency,
in their resp. rotating frames, (Zeeman splittings are equal, in
both rotating frames) then, exchange of rotating frame Zeeman
energy will take place between both spin systems, and an internal
equilibrium is reached in the double rotating frame (characterized
by an inverse temperature β_1).

Formally, the full Hamiltonian describing the I-S system,
subjected to two strong r.f. fields H_{1I} and H_{1S}, at
frequencies ω_{oI} and ω_{oS} has the form:

$$H = H_o + H_{dII} + H_{dSS} + H_{dIS} + H_{1I}(t) + H_{1S}(t)$$

It is then appropriate to transform to a rotating frame, in
which the r.f. fields are stationary (Redfield, 1955). In
this case we need a double rotating frame induced by the
transformation:

$$R = \exp\left[-it(\omega_{oI} I_z + \omega_{oS} S_z)\right]$$

In this frame, the Hamiltonian is transformed to

$$H_R = H_d^o + H_{1I} + H_{1S} + \text{time-dep. terms}$$

$$H_{1I} = -\gamma_I \hbar H_{1I} \sum_i I_{ix}$$

$$H_{IS} = -\gamma_S \hbar H_{IS} \sum_i S_{ix}$$

The phase of the rotation is chosen to put H_{1I} and H_{1S} along
the x-axis in the I and S rotating frames.
Thermodynamics can then be applied in the rotating frame,
since the Hamiltonian is effectively time independent. The
two terms H_{1I} and H_{1S} are considered as reservoirs of Zeeman
energy. H_{1I} do not commute with either H_{dII}^o or H_{dIS}^o, and as
long as $H_{1I} \neq 0$, energy can be transferred between the Zeeman
reservoir H_{1I} and the secular dipolar reservoirs H_{dII}^o or H_{dIS}^o.
The same reasoning on the S spins, leads to the conclusion,
that H_{dIS}^o provides a coupling mechanism between H_{1I} and H_{1S},
so that they can exchange energy, at a rate determined by
the cross relaxation time T_{IS}, and ultimately approach a

state of full internal equilibrium in the rotating frame described by a temperature β_1.

The result of this "mixing" is :

1) a dramatic cooling of the S spin system (due to its small heat capacity compared to that of the I spin system), resulting in an ordering of the S spins in the rotating frame in the form of a magnetization M_S along H_{1S}, their effective field.

2) a slight heating of the I-spin reservoir, resulting in a small decrease of M_I.
The r.f. field H_{1I}, needed to realize the Hartmann-Hahn condition, is in this case, delivered by the r.f. field which locks the magnetization M_I in the rotating frame.

We can now calculate the magnetization of the S spin system after such a thermal mixing.
Since the total spin energy must be conserved, we have, by writing the energy before and after the contact:

$$\beta_1 C_I H_{1I}^2 + \beta_1 C_S H_{1S}^2 = \beta_0 C_I H_{1I}^2$$

where we neglected the energy $\beta_L C_S H_{1S}^2$ of the S-spin system before the mixing.
With (6) we find

$$\beta_1 = \beta_0 \, [1+\varepsilon]^{-1} \tag{7}$$

where $\varepsilon = (C_S/C_I) \, (\gamma_I^2/\gamma_S^2) = [N_S S(S+1)]/[N_I I(I+1)]$ (8)

For ^{13}C and ^{1}H, $\varepsilon \simeq 0.01$.
The S-magnetization following this single contact is:

$$M_S^{(1)} = \beta_1 C_S H_{1S}$$

Using eqns. (5), (6) and (7), we find:

$$M_S^{(1)} = (\gamma_I/\gamma_S)(1-\varepsilon)\beta_L C_S H_o \tag{9}$$

We can compare this to the magnetization we would have obtained by letting the S spin system equilibrate with the lattice in an external field H_o (eq. 1):

$$M_S^{(o)} = \beta_L C_S H_o$$

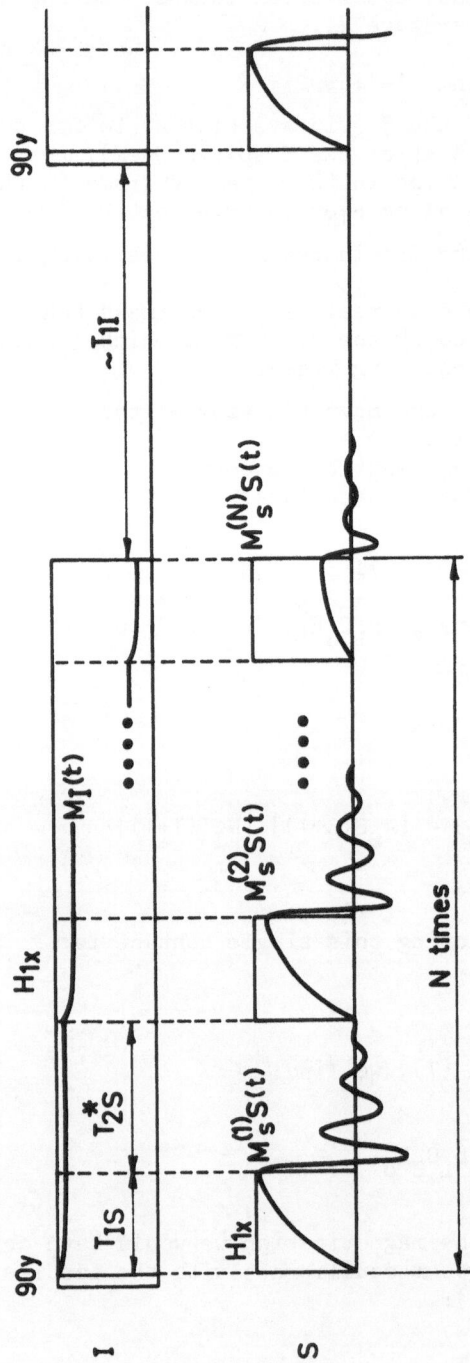

Fig. 2. Direct observation high-resolution double resonance ("Proton Enhanced Nuclear Induction Spectroscopy"). N contacts are made in a single experiment, the S signals being coherently added. The experiment is repeated at a rate $\sim 1/T_{1I}$.

In one single cross polarization, we have gained a factor

$$(1-\varepsilon)\gamma_I/\gamma_S$$

This is a factor 4 for ^{1}H and ^{13}C, and approx. 10 for ^{1}H and ^{15}N. (A gain of 4 in sensitivity, is a gain of 16 in time!). Notice that in the calculation we neglected the dipolar reservoirs, assuming them much smaller than the rotating frame Zeeman reservoirs (i.e., H_1 much larger than the local field). If this is not the case one gets a smaller sensitivity enhancement.

If, in a further step, we now suddenly remove (turn-off) the r.f. field H_{1S} (and at the same time "disconnect" the two spin reservoirs), the decay of $M_S^{(1)}$ to zero will be observed and the Zeeman order of the S spin system in the rotating frame is destroyed. The rotating frame temperature of the S system being again infinitely high, the whole process of (-MIX-OBSERVE-) can be repeated many times, until the I reservoir is completely warmed up and its magnetization is depleted. After the nth step, we find for the S magnetization:

$$M_S^{(n)} = (\gamma_I/\gamma_S)(1-\varepsilon)^n M_S^{(o)} \tag{10}$$

The signals resulting from these successive contacts are then coherently added for signal averaging and will ultimately be subjected to Fourier transformation to obtain the S spectrum. As the H_{1I} (spin-locking) field is still on, not only during the contacts, where it satisfies the Hartmann-Hahn condition, but also in between, during the observation period, then this same H_{1I} field will decouple the I spins, and the S signal after each contact will give us a high-resolution spectrum. The complete experiment, shown in figure 2, was called "Proton Enhanced Nuclear Induction Spectroscopy" (Pines, Gibby and Waugh, 1972).

It is now clear that this multiple-contacts (or even single contact) double resonance method, gives us by direct observation of the S spins free induction decay, a high resolution signal, with a tremendous gain in sensitivity.

Sensitivity

An extensive calculation of the gain in sensitivity was carried out by Pines and Waugh (1973), the results of which are now given here.
The gain in sensitivity for a cross-polarization experiment, where the number of contacts have been optimized for maximum S/N power ratio (by taking N too high, one exhausts the I magnetization and accumulates noise) is given by

$$(S/N)_{CP} \, / (S/N)_{FID} \, = (0.41/\varepsilon)(\gamma_I/\gamma_S)^2 \qquad\qquad (11)$$

For ^{1}H and ^{13}C this is about 650,
for ^{1}H and ^{15}N about 12,000 .

If one further takes into account the rate at which the signal comes out of the noise in each of those experiments, by writing that the cross polarization experiment is repeated after a time T_{1I}, and the FID after a time T_{1S}, then the overall gain in sensitivity is:

$$G_{CP} = (0.41/\varepsilon)(\gamma_I/\gamma_S)^2(T_{1S}/T_{1I}) \qquad\qquad (12)$$

As in most cases $T_{1S} > T_{1I}$, this is even a more favorable situation.
One calculates that this is 0.41 the gain one obtains in an adiabatic transfer (the most efficient process possible), of all the polarization from the I to the S reservoir. Nevertheless, such gains are in practice somewhat illusory. Indeed the I-magnetization cannot be maintained indefinitely along H_{1I}, but decreases in a time $T_{1\rho}$, characterizing the decay of M_I along H_{1I}. The fractional decrease of M_I which was determined by ε, the ratio of the S and I heat capacities, is now determined by the ratio $T_{2S}^*/T_{1\rho I}$, where $1/T_{2S}^*$ sets the spectral resolution. Replacement of ε by $T_{2S}^*/T_{1\rho I}$ is a much less favourable situation.

A VARIATION OF THE BASIC METHOD. THE SINGLE CONTACT TOTAL TRANSFER OF POLARIZATION

In the direct observation high resolution double resonance experiment, we clearly distinguished 4 steps:

"prepare", where the I spins polarize to full magnetization;
"hold", during which period the I-spin "order" is maintained (here, by spin-locking M_I in the I rotating frame);
"mix", where the order is transferred from the I to S spin system (here, by transfer of Zeeman order from the I rotating frame to the S rotating frame);
"observe", the S signal is observed and recorded while the I system is decoupled.

There are of course several variations of this basic schema. In the light of recent results, one variation (Pines, Waugh, 1973; Pines, 1973) proves to be very rewarding. Instead of spin-locking the I-magnetization along H_{1I} in the I-rotating frame, by which we transfer Zeeman order into the rotating frame, we will now transfer Zeeman order into the dipolar reservoir. This can be done in various ways: Adiabatic Demagnetization in the Rotating Frame (ADRF) (Slichter, 1961) following a spin-locking, ADRF by adiabatic fast passage and removal of H_1 at resonance (Anderson, 1962), or by the two-pulse transfer technique (Jeener, 1967).

Assume we begin again with the I spins polarized at the lattice temperature. We now perform an ADRF: M_I is locked along H_{1I} as in the previous method. H_{1I} is then adiabatically reduced to zero, leaving the magnetization M_I in the local dipolar field in the rotating frame H'_L, given by Goldman (1970)

$$H'_L{}^2 = (3\gamma_I^2)^{-1} < \Delta\omega^2 >_{II}$$

The I spin system being isolated, this results in a cooling of the spin system, to an inverse temperature

$$\beta_o = \beta_L (H_o/H'_L) \tag{13}$$

Keeping the I-system in its dipolar state, does not require any continuous r.f. irradiation any more. This is an important advantage, as continuous irradiation can lead to unacceptable power dissipation in the probe and heating of the sample.

In order to "mix" the I and S spin states we now turn on the H_{1S} at resonance, such that

$$\gamma_S H_{1S} = \alpha\gamma_I H'_L \tag{14}$$

Energy exchange now takes place between the I spin dipolar reservoir H^o_{dII} and the S spin rotating frame Zeeman reservoir H_{1S} via H^o_{dIS} reaching a common temprature β_1.
As the total energy is conserved, we can, like in the previous section, equate the energy before and after mixing, yielding for the final temperature:

$$\beta_1 = \beta_o [1 + \epsilon\alpha^2]^{-1} \tag{15}$$

and for the magnetization,

$$M_S^{(1)} = (\gamma_I/\gamma_S)(\alpha/1+\alpha^2\epsilon) M_S^{(o)} \tag{16}$$

For $\alpha = 1$, the Hartmann-Hahn condition is exactly met and we have the same results as for the spin-locking case.

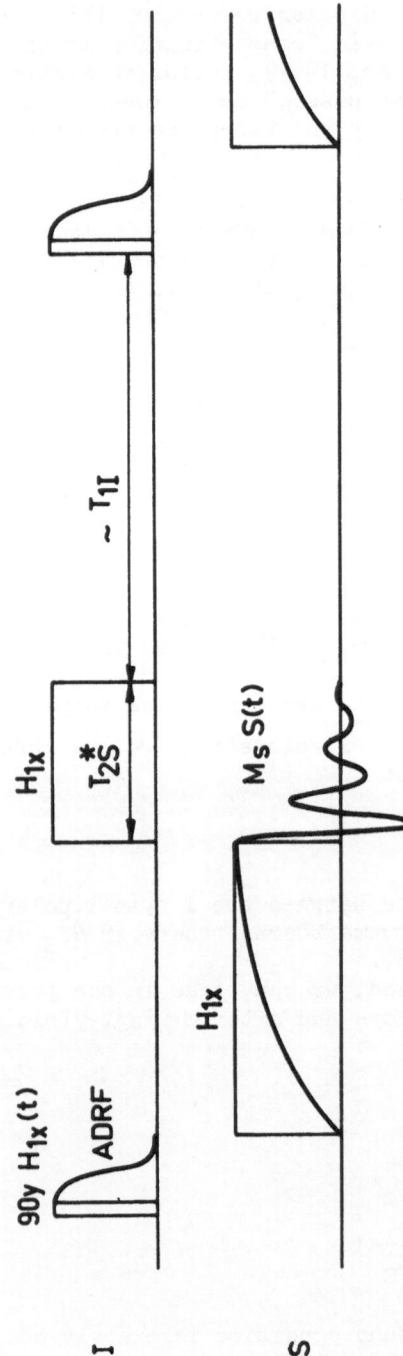

Fig. 3. A most rewarding version of the direct observation high resolution double resonance method. In a single experiment and one single contact, one transfers the whole I spin dipolar order to the S reservoir, with

$$H_{1S} = (\sqrt{\varepsilon})^{-1} H'_{LI}$$

I-r.f. irradiation is only necessary for decoupling.

Much more interesting is the case where $\alpha \gg 1$ (unmatched H-H condition). It is readily seen from eq. (16) that we then obtain a large S signal in a single shot (contact). The maximal value of $M_S^{(1)}$ occurs for $\alpha\sqrt{\epsilon} = 1$, or in other words, for a field H_{1S}, such that

$$H_{1S} = (1/\sqrt{\epsilon})H_L'$$

$M_S^{(1)}$ is then

$$M_S^{(1)} = (\gamma_I/\gamma_S)(1/(2\sqrt{\epsilon}))M_S^{(o)} \tag{17}$$

This is half the polarization we can get from fully adiabatic transfer (now, in one single contact). The "observe" phase is unchanged; the S-signal is recorded in the presence of a I-decoupling field. The whole sequence is depicted in figure 3. An important advance of this total cross-polarization over the multiple contact experiment, is that decoupling (and observing) occurs only once, thus reducing the total power requirements. Although this method looks very profitable, we have to make one restriction.

From the work of Mc Arthur (1969) we know that the cross-relaxation time T_{IS} exponentially increases when α increases, and may even become prohibitively long. H_{1S} is then long, and sample heating may again become a problem. At the same time, one has to keep T_{1S} shorter than T_{1DI}, the dipolar spin relaxation time of the I system. A compromise can then be made, between r.f. irradiation amplitude, mixing time, and loss in sensitivity.

INSTRUMENTAL REQUIREMENTS

We now describe the basic features of the spectrometer required to perform these direct observation double resonance experiments (Fig. 4).

Two synthesizers provide the basic r.f. for the I and S irradiation. In the configuration shown, and to set some numbers, the magnetic field has a value of 22.8 kG, so that for ^1H, the corresponding Larmor frequency is 97.2 MHz and for ^{13}C, 24.4 MHz. Superheterodyne operation (30 MHz) allows for minimum changes when switching to other nuclei (or frequencies).

All phase shifting necessary for spin-locking and ADRF sequences are done at the intermediate frequency, along with the main amplification of the precession signal and the detection. Gating of the various r.f. channels at the intermediate frequency is

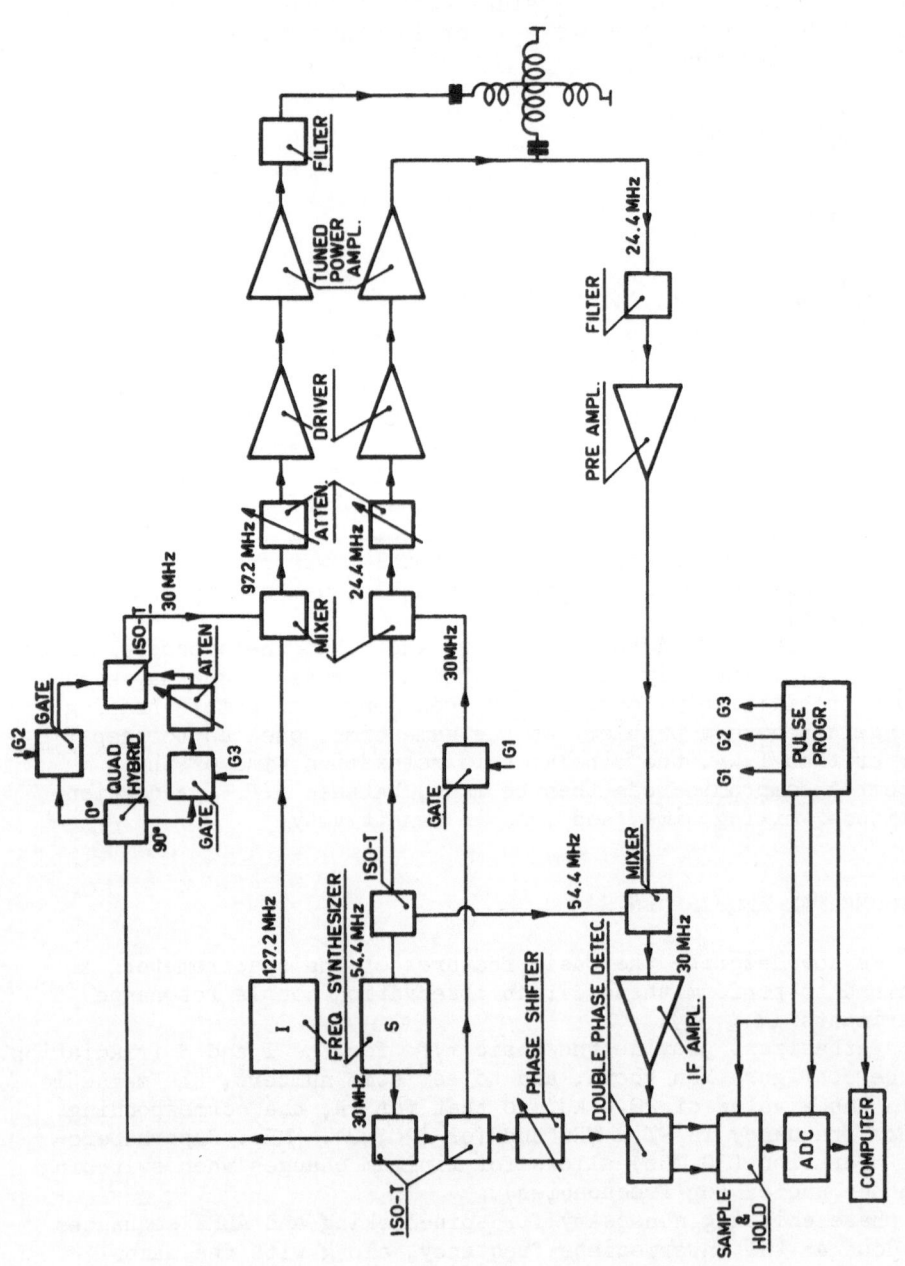

Fig. 4. Double resonance spectrometer, for direct high resolution observation of rare nuclei (S).

done by the pulse programmer which provides the variable pulse
widths and delays, and controls the timing of the data acquisition
system (sample-and-hold, ADC).
After mixing, the r.f. signals are then amplified in tuned power
amplifiers and sent to the sample coils. The S spin free precession
signal is amplified, mixed and further reamplified at the
intermediate frequency. After phase detection signals from the
quadrature phase detector are digitized and further processed.

Filters

A serious problem in the detection of the S-signal is the
leakage of the strong I r.f. frequency which produces large
bias voltages at the preamplifier imput. This is reduced by
using crossed sample coils, and by insertion of a multistage
filter in the output of the I-channel power amplifier and in the
S receiver chain. This is essential for the detection of the
small S signals while decoupling on the I frequency with a strong
r.f.

Power

It is interesting to point out that the requirement on the
power necessary to average out the strong dipolar S coupling
of the I spins with the S spins, is not so stringent as in the
multiple-pulse method. Indeed, in double resonance, decoupling
is continuous, so that multiple pulse, with a duty factor δ,
will require $1/\delta$ times more power to produce the same effective
H_1 field.
A couple of hundred watts are then sufficient to efficiently
decouple the I spins.

Q-factor

Furthermore, in multiple-pulse experiments one has to keep
the Q factor of the coil reasonably low, as the magnetization
has to be sampled in between closely spaced pulses (of the order
of 10 µs) so that receiver dead time due to high -Q probe
ringing is an important problem. In double resonance experiments
this problem does not exist and one can easily use high Q coils,
improving the sensitivity of the signal and making transmitter
power considerations easier to meet (a few hundred watts). Coils
with Q factors of 100 are currently used in these experiments.
It is clear that the spectrometer does not involve any highly
critical part or adjustment, and even the pulse programmer can
be kept very simple if one has only one specific application in
mind. In general, any spectrometer can, with the adjunction of
a second channel (with only moderate r.f. power) be used for

direct observation of double resonance. The modifications being
nowadays made rather easy by the use of commercially available
digital and linear integrated circuits and solid-state-devices.

We confined ourselves to a description of the direct
observation double resonance method (without entering in all
its various modifications) because of its relative simplicity
and of the tremendous flux of new and accurate results already
obtained in a preliminary stadium. These results and further
applications will be discussed in the next talks.
Modifications of the normal Hartmann-Hahn indirect observation
method have been proposed for high resolution (Mansfield, 1971,
1973; Bleich, 1971). An obvious advantage is the observation
of the abundant I-spins giving large signals at a usually much
higher frequency than for S spins. The achievable sensitivity
of this method is calculated to be roughly the same as for the
direct observation method, giving the latter one some advantages
due to the relative simplicity, ease of adjustment and absence
of distorsions. For $T_{1\rho I}$ below 100 ms, the sensitivity of the
indirect method is better, mainly for S nuclei more diluted
than ^{13}C. However the indirect detection method involves a
mapping of the S spectrum, point-by-point. If irrespective of
optimum sensitivity considerations, the time required to obtain
the complete S spin spectrum, is roughly T_{1I} in the direct
observation method, it will be $T_{1I}F_ST_{2S}^*$ in the indirect method
(F_S, the highest frequency in the S spectrum; $1/T_{2S}^*$, the
spectral resolution and $F_ST_{2S}^*$, the number of points in the
spectrum). The direct observation method is obviously the
quickest, as it delivers immediately a distorsion-free spectrum
and, even if it is not in all circumstances the most sensitive
one, it will at least not put stringent (sometimes intolerable)
requirements on the stability of the spectrometer.

REFERENCES

A.G. ANDERSON and S.R. HARTMANN, Phys.Rev. 128, 2023 (1962)

E.R. ANDREW, "Progress in NMR Spectroscopy" 8, Pergamon Press,
New York (1971)

H.E. BLEICH and A.G. REDFIELD, J.Chem.Phys. 55, 5405 (1971)

M. GOLDMAN, "Spin Temperature and Nuclear Magnetic Resonance in
Solids", Oxford Univ. Press, London (1970)

S.R. HARTMANN and E.L. HAHN, Phys.Rev. 128, 2042 (1962)

J. JEENER and P. BROEKAERT, Phys.Rev. 157, 232 (1967)

P.C. LAUTERBUR, Phys.Rev.Lett. 1,343 (1958)

F.M. LURIE and C.P. SLICHTER, Phys.Rev. 133, A1108 (1964)

P.M. MANSFIELD, "Progress in NMR Spectroscopy" 8, Pergamon Press, New York (1971)

P. MANSFIELD and R.K. GRANNELL, J.Phys. C4, L197 (1971); ibid. C5, L226 (1972); Phys.Rev. B8, 4149 (1973)

A. PINES, W.K. RHIM and J.S. WAUGH, J.Chem.Phys. 54, 5439 (1971)

A. PINES, M.G. GIBBY and J.S. WAUGH, J.Chem.Phys. 56, 1776 (1972); ibid. 59, 569 (1973)

A. PINES, "Proceedings of the First Specialized Colloque Ampere" Krakow 1973, 165 (1973)

A. PINES and T.W. SHATTUCK, Chem.Phys.Lett. 23, 614 (1973)

A.G. REDFIELD, Phys.Rev. 98, 1787 (1955)

L.R. SARLES and R.M. COTTS, Phys.Rev. 111, 853 (1958)

C.P. SLICHTER and W.C. HOLTON, Phys.Rev. 122, 1701 (1961)

I. SOLOMON, C.R.Acad.Sci. 248, 92 (1950)

J.H. VAN VLECK, Phys.Rev. 74, 1168 (1948)

HIGH RESOLUTION DOUBLE RESONANCE IN SOLIDS

RECENT DEVELOPMENTS AND APPLICATIONS

J. Tegenfeldt

University of Uppsala

Uppsala, Sverige

SOME INTRODUCTORY REMARKS

Double resonance techniques are useful among other things
for the observation of high-resolution spectra in solids when
(1) the spin species investigated (S in the following) has
a low concentration and (2) there is a more abundant spin
species (I) present in the sample. The various methods may
be classified into two groups:

(a) Indirect detection methods. In these the S-spins are
    observed indirectly through their influence on the I-spin
    signal [1 - 3].

(b) Direct detection methods. Here the S-spin free induction
    decay is observed directly and the I-spins are used merely
    as a source of polarization [4].

I will concentrate on the direct detection methods for the
simple reason that they so far have produced more results.

The basics of this type of experiment have been discussed
already by Van Hecke in one of the preceding lectures, and I
will merely add some, not necessarily very recent, refinements
and special tricks that may be employed to improve whatever
signal is observed and I will also talk about some applications
of these methods.

In order to establish a basic, schematic frame-work for
the following discussion, let us take a look at the fundamental
cycle of events that make up these experiments. This is shown
in figure 1.

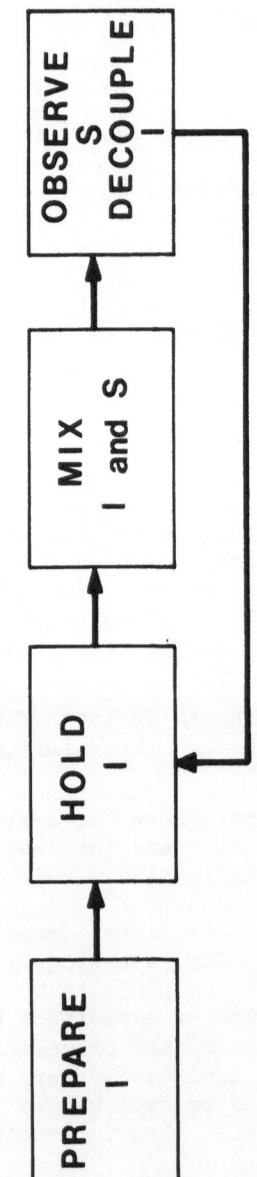

Fig. 1. The sequence of events in a typical double resonance experiment.

In words: the I spins are polarized in a strong magnetic field
and cooled down to a very low temperature in some reference frame
and held there, and then contact is established with the S
system causing a transfer of polarization to that system.
Finally, one observes the S magnetization that has been created,
while at the same time the effect of the I-S dipolar interaction
is eliminated through irradiation of the I spins at their
resonance frequency. The entire cycle is then repeated and the
signal accumulated until an adequate S/N ratio has been reached.

Also, let me remind you about what makes this way of
detecting the rare S spin superiour to a single resonance
experiment. The factors may be summarized as follows:

1. The actual S spin magnetization, M_S, obtained in a single
   cross polarization via the I spins, is larger than the
   magnetization obtained by direct polarization in the static
   field. The enhancement is $(\gamma_I/\gamma_S)(1 - \epsilon)$ when the Hartmann-
   Hahn condition [5] is fulfilled. The quantity ϵ is the
   heat capacity ratio between the S and I systems, usually
   a very small quantity.

2. The time between two Bloch decays is for the direct
   polarization of the order of $T_{1S} \gg T_{IS}$, the cross relaxation
   time determining the delay required between two mixings in
   a multiple contact experiment. Of course one has to
   repolarize the I spins once in a while, but even in a single
   contact experiment, however, there is often a saving in time
   since the delay necessary to repolarize the I spins is
   governed by T_{1I}, which is normally shorter than T_{1S}.

After these introductory remarks, let me now go over to a
discussion of some variations in the basic scheme, that can be
useful under a variety of circumstances. To begin with let us
look at the first step of figure 1, preparation of the I spins.

I SPIN PREPARATION RECIPES

An important factor in deciding if a double resonance
experiment will yield an acceptable result within reasonable
time, is the time consumed in the first step, preparation of
the I spin system. Here T_{1I}, the spin-lattice relaxation time
for the I spins, is important. This is because one has to wait
for a period of the order of a few T_{1I} to restore the I spin
magnetization between two cycles. For this reason a short
T_{1I} is desirable. Various ways can be thought of that may be
used to overcome the problem with a long T_{1I}:

(a) Shortening of T_1 by doping the sample with paramagnetic
     substances. This is often the most convenient way, but not
     always possible and sometimes undesirable if relaxation
     properties are to be studied.

(b) Using several samples, polarizing them outside the probe and
     switching sample when polarization is running out for the
     current sample. This creates some tricky practical problems
     and puts severe requirements on the patience of the
     experimenter.

(c) In samples containing a third nuclear species (denoted X,
     abundant) with a short T_1, and with spin >1/2 and a
     quadrupole coupling of suitable size, a special trick,
     level crossing in the laboratory frame, may be used to make
     a rapid transfer of polarization from the easily polarized
     X spins to the I spins [6, 7]. The procedure for the
     preparation of the I spin magnetization is the following:
     (1) Polarize the X spins in a strong magnetic field;
     (2) Lower the magnetic field until the quadrupole split
          levels of the X spins cross the Zeeman levels of the I
          spins, thereby causing a rapid transfer of polarization
          to the I spins.

Examples of suitable X nuclei are ^{35}Cl and ^{37}Cl, both
usually having a suitable quadrupole splitting when covalently
bonded in organic compounds (order of 30 MHz). The technique
is referred to as cascaded enhancement nuclear induction
spectroscopy, CENIS [8].

Figure 2 shows the result of the application of this
technique in a crystal of p-dichlorobenzene, for which T_{1I} is
several hours at room temperature [8]. A rather enormous gain
in sensitivity is apparent, compared to a direct polarization
of the I spins.

After polarizing the I spins we want to prepare a cold
I spin reservoir to use as a means for cooling the S spins
and create a S spin magnetization. In one of the preceding
lectures, one way of doing this was described: spin-locking
the I spins along the effective field and mixing with I spin
order as Zeeman order in the rotating frame, and repeating
this mixing several times per I spin polarization. I will
discuss and alternative procedure, mixing with the I spin order
transferred to dipolar order through adiabatic demagnetization
in the rotating frame (ADRF), and with a single, long contact
per I spin polarization [4].

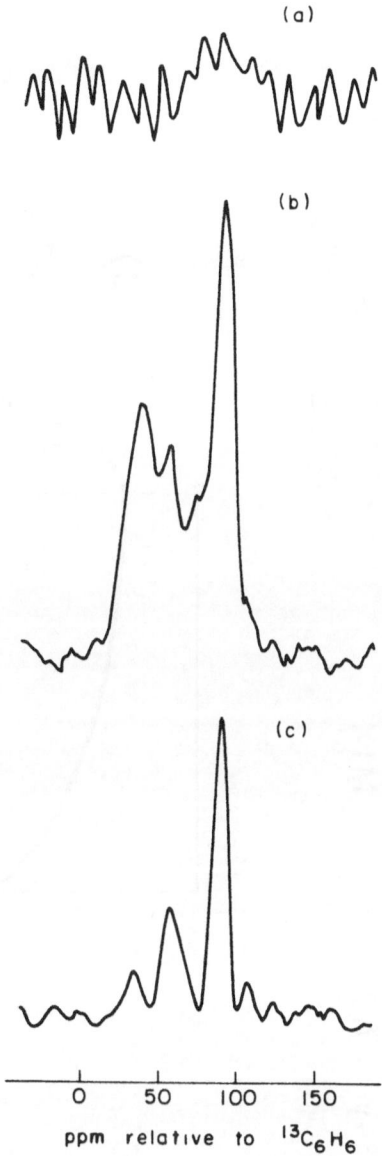

Fig. 2. Natural abundance ^{13}C spectrum of a single crystal of p–dichlorobenzene [8]. (a) Spectrum obtained after direct polarization of the protons for 5 hours. (b) Spectrum obtained using the CENIS technique. (c) Same as (b) but with shorter mixing time.

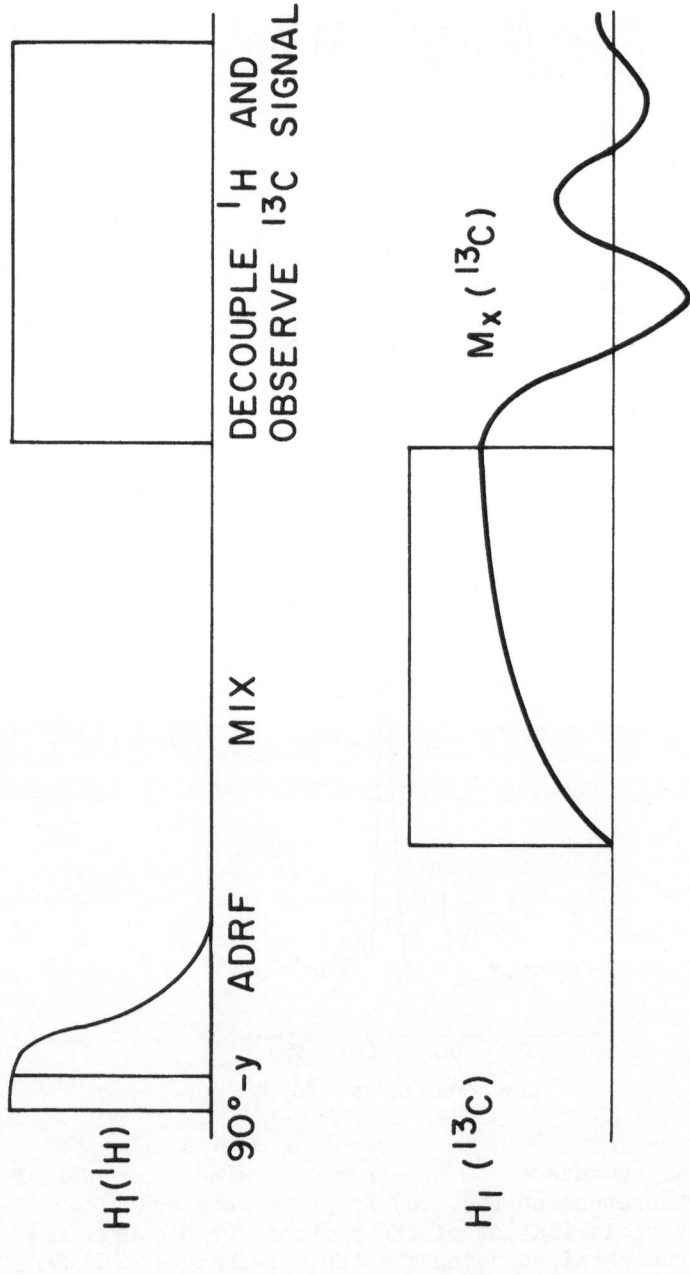

Fig. 3. One version of the double resonance high resolution experiment involving mixing with the I spin dipolar state after adiabatic demagnetization of the I spins in the rotating frame.

The sequence of events for this experiment is shown in
figure 3. The first step after polarizing the I spins is to
spin-lock M_I along the appropriate effective field and then
adiabatically to reduce H_{1I} to zero. This results in a cooling
of the I spin system through the transfer of Zeeman order to
dipolar order in the local field H'_L in the rotating frame,
defined by

$$H'^2_L = \frac{1}{3} M_2$$

for pure dipolar interactions. M_2 is the van Vleck second moment
of the I spin absorption line. The inverse temperature β_o
in the rotating frame after demagnetization is related to
the inverse laboratory temperature by

$$\beta_o = \beta_L \, (H_o/H'_L)$$

This now leaves us with an ordered, cold spin system, for
which the order can be maintained for a time of the order of
T_{1D}, without the presence of an RF field H_{1I} — this of course
has the advantage that less power is fed into the probe
during the mixing and hold periods, with less probe heating
as a result.

After having prepared the I spins we turn to the mixing
step, applying an RF field H_{1S} at the resonance frequency
of the S spins and with a magnitude such that

$$\gamma_S H_{1S} = \alpha \gamma_I H'_L$$

With $\alpha = 1$, this corresponds to the Hartmann-Hahn condition,
but we will investigate what happens when $\alpha > 1$ (mismatched
Hartmann-Hahn condition), which is actually a more interesting
case.

If we neglect spin-lattice relaxation during mixing, the
total spin energy is conserved and one can derive the following
expression for the final, common inverse spin temperature, β_1

$$\beta_1 = \beta_o \, (1 + \varepsilon\alpha^2)^{-1}$$

This gives the S spin magnetization after mixing $M_S^{(1)}$ in
terms of the magnetization $M_S^{(o)}$ obtainable by direct
polarization in H_o

$$M_S^{(1)} = (\gamma_I/\gamma_S) \, \alpha M_S^{(o)}/(1 + \alpha^2\varepsilon)$$

With $\alpha = 1$ we reach the same enhancement as for mixing with the magnetization M_I spin-locked and with the Hartmann-Hahn condition fulfilled. However, as α is increased, $M_S^{(1)}$ increases up to a maximum for $\alpha^2 \varepsilon = 1$. This maximum is

$$M_{S,\text{max}}^{(1)} = (\gamma_I/\gamma_S) \, M_S^{(o)}/2 \, \sqrt{\varepsilon}$$

This is considerably more than what is achieved in a mixing with $\alpha = 1$; in fact it is 50 % of the maximum transferable polarization - i.e. what one obtains after a completely adiabatic transfer of polarization from the I to the S spin system.

However, you never get anything without paying for it somehow; the price you have to pay in this case is a rapid decrease in crossrelaxation rate, T_{IS}^{-1}, as one leaves the Hartmann-Hahn condition. This is illustrated in figure 4, which shows some results of cross relaxation rate measurements for ^{43}Ca-^{19}F double resonance in CaF$_2$ [9].

Sometimes, however, the ADRF mismatched Hartmann-Hahn technique works very well – an example is given in figure 5. This is a ^{13}C spectrum of calcium formate with ^{1}H as the abundant spin [10]. Using the spin-lock version and the Hartmann-Hahn condition, the spectrum looked essentially like pure noise for the same total time spent on data collection.

CROSS POLARIZATION DYNAMICS

While dealing with the mixing step, let us consider some features of the dynamics of this process. The hamiltonian of a system containing two spin species, may be written

$$H = H_I + H_S + H_{IS} + H_{rf}(t)$$

Here H_I contains the Zeeman interaction of the I spins with the static field H_o and the dipolar interaction between the I spins; H_S contains the same interactions for the S spins and H_{IS} is the dipolar interaction between the I and the S spins. The term $H_{rf}(t)$ describes the interaction with the radiofrequency field. Transforming to a rotating frame with transformation

$$R = \exp\left[-it(\omega_{oI} I_z + \omega_{oS} S_z)\right]$$

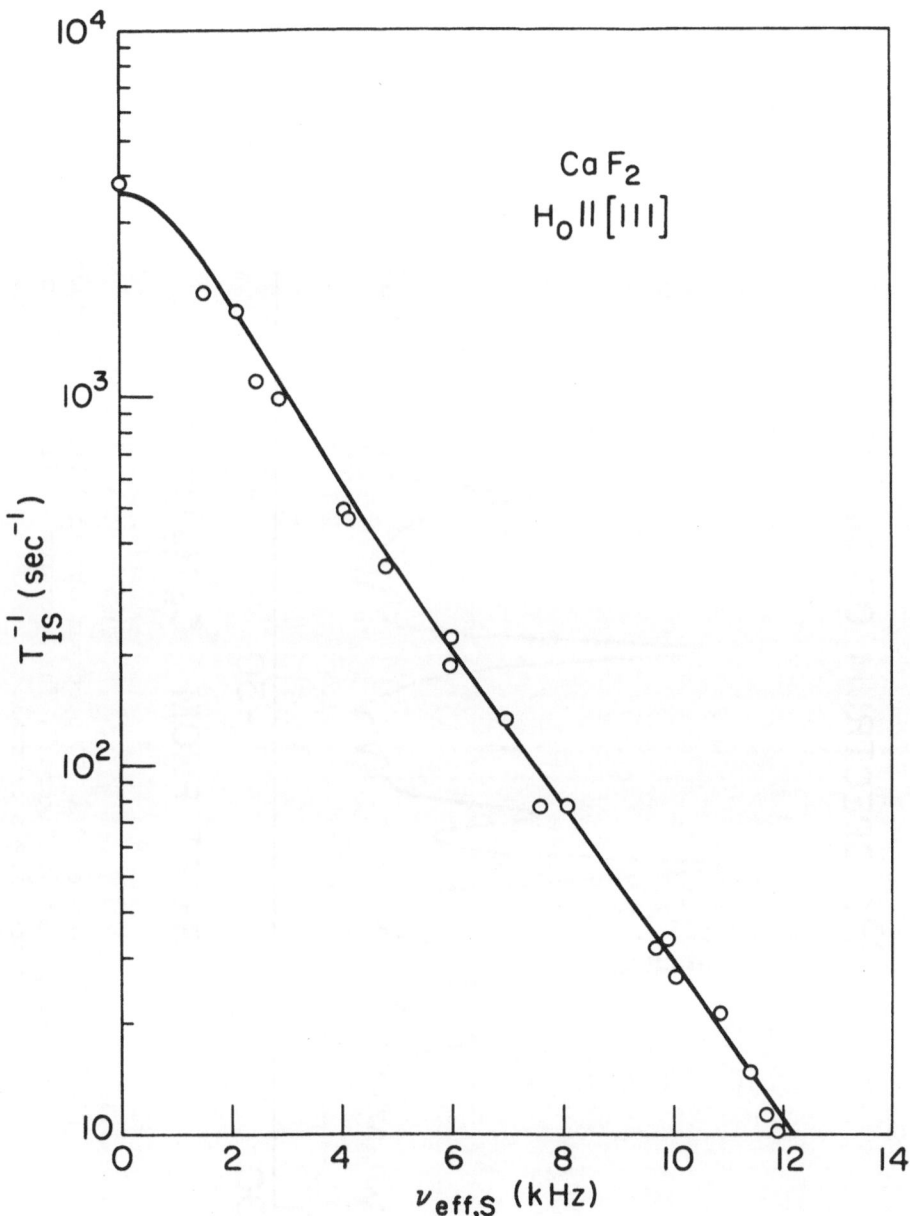

Fig. 4. Cross relaxation rates as a function of the ^{43}Ca
        rotating frame effective field for ^{43}Ca – ^{19}F mixing in
        CaF$_2$ with the I spins (^{19}F) in a dipolar state. The
        experimental points are taken from ref. 9 and the
        calculated curve from ref. 11.

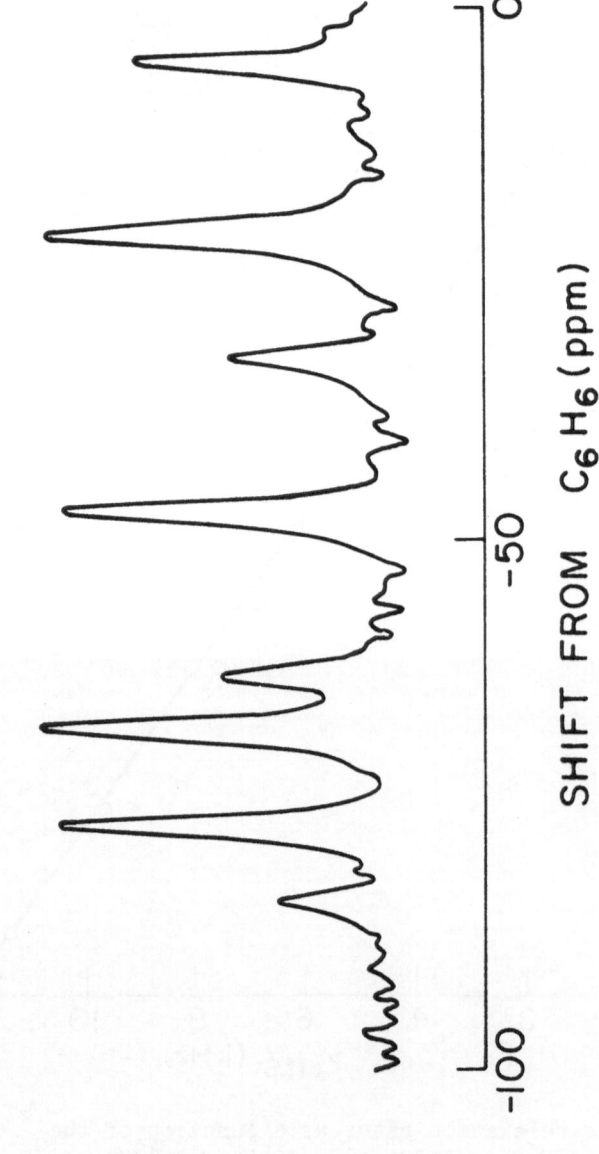

¹³C SPECTRUM OF Ca(HCO₂)₂

SHIFT FROM C₆H₆ (ppm)

Fig. 5. An example of a ¹³C spectrum obtained using the sequence
depicted in figure 3. The spectrum is from a calcium
formate single crystal.

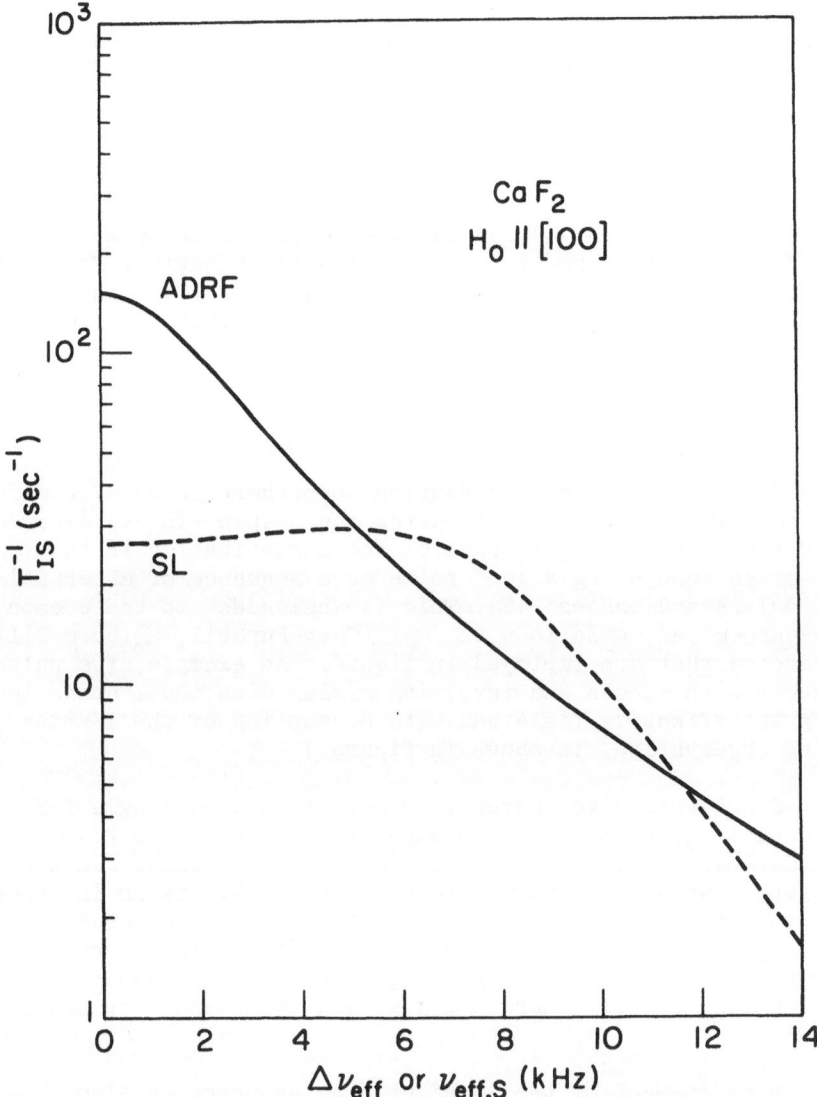

Fig. 6. Theoretically calculated cross relaxation rates for
^{43}Ca-^{19}F mixing in CaF$_2$ [11]. The abscissa is the
effective field in the rotating frame for the S spins
(^{43}Ca) for the curve labelled ADRF (mixing with the I
spins in a dipolar state) and the difference between
the I and S spin effective fields for mixing with the
I magnetization spin-locked (SL).

and neglecting oscillatory terms in the transformed Hamiltonian
one obtains a time independent hamilton operator

$$H_R = H_1 + H_2 + H_p$$

where H_1 contains I spin operators, H_2 S spin operators and H_p
contains interaction between I and S spins. Treating H_p
as a perturbation that couples H_1 and H_2 one may in principle
calculate the cross relaxation rates T_{IS}^{2-1}. This has been done
for CaF_2 (I = ^{19}F, S = ^{43}Ca) and the results display a
considerable difference in cross relaxation behaviour for mixing
in the two extreme cases: (1) after an ADRF or (2) with the I
spins locked along the effective field [11]. This is illustrated
in figure 6.

ECHOES

        If we look at the observation step, there is one variation
that sometimes may be useful. After the S spin FID has decayed,
it is possible to restore part of the magnetization in the
form of an echo using a 180° pulse or a sequence of alternating
180° pulses and echoes [12]. This is analogous to the common
spin-echoes and variations thereof (Carr-Purcell, Meiboom-Gill
sequences) that are employed in liquids. An example of a pulse
sequence with echoes and involving mixing with the I spins locked
along the effective field and with decoupling of the I spins
during observation, is shown in figure 7.

        Since this gives several signals from one, single I-S
contact, it is tempting to conclude that this should give a
considerable improvement of the signal over a single FID per
contact. However, we have to remember that this is an inhomogeneous
echo and only a portion of the magnetization is restored. The
echo amplitude is thus successively decreasing in the presence
of constant noise and trying to squeeze too many echoes out of
one I-S contact will therefore just add noise and very little
signal.

        A refinement of the straight echo sequence is also shown
in the figure: In the middle of one of the echoes, H_{1S} is
turned on again, and since some magnetization has been restored
along the S spin effective field already before H_{1S} is turned on,
the mixing starts out with $M_S > 0$ and therefore the mixing
requires less time [13].

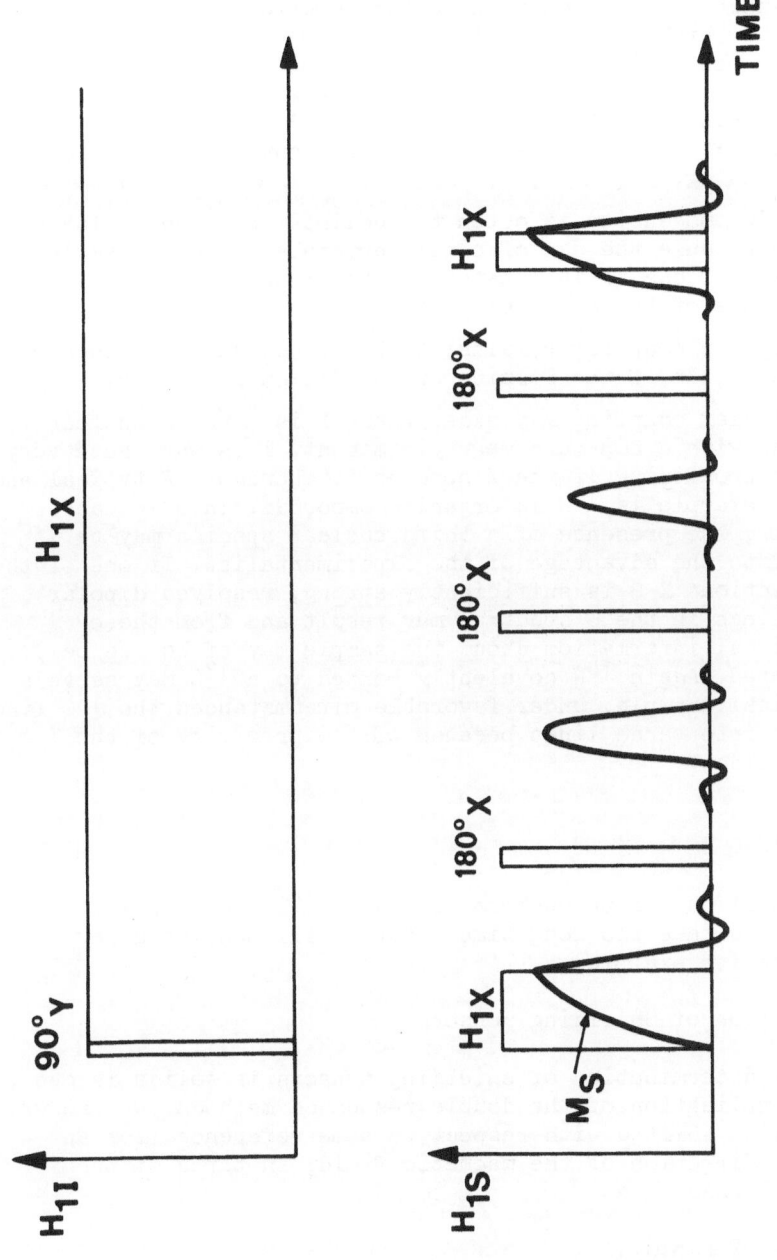

Fig. 7. Double resonance high resolution experiment including
z-restored spin echo pulse sequence [13].

RESOLUTION

Before I go on with the discussion of applications there
is one aspect I should mention. Although we call these
experiments "high resolution," we must remember that the
attainable resolution by no means approaches what one can
accomplish in liquids. Typically the line width in, say, $^{13}C-^{1}H$
double resonance spectra will be of the order of a ppm in
fields of 10-12 kG. There are several reasons for the residual
line-broadening.

1. Field inhomogeneity is often the dominant broadening factor
   partly because the use of single crystals prevents sample
   spinning if one is interested in the orientational
   dependence of the chemical shift.

2. S-S dipolar coupling remains; this is, however, not much of
   a restriction if the S spins are sufficiently dilute.

3. X-S dipolar coupling may exist where X is a third nuclear
   species with a non-zero magnetic moment. This may cause very
   severe broadening if the X species is abundant. A typical and
   common example is ^{14}N in organic compounds. In some cases,
   however, the presence of a third nuclear species may be
   turned to the advantage of the experimentalist. If one of the
   interactions X-S is sufficiently strong, resolved dipolar
   splittings in the S spectrum may result and from these
   additional information about the sample may of course be
   extracted. Again ^{14}N covalently bonded to a ^{13}C may serve as
   a typical example: under favorable circumstances the ^{13}C line
   splits into three lines because of the proximity of the I = 1
   ^{14}N nucleus.

APPLICATIONS

I will not go through an exhaustive list of applications
- this would take too long time - but I will mention a few
representative examples.

Determination of Shielding Tensors

The determination of shielding tensors in solids is one
obvious application of the double resonance methods. We can write
the chemical shift σ with respect to some reference, for an
arbitrary direction of the magnetic field, in terms of the
shielding tensor $\underline{\sigma}$ as

$$\sigma = \sum_{i,j} n_i n_j \sigma_{ij}$$

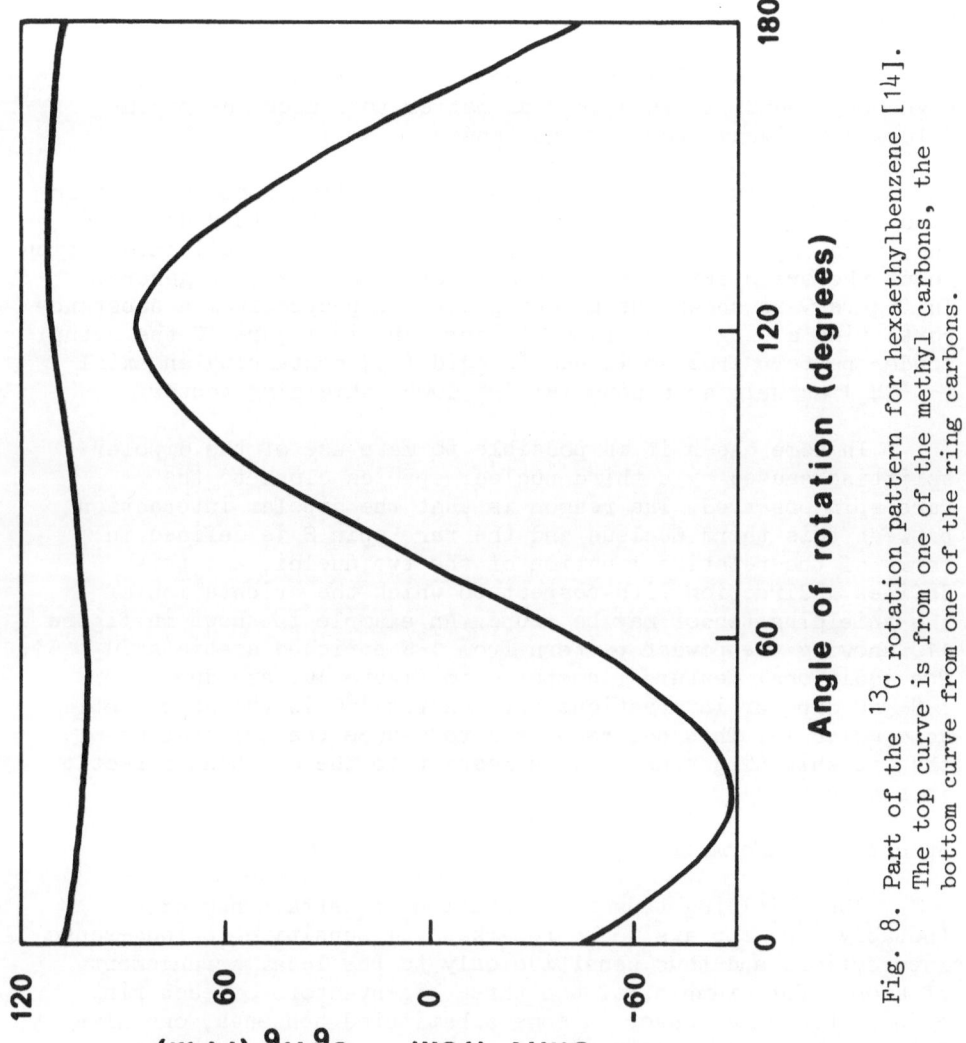

Fig. 8. Part of the ^{13}C rotation pattern for hexaethylbenzene [14].
The top curve is from one of the methyl carbons, the
bottom curve from one of the ring carbons.

The vector $\underline{n}$ is a unit vector along the magnetic field. The
components of $\underline{\sigma}$ are usually found by recording σ as a function
of the rotation angle for a rotation of a single crystal about
three mutually perpendicular axes. A typical rotation pattern
is shown in figure 8. The figure shows part of the data from a
single crystal of hexaethylbenzene (S = ^{13}C, I = ^{1}H) [14].
Note the period of 180°.

Once the six independent components of the symmetric $\underline{\sigma}$
have been found, it is a trivial matter to deduce the eigenvalues and eigenvectors of the tensor.

From a powder it is also possible to find some information
about the shielding tensor: it is still possible to deduce
the three eigenvalues of the tensor, but of course all information
about the orientation of the eigenvectors is lost, in general.
In figure 9 is shown the powder pattern expected from a substance
containing a single shielding tensor, and in figure 10 the actual
powder pattern from solid acetic acid [15] containing an axial
($-^{13}$CH$_3$) as well as a nonaxial ($-^{13}$COOH) shielding tensor.

In some cases it is possible to make use of the dipolar
splitting caused by a third nuclear species close to the
rare spin observed. The reason is that the dipolar interaction
between this third nucleus and the rare spin S is defined in
terms of the relative location of the two nuclei, and this
defines a direction with respect to which the orientation of
the shielding tensor may be found. An example is shown in figure
11, showing the powder pattern from ^{13}C enriched acetic acid [15].
The additional features, compared to figure 10, are due to
^{13}C-^{13}C dipolar interactions between two ^{13}C in the same acetic
acid molecule. This has been used to deduce the orientation of
the ^{13}C shielding tensors with respect to the C-C bond direction
in acetic acid.

Structural Information

The shielding tensor orientation of certain nuclei
(notably ^{13}C) appears to be remarkably insensitive to long-range
interactions and thus sensitive only to the local arrangement
of atoms. For example, of the three eigenvectors of each ring
carbon shielding tensor in some substituted benzenes, one always
seems to be very closely normal to the ring. One can use this
property to find the orientation of this plane in a crystal
or the relative orientation of planar molecules.

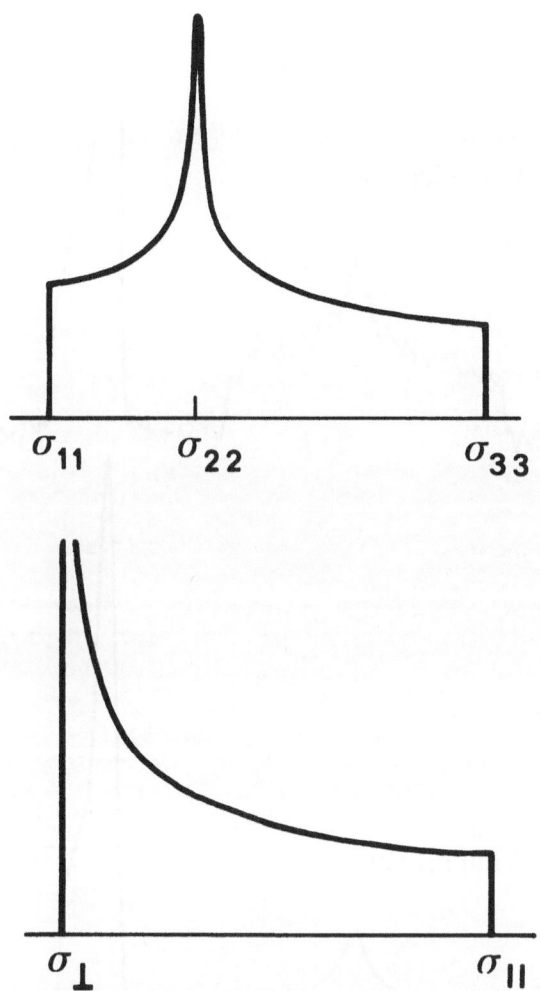

Fig. 9. Theoretical powder patterns for a non axial (top) and
      an axial (bottom) shielding tensor.

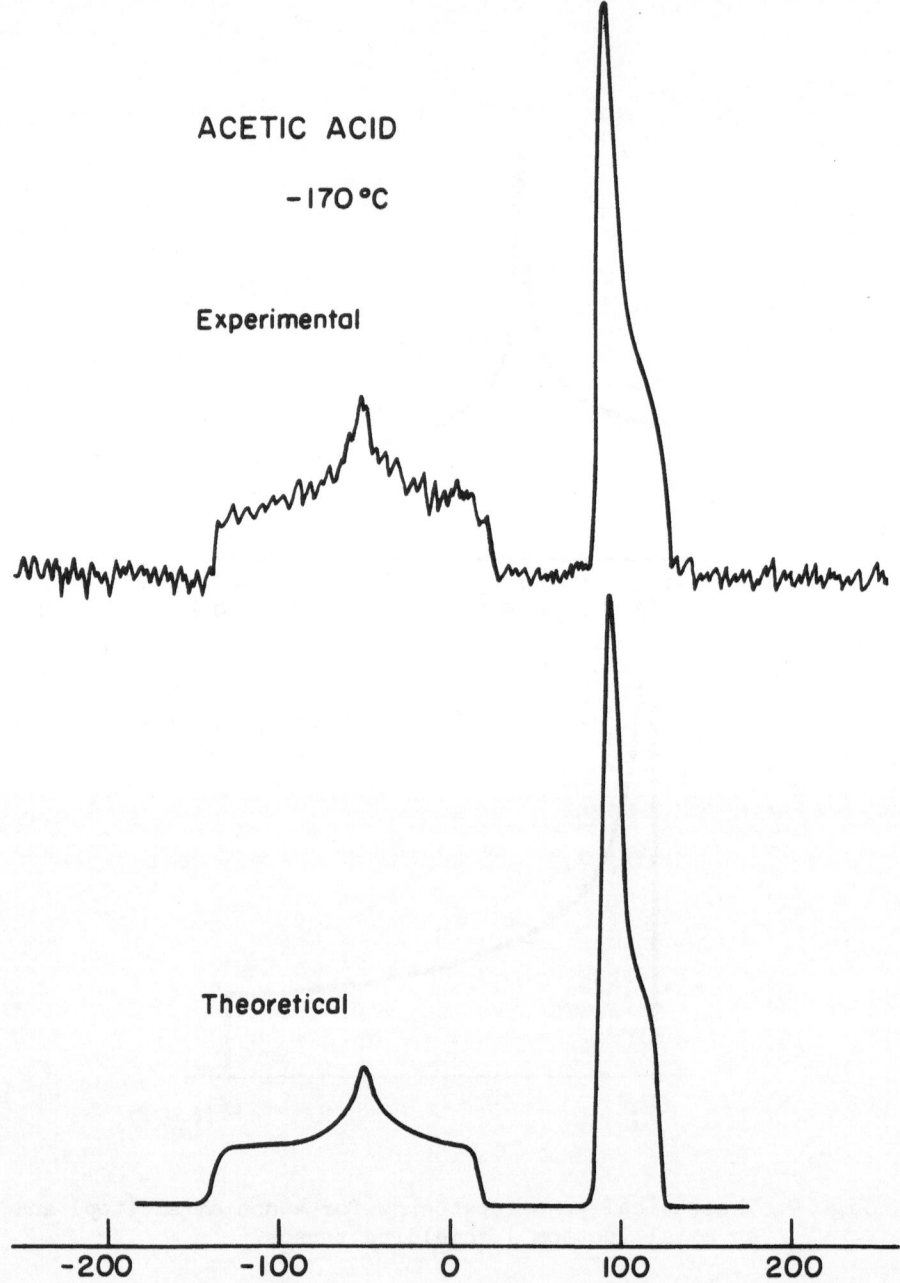

Fig. 10. Powder pattern from natural abundance ^{13}C in acetic acid [15].

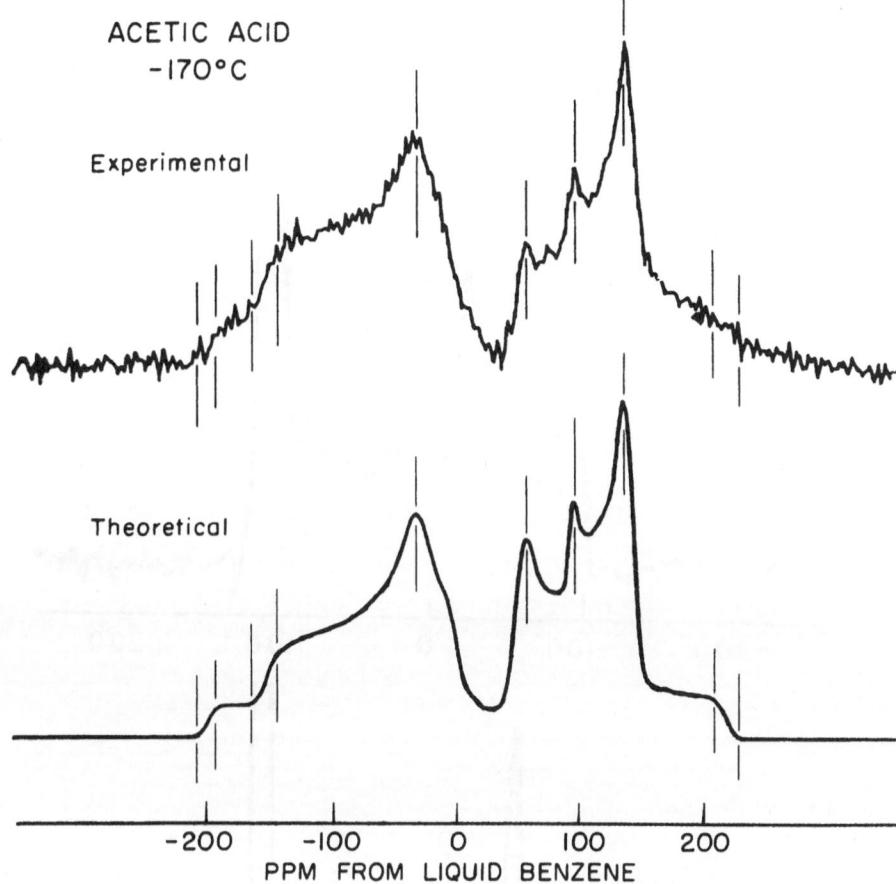

Fig. 11. Powder pattern from acetic acid, doubly enriched by
         45 % in ^{13}C [15].

Molecular Motions

     A sufficiently rapid rotation of the molecule observed,
leads to an averaging of the shielding tensor. In the extreme
case, when the motion is isotropic, only the isotropic chemical
shift $(1/3)$Tr$\underline{\sigma}$ is observed, just as in a liquid. An example is
adamantane.

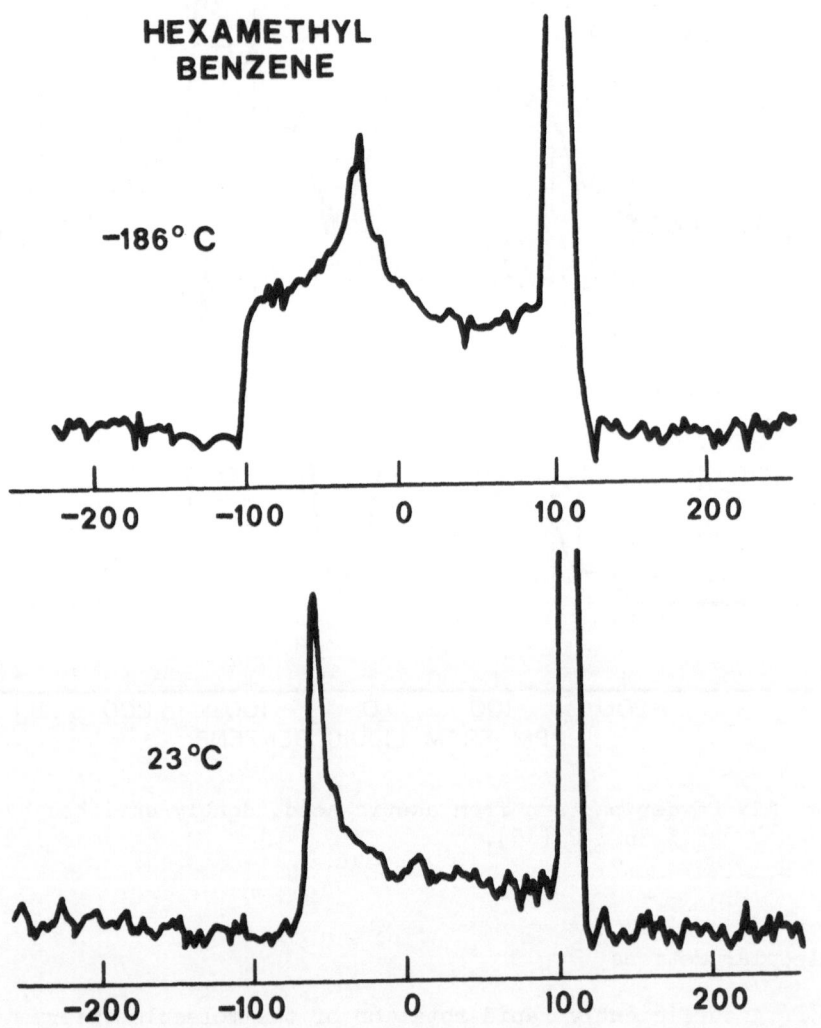

Fig. 12. Powder pattern from natural abundance ^{13}C in hexamethyl-
         benzene [16].

In other cases the motion is strongly anisotropic, as
for example in hexamethylbenzene, which rotates about the normal
to the molecular plane at room temperature, giving axially
symmetric tensors [16]. That this is actually a motional effect
can be shown by freezing out the motion by lowering the temperature
sufficiently: the axial symmetry of the ring carbon tensors
is lost. This is shown in figure 12.

Diffusion

For the study of diffusion in solids one may take
advantage of the possibility of making inhomogeneous echoes
using double resonance methods, as described above. This is
analogous to the use of echoes in studying diffusion in liquid
systems, using conventional pulse techniques.

The full recovery of the magnetization in an inhomogeneous
echo experiment depends on the absence of diffusion; any
translational diffusion of the molecules studied will destroy
the magnetization irreversibly and the amplitude of the echo
will decrease. By doing the experiment with a known magnetic
field gradient, quantitative information about this diffusion
may be obtained [12].

Large Molecules, Biological Systems

The methods may be useful for looking at large, organic
molecules, for example molecules occurring in biological systems.
One reason is that in many biological systems long rotational
correlation times, for the molecules of interest, result in a
residual dipolar coupling, that will smear out spectra
obtained by using conventional NMR techniques. Of course, part
of this residual dipolar coupling may be eliminated by decoupling,
as described earlier in these lectures. Furthermore, it is
precisely this residual dipolar coupling that makes possible
sensitivity enhancement through crosspolarization. The technique
has been applied to model lipid systems producing partly
resolved multiline spectra [17], whereas conventional NMR does
not give resolved spectra for these systems.

REFERENCES

1. P. MANSFIELD and P.K. GRANNELL, J.Phys. C4, L197 (1971)
                                 ibid. C5, L226 (1972)

2. P.K. GRANNELL, P. MANSFIELD and M.A.B. WHITHAKER, Phys.Rev. B8,
   4149 (1973)

3. H.E. BLEICH and A.G. REDFIELD, J.Chem.Phys. 55, 5405 (1971)

4. A. PINES, M.G. GIBBY and J.S. WAUGH, J.Chem.Phys. 56, 1776 (1972)
   ibid. 59, 569 (1973)

5. S.R. HARTMANN and E.L. HAHN, Phys.Rev. 128, 2042 (1962)

6. D.E. WOESSNER and H.S. GUTOWSKY, J.Chem.Phys. 29, 804 (1958)

7. M. GOLDMAN, Compt.Rend. 246, 1038 (1958)

8. D.E. DEMCO, S. KAPLAN, S. PAUSAK and J.S. WAUGH, to be published

9. D.A. McARTHUR, E.L. HAHN and R.E. WAHLSTEDT, Phys.Rev. 188, 609 (1969)

10. J.L. ACKERMAN, J. TEGENFELDT and J.S. WAUGH, J.Amer.Chem.Soc., in press

11. D.E. DEMCO, J. TEGENFELDT and J.S. WAUGH, submitted to Phys.Rev.

12. A. PINES and T.W. SHATTUCK, Chem.Phys.Letters 23, 614 (1973)

13. J.S. WAUGH, J.Mol.Spectry 35, 298 (1970)

14. S. PAUSAK, J. TEGENFELDT and J.S. WAUGH, J.Chem.Phys. 61, 1338 (1974)

15. S. KAPLAN, Ph.D.Thesis, M.I.T. (1974)

16. A. PINES, M.G. GIBBY and J.S. WAUGH, Chem.Phys.Letters 15, 373 (1972)

17. J. URBINA and J.S. WAUGH, Ann.N.Y.Acad.Sci. 222, 733 (1973)

WIDE LINE DOUBLE RESONANCE AND RELAXATION IN THE
ROTATING AND LABORATORY FRAMES

R. Blinc

University of Ljubljana, Institut Jozef Stefan

Ljubljana, Jugoslavia

DOUBLE RESONANCE SPECTROSCOPY

Double resonance is a trigger detection method which uses the "strong" NMR signal of the A spin system to detect the "weak" nuclear resonance or nuclear quadrupole resonance of the B spins. The method is based on the fact that, if certain conditions are met, the effect of a radiofrequency perturbation of the B system can be via A - B dipole-dipole coupling transferred to the A system, integrated and detected as a change in the NMR signal of the A spins. The transfer of the r.f. perturbation will be maximized if the separation between the energy levels of the B spins (in any given representation in which the B spins are quantized) exactly matches the separation between the energy levels of the A spins in the representation in which these are quantized. Depending on the representation in which the resonance transfer of the r.f. perturbation is performed we may distinguish

  i) double resonance in the rotating respectively dipolar frame
 ii) double resonance in the laboratory frame
iii) double resonance between the laboratory and the rotating frame.

Each double resonance experiment involves three major steps:
a) first, the A spins are ordered
b) second, the B spin system is transformed into a state of maximum possible disorder and is coupled to the A system
c) third, the remaining order in the A spin system is detected.

A simple thermodynamic analogy is as follows:

The A and B systems represent two thermal reservoirs, a large
and a small one, which are connected with a heat conducting link.
The B spin system can be heated externally by irradiation with
a resonant rf field, but its heat capacity is too small to
allow a direct measurement of the resulting change in its
temperature. If, however, the B system is continuously heated
for a long enough time, and the thermal conductivity of the
link to the A system is sufficiently high, the temperature of
the A system will make a detectable change.

There are several quantities which can be measured in a
double resonance experiment:
a) the fact that the temperature of the A system has changed shows
   that resonant rf absorption in the B spin system has occured.
   In this way the S spin NMR spectrum can be measured.
b) in addition, one can measure the ratio of the heat capacities
   of the two systems, the thermal conductivity of the link, and
   the time the two thermal reservoirs remain isolated from the
   surrounding lattice reservoir, i.e. the spin-lattice
   relaxation times.

The basic physics of a double resonance experiment in the
rotating frame may be perhaps understood in the following way:

The Hamiltonian of a spin system - for instance A - in the
rotating frame is time independent. At exact resonance it is
equivalent to the Hamiltonian of a spin system in a static magnetic
field with the rf field $H_1 \parallel x$ replacing the static field $H_o \parallel z$.
The direction (x) of H_1 thus determines the axis of quantization.
A field, which oscillates perpendicular to the direction of
quantization (x) with a frequency which - multiplied by Planck's
constant - matches the energy level separation in the rotating
frame will induce resonance transitions in the A-system. Such
a field is created by the magnetization of the B spins if
irradiated with another rf field of appropriate frequency and
amplitude, which satisfies the double resonance condition.

We shall in the following limit ourselves to the special
case where the spin-lattice relaxation times of both species
are long compared to the spin-spin relaxation times involved.

The experimental procedure is illustrated in figure 1. The
two spin species are first polarized in a strong external magnetic
field $H_o \parallel z$. Then we apply along the x-direction a radiofrequency
field pulse $H_1(t)$ which is resonant ($\omega_A = \gamma_A H_o$) with respect to
the A spins and has an amplitude H_{1A} and a pulse length t

$$\gamma H_{1A} t = \pi/2 \qquad\qquad\qquad\qquad\qquad (1)$$

so that it turns the A - magnetization from the z into the y
direction. Next we change the phase of the above r.f. field by 90

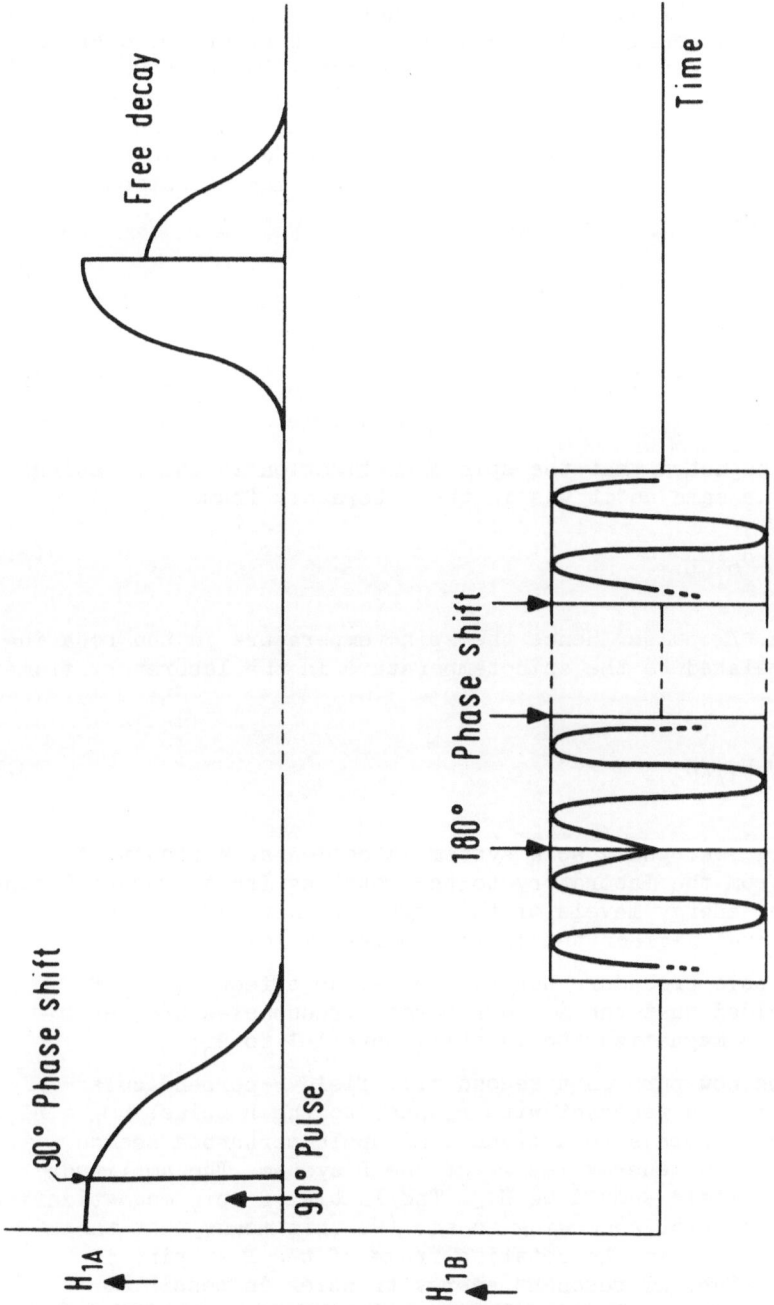

Fig. 1. Pulse sequence for double resonance in the dipolar respectively rotating frame.

degrees so that the r.f. field and the A magnetization are parallel
to each other. For an uncoupled system we would now have a
stationary situation, i.e. the A magnetization and the r.f. field
are locked and would remain parallel to each other for ever. In
view of the coupling of the A spins to the lattice, however, this
state decays with a time constant $T_{1\rho}$, which we call the spin-
lattice relaxation time in the rotating frame. For times short
compared to $T_{1\rho}$ a spin-temperature Θ can be defined in the
rotating frame which is related to the spin magnetization in
this frame by

$$M = CH_{1A}/\Theta \qquad\qquad (2.a)$$

with C being a Curie constant

$$C = \gamma_A^2 \hbar^2 \, A(A+1)N/3k \qquad\qquad (2.b)$$

It should be noted that the spin magnetization in the rotating
frame is the same as it was in the laboratory frame

$$M = C \, H_o/\Theta_L \qquad\qquad (3.a)$$

before the $\pi/2$ pulse. Hence the spin-temperature in the rotating
frame is related to the spin-temperature in the laboratory frame
Θ_L by

$$\Theta = \Theta_L H_{1A}/H_o \qquad\qquad (3.b)$$

and $\Theta \ll \Theta_L$, i.e. the A spin system is cooled as a result of
transfer from the laboratory to the rotating frame. The splitting
between the energy levels of this sytem is now γH_{1A} if we can
for the moment neglect the local dipolar fields.

The above procedure has had so far no effect on the B
spins provided that the A and B Larmor frequencies are far
apart. The B magnetization is still parallel to H_o.

Let us now turn on a second r.f. field – perpendicular
to H_o – which is resonant with respect to the B spins: $\hbar\omega_B = \Delta E_B$.
Here ΔE_B corresponds to a given quadrupole perturbed Zeeman
splitting of the energy levels of the B system. The amplitude
of the r.f. field should be H_{1B}. The initial B spin magnetization
along H_{1B} is zero. According to eq. (2) this means that the
spin temperature in the rotating frame of the B – spins is
infinitely high. If resonant energy transfer is possible,
i.e., if the energy level splittings in the two rotating frames

$\Delta E_{A'}$ and $\Delta E_{B'}$ are matched:

$$\Delta E_{A'} = \Delta E_{B'}$$

or

$$\gamma_A H_{1A} = \gamma_B H_{1B} \qquad (4.a)$$

with

$$\hbar \omega_B = \Delta E_B \qquad (4.b)$$

the spin temperatures of the A and B system will become equal
rather soon because of dipolar interactions, and a finite B
magnetization along H_{1B} is created. In order that the heat flow
from the B to the A system does not stop, we invert the phase
of H_{1B} for 180°. After that the B magnetization and the field
H_{1B} have opposite directions so that the B spin system has a
negative spin temperature (see eq. 2.a). Since negative temperatures
are as a matter of fact, "hotter" than positive temperatures,
the heat flow from the B to the A system continues. The procedure
is repeated for a time of the order of $T_{1\rho}$.

The requirement of having two simultaneous strong r.f. fields
during the double resonance experiment can be avoided by performing
an adiabatic demagnetization in the rotating frame of the A spins
(Fig. 1). The A spins are in this case spin locked in their
dipolar frame and the double resonance condition (4.a) has to
be replaced by

$$\gamma_A H_{loc} \approx \gamma_B H_{1B} \qquad (4.c)$$

where H_{loc} is the local dipolar field acting on the A spins. The
double resonance conditions (4.a) and (4.c) are strictly valid
only for nuclei with no quadrupole interactions. Rigorously
one should of course match the rotating frame energy level
splitting of the A system to the rotating frame energy level
splitting of the B system.

Since the heat flow from B to A does not change the total
energy of the combined system, it can be easily shown that after
n 180° phase changes (Fig. 1) the common spin temperature Θ_n
equals

$$(1/\Theta_n) = (1/\Theta_o)/(1+\varepsilon)^n \approx (1/\Theta_o)\exp(-n\varepsilon) \qquad (5)$$

where $\varepsilon \ll 1$ is the ratio of the magnetic heat capacities of the two systems.

After that the A spins are adiabatically remagnetized and the free precession signal of the A spins is measured as:

$$M \overset{\sim}{\sim} M_o \exp(-n\varepsilon) \tag{6}$$

if spin-lattice relaxation effects can be neglected. In practice ΔE_B is not known and the frequency ω_B as well as the amplitude H_{1B} have to be adjusted to satisfy both the "resonance" (4.b) and the "double resonance" (4.c) conditions. Normally H_{1B} is fixed and ω_B is varied. The resonant value of ω_B is detected as this value where the B and – through it – the A spin system is heated resulting in a decrease in the free precession signal of the A spins.

In the above case (i) both spin systems are quantized in their respective rotating frames, i.e. frames which rotate around the direction of the external magnetic field with the corresponding Larmor frequencies. Since the measurement is performed in a large external magnetic field single crystalline samples are required.

This difficulty does not arise in case (iii) where the A spin system is quantized in the local field and the B system in the rotating frame. The experiment is performed in the following way (Fig. 2). The two spin species A and B are first polarized in a strong external magnetic field and then adiabatically demagnetized to zero external field by moving the sample out of the magnet. The spin temperature of the A spin system decreases in this process. The B spin system is now irradiated with a strong r.f. field H_{1B}. If the frequency of this field corresponds to the given splitting of the quadrupole energy levels of the B spins, the populations of these two levels equalize, resulting in an infinite spin temperature of the B system. If resonant energy transfer is possible, i.e. if the A precession frequency in the local field H_L equals to the B precession frequency in the field H_{1B}

$$(\gamma_A H_L)_A = \gamma_B H_{1B}$$

and if the dipolar coupling between the A and the B system is non-zero, the resulting heat flow from the "hot" B to the "cool" A system will soon equalize the two spin temperatures. In order that the heat flow from B to A does not stop, we invert the phase of H_{1B} for 180° after a time of the order of the A B cross relaxation time. After that the B spin system has a negative spin temperature and – since negative temperatures are "hotter"

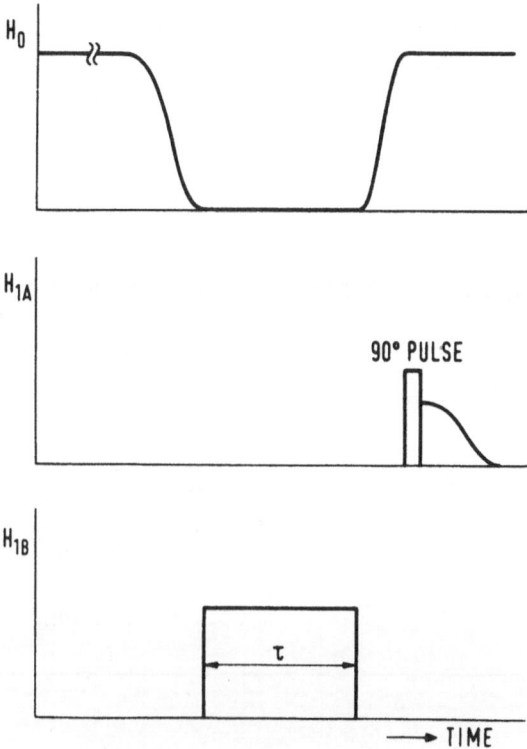

Fig. 2. Pulse sequence and variation of the Zeeman magnetic field
        with time for double resonance between the laboratory
        and the rotating frames.

than positive ones – the heat flow from the B to the A system
continues. This procedure is repeated for a time of the order
of the dipolar spin-lattice relaxation time $(T_{1D})_A$. Finally the
sample is moved back into the magnet and the remaining A
magnetization is determined by measuring the A free induction
decay signal following a 90° pulse. The sensitivity of this method
is comparable to the one of the double resonance in the rotating
frame technique, and polycrystalline samples can be used.

An example of the use of this method is the determination
of the O^{17} NQR spectra in KH_2PO_4 via proton O^{17} double resonance.
The experimental procedure is as follows:

The proton system is first cooled by adiabatic demagnetization
into a zero magnetic field and then heated by thermal contact
with the O^{17} system. The O^{17} system is heated when the search
frequency $\nu = \omega/2\pi$ of the radiofrequency field H_1 equals to the
splitting between any two pairs of the pure quadrupole energy
levels of O^{17}:

Table 1. O^{17} NQR frequencies, quadrupole coupling constants and
asymmetry parameters in KH_2PO_4

Paraelectric phase	Ferroelectric phase
T = 190 K	T = 77 K
ν_I = 2.47 MHz e^2qQ/h = 5.16 MHz	ν_I = 2.95 MHz e^2qQ/h = 5.96 MHz
ν_{II} = 1.47 MHz	ν_{II} = 1.65 MHz
ν_{III} = 1.00 MHz η = 0.55	ν_{III} = 1.30 MHz η = 0.72
	ν_I = 2.20 MHz e^2qQ/h = 4.85 MHz
	ν_{II} = 1.45 MHz
	ν_{III} = 0.75 MHz η = 0.18

$$(h\nu)_{res} = (\Delta E)_a$$

To achieve resonance energy transfer the amplitude of the radiofrequency field H_1 must satisfy the Hahn double resonance condition:

$$(\gamma H_{loc})_H = [(\gamma H_1)^2_{O17} + (\Delta\omega)^2]^{\frac{1}{2}} \quad ;\Delta\omega = \omega_{res} - \omega$$

The proton system is finally adiabatically remagnetized and the remaining proton magnetization measured as a function of the search frequency ν.

The results are shown in figure 3, figure 4 and table 1.

The above technique has one important limitation. The magnetic dipole-dipole coupling of integer spin nuclei to neighbouring nuclei of half integer spin is quenched in zero magnetic field and double resonance cannot be applied in the above form.

This difficulty is avoided in a level crossing experiment where heat transfer between the A and the B spin species takes place only when the resonance frequencies of these two species

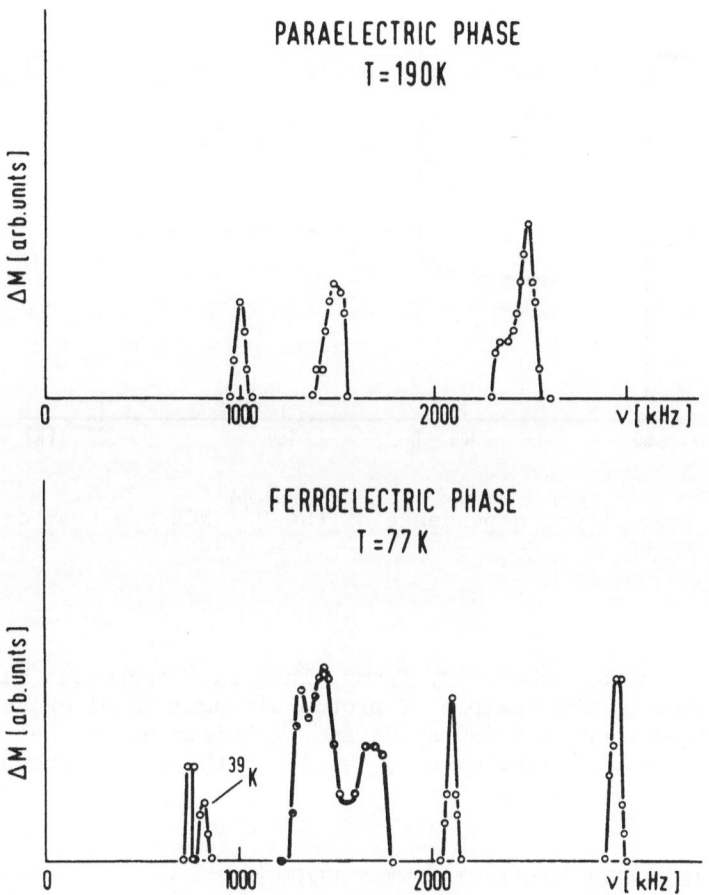

Fig. 3. O^{17} NQR spectrum of KH_2PO_4 above and below T_c.

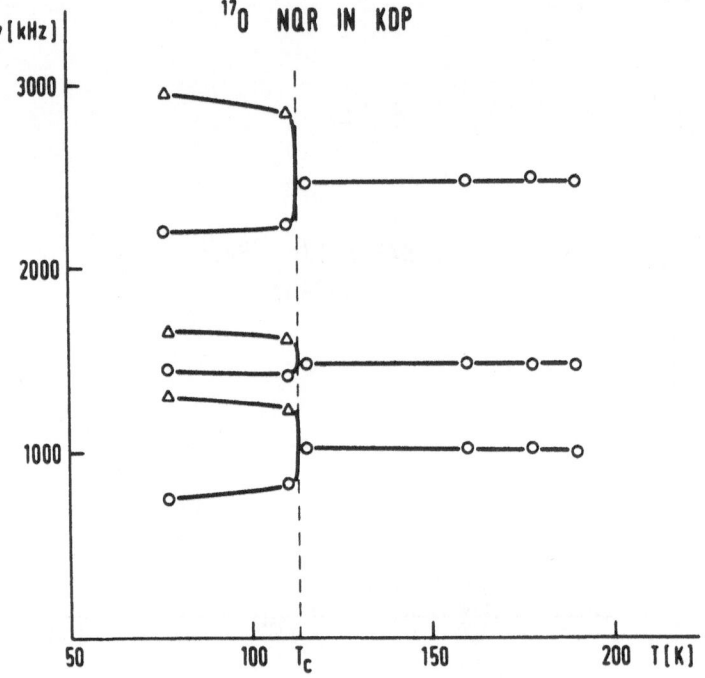

Fig. 4. Temperature dependence of the O^{17} NQR spectrum of
        KH_2PO_4.

are equal in the laboratory frame. This last technique has been
illustrated on the example of proton-nitrogen level crossing
double resonance in amino acids and nucleic acids by the
Ljubljana group (Blinc et al., 1972) and the Oxford group
(Edmonds et al., 1972).

DOUBLE RESONANCE RELAXATION MEASUREMENTS

Double Resonance Relaxation Measurements in the Rotating Frame
for $T_{1\rho B} \lesssim T_{1\rho A}$

     The double resonance spin-lattice relaxation problem
is equivalent to the thermodynamic problem of energy exchange
between two heat reservoirs, A and B, the thermal coupling of
which to the surrounding heat bath is characterized by time
constants $T_{1\rho A}$ and $T_{1\rho B}$, respectively. When thermal contact

between the two reservoirs is established, it is characterized by a time constant τ_{AB} which is short as compared to $T_{1\rho A}$ or $T_{1\rho B}$. The energy of a given reservoir can be described by a spin temperature T_s

$$E_A = C_A \beta_A \ , \ E_B = C_B \beta_B \ , \beta = 1/kT_s \tag{7}$$

which is in thermal equilibrium equal to the lattice temperature.

$T_{1\rho B}$ is determined by observing E_A respectively β_A as a function of time of thermal contact to the B reservoir. The mathematical problem is the determination of the time development of the spin temperature of the A system, β_A, when in thermal contact with the B system. The change in energy respectively spin temperature is described by

$$d\beta_A/dt = -\beta_A/T_{1A} -(\beta_A-\beta_B)/\tau_{AB} \ ; \ d\beta_B/dt = -\beta_B/T_{1B} -(\beta_B-\beta_A)/\tau_{BA} \tag{8,9}$$

with

$$\tau_{BA} = \tau_{AB}/\epsilon \tag{10.a}$$

where ϵ is the ratio of the magnetic heat capacities of the two systems:

$$\epsilon = C_B/C_A \tag{10.b}$$

For given initial conditions $\beta_A \overset{\sim}{=} \beta_A(0)$ and $\beta_B = 0$ we find

$$\beta_A(t) = \beta_A(0) \{\exp[-t/\epsilon\tau_{AB}]+(1-\epsilon) \exp[-t/(1+\epsilon)T_{1\rho A}].$$

$$\exp[-t/(1+\epsilon)T_{1\rho B}]\} \tag{11}$$

provided that $\epsilon\tau_{AB} \ll T_{1\rho A}, T_{1\rho B}$.

If there is no thermal contact between the two systems ($\tau_{AB}^{-1} = 0$) the spin temperature of the A-system on the other hand varies as

$$\beta_A^{(0)}(t) = \beta_A^{(0)}(0) \cdot \exp(-t/T_{1\rho A}) \tag{12}$$

As the magnetization of the A system M_A is proportional to

β_A, all time constants of the problem can be determined by observing the time development of β_A via the corresponding free induction decay signals.

The actual experiment is performed as follows. First a low spin temperature is generated in the A spin system by adiabatic demagnetization in the rotating frame (ADRF) [3, 4]. Then a B rf pulse of variable length is turned on for a time t_{Brf} to establish thermal contact between the two spin systems by equalizing the transition frequencies in the two rotating frames:

$$\omega_{eff}(A) = \omega_{eff}(B) \tag{13}$$

where $\omega_{eff} = \gamma H_{eff}$.

The remaining A spin magnetization is finally measured via a free induction signal (or a "dipolar" signal after a 45° pulse) at a fixed time $t \approx T_{1A}$ thus making the term $\exp[-t/(1+\epsilon)T_{1\rho A}]$ in eq. (11) constant. The logarithm of the A signal amplitude as a function of t_{Brf} is a straight line with a slope $[\epsilon/(1+\epsilon)](1/T_{1\rho B})$ from which T_{1B} can be determined.

Double Resonance Relaxation Measurements in the Rotating Frame for $T_{1B} > T_{1A}$

If T_{1B} is long as compared to T_{1A}, another procedure [5] has to be used for the determination of T_{1A}. First the A spin system is cooled by ADRF. Then thermal contact between the A and B system is established by applying a large effective field

$$H_{eff} = (1/\gamma) \, [(\Delta\omega)^2 + (\gamma H_1)^2]^{1/2} \tag{14}$$

(which is off resonance by $\Delta\omega$) to the B system, thus cooling it down too. After that $H_{eff}(B)$ is turned off so that the thermal contact is disconnected and the B magnetization relaxes along the laboratory field H_0 with the time constant T_{1B}. The A system is now first saturated by a series of 45° pulses and then remagnetized by reconnection to the B system. The time constant T_{1B} is then determined by measuring the magnetization of the A system as a function of the time t between the two thermal contacts (Fig. 5).

The time development of the magnetization of the A system is here given by:

$$M_A(t) \approx M_A(0) \cdot \epsilon \cdot \exp(-t/T_{1B}) \tag{15}$$

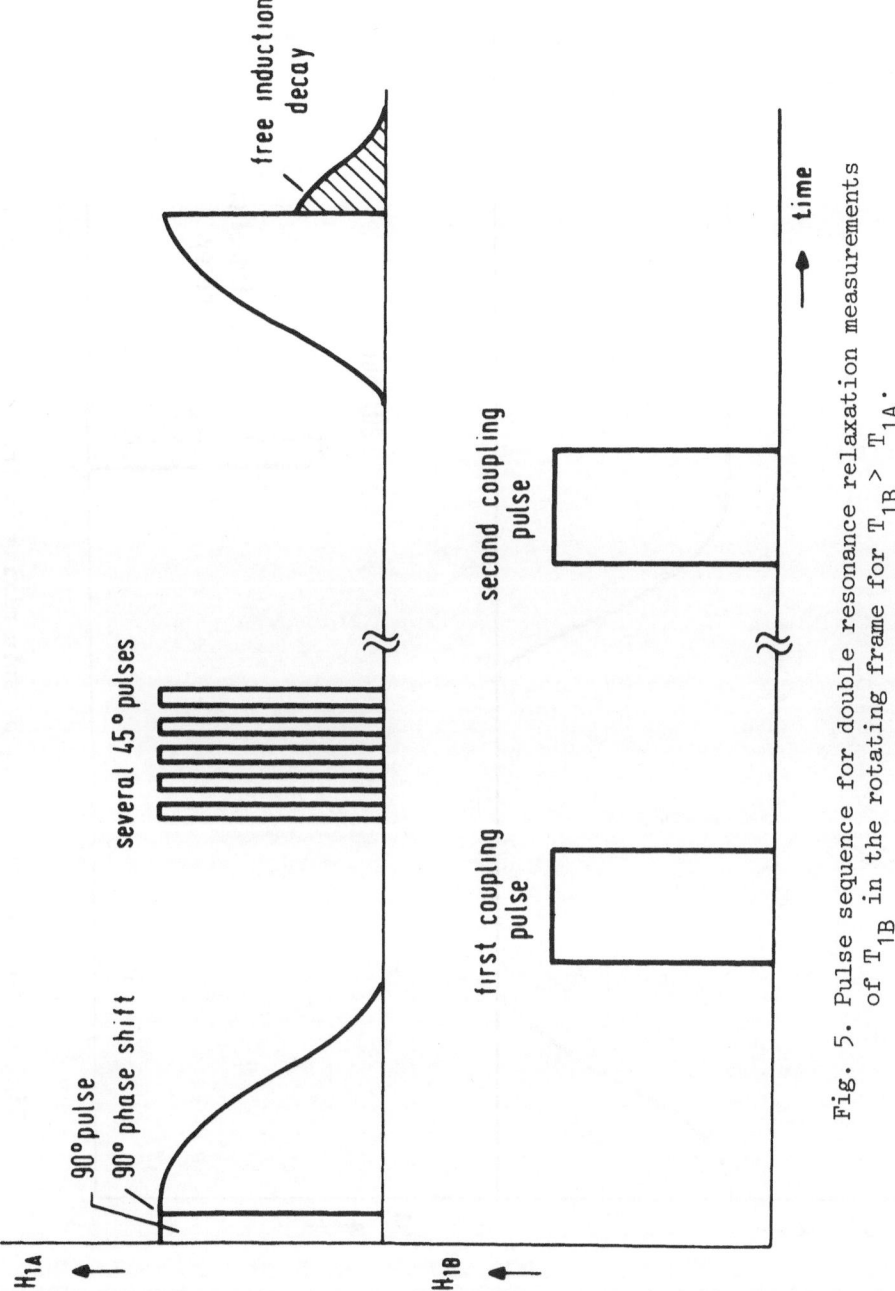

Fig. 5. Pulse sequence for double resonance relaxation measurements of T_{1B} in the rotating frame for $T_{1B} > T_{1A}$.

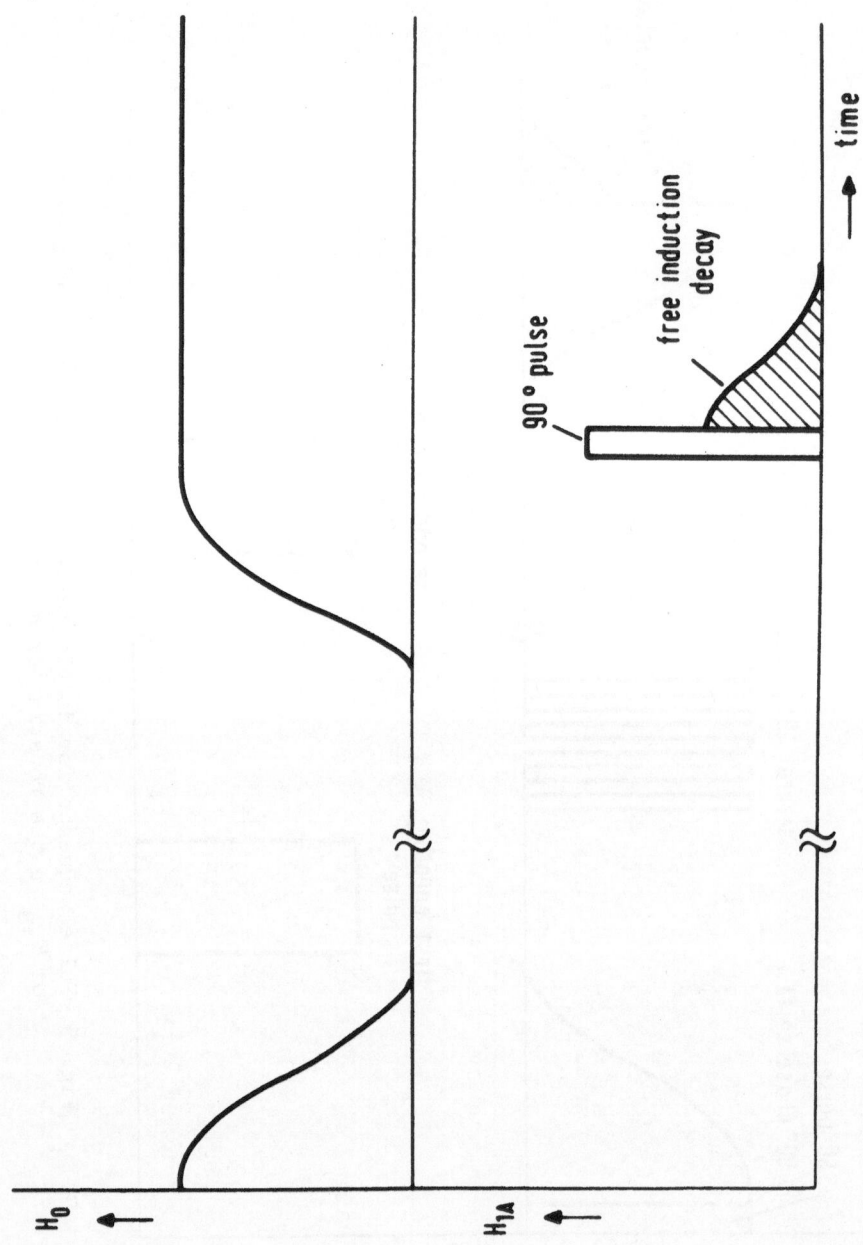

Fig. 6. Measurement of T_{1B} by spin mixing in the laboratory frame.

Determination of Quadrupolar Relaxation Times in Powdered Samples
by Level Crossing in the Laboratory Frame

A disadvantage of the double resonance methods in the
rotating frame lies in the fact that for the determination of the
NQR spectra and relaxation times of nuclei with a non-zero
quadrupole coupling single crystals are required.

This problem can be avoided by multiple level crossing
and spin mixing in the laboratory frame [4].

The A system is first magnetized in a strong external
magnetic field, and then adiabatically demagnetized to zero
field by moving the sample out of the magnet. At an intermediate
field the Zeeman splitting of the A system and the Zeeman perturbed
electric quadrupole splitting of the B system are equal, and
resonance energy transfer takes place. The system is kept
for a time t in zero field and then adiabatically remagnetized
resulting in a second level crossing.

T_{1B} is simply determined by measuring the A signal after
remagnetization as a function of the time the sample stays
in zero field (Fig. 6):

$$M_A(t) = A \exp(-t/T_{1A}) + B \exp(-t/T_{1B})$$

The relaxation time T_{1B} can be assigned to a given spin
species by saturating the energy levels of the corresponding
species or by an inspection of the relative magnitude of the
two coefficients, A and B.

REFERENCES

1. S.R. HARTMANN and E.L. HAHN, Phys.Res. 128, 2042 (1962)
   F.M. LURIE and C.P. SLICHTER, Phys.Rev. 133A, 1108 (1964)

2. R. BLINC, M. MALI, R. OSREDKAR, A. PRELESNIK, J. SELIGER, I. ZU-
   PANCIC and L. EHRENBERG, J.Chem.Phys. 57, 5087 (1972)
   D.T. EDMONDS and P.A. SPEIGHT, J.Mag.Res. 6, 265 (1972)

3. D.A. McARTHUR, E.L.HAHN and R.E. WALSTEDT, Phys.Rev. 188, 609
   (1969)

4. R.E. SLUSHER and E.L. HAHN, Phys.Rev. 166, 332 (1960)

5. E.L. HAHN, "Magnetic Resonance and Relaxation", R. Blinc, Ed.,
   14, North-Holland Publishing Co. (1967)

6. D. STEHLIK, "Pulsed Magnetic and Optical Resonance", R. Blinc,
   Ed., 63, Published by J. Stefan Institute (1972)

NUCLEAR MAGNETIC DOUBLE RESONANCE BASED ON STRONG RF MAGNETIC

FIELD INDUCED COUPLING BETWEEN SPIN SYSTEMS[*]

R. Blinc

University of Ljubljana, Institut Jozef Stefan

Ljubljana, Jugoslavija

INTRODUCTION

Nuclear double resonance as introduced by Hartmann and Hahn [1] and later by Lurie and Slichter [2] is a high sensitivity technique which detects the weak NMR or NQR signal of type B nuclei via their effect on the A spin system which exhibits a strong NMR signal. The technique is based on the existence of nuclear dipole-dipole coupling between the two spin systems in a frame of reference in which the energy levels of the A and B spins are equally spaced, which makes entropy transport possible.

Double resonance in the laboratory frame of one spin system and in the rotating frame of another spin system has been studied by Slusher and Hahn [3]. It represents an extension of the rotating frame nuclear double resonance [1, 2] to the zero magnetic field region.

Dipolar coupling between integer nuclear spins (B spin system) in an asymmetric electric field gradient and any other nonresonant spins (A spin system) is however highly reduced in zero magnetic field [4]. This effect is usually called spin quenching. The level crossing technique [5, 6, 7] where the A and B spin systems couple in a nonzero magnetic field is therefore used to detect pure NQR spectra of integer spin nuclei by nuclear

---

[*] This lecture is based on a paper by J. Seliger, R. Blinc, M. Mali, R. Osredkar

double resonance.

   It is the purpose of this lecture to show that the two
spin systems couple even though spin quenching occurs when a
strong RF magnetic field is applied to the sample with a
frequency which is either (a) near to one of the quadrupole
transition frequencies of the B nuclei when the sample is in
zero magnetic field and the A spins exhibit no quadrupole
coupling, or (b) equal to $\omega_A \pm \omega_B$ when the Zeeman or quadrupole
coupling of the A and B nuclei is nonzero. Here ω_A is one
of the transition frequencies between the energy levels of the
A nuclei, and ω_B is one of the transition frequencies between
energy levels of the B nuclei.

   This effect is one of the forms of the well known solid
effect [8]. It was already observed by Abragam and Proctor [9]
in LiF, by Landesman [10] in paradichlorbenzene and by Koo [6]
and Edmonds et al [7] in zero magnetic field nuclear double
resonance detection of ^{14}N nuclei.

   A schematic representation of the RF magnetic field induced
coupling between the A and B spin systems as well as a
schematic representation of the coupling between the A and B
spin systems in the case of rotating frame nuclear double
resonance is shown in figure 1.

   This RF magnetic field induced coupling between spin systems
enables one to detect resonance frequencies of the B nuclei
via the signal of the A nuclei. A new nuclear double resonance
technique is proposed, based on this coupling, which is of
particular importance in detection of pure NQR spectra of
integer spin nuclei, since it can be used in many cases when
the level crossing technique fails.

   The double resonance lineshapes obtained by this new technique
are evaluated for the case of ^{1}H–^{14}N double resonance in zero
static magnetic field, as well as for the case of nuclear double
resonance between two purely magnetic spin systems in a strong
static magnetic field and for the case of nuclear quadrupole
double resonance.

   A theoretical estimate of the sensitivity of this new
technique is presented together with some experimental results.

THE ORIGIN OF THE RF MAGNETIC FIELD INDUCED COUPLING BETWEEN
SPIN SYSTEMS

   In this section we are going to describe the physical
origin of the RF magnetic field induced coupling between spin
systems.

   In the following description we shall limit ourselves to

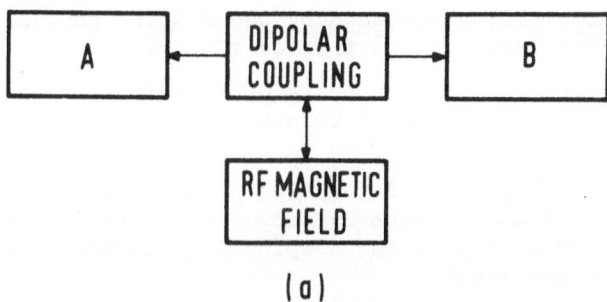

(a)

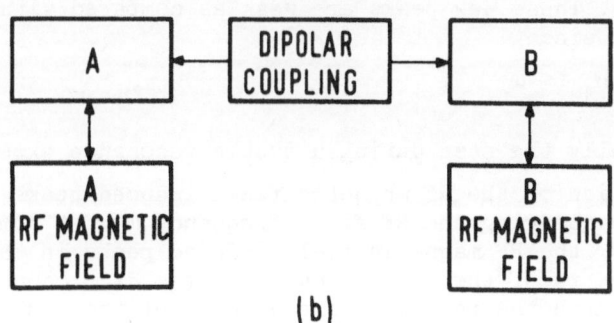

(b)

Fig. 1. Dipolar coupling between the A and B spin systems in
the case of solid effect (a) and in the case of
rotating frame nuclear double resonance (b).

the case of well resolved spectra of the A and B nuclei, i.e. to the case when the energy differences between Zeeman, quadrupole or mixed energy levels of both nuclei are much larger than the resonance linewidths. The case of coupling between purely magnetic A spins and quadrupole B spins in zero magnetic field will be treated at the end of this section.

The A spins precess in static magnetic field or in local electric field gradients at the A spin sites or in both with their resonance frequencies ω_A. The frequency spectra of the dipolar magnetic fields produced by the A spins consist therefore of discrete peaks at the frequencies ω_A.

Similarly the frequency spectra of the dipolar magnetic fields produced by the B spins consist of discrete peaks at the B spin resonance frequencies ω_B.

The frequency spectra of the dipolar magnetic fields produced by the A and B spins are shown in figure 2.a. Only one resonance frequency of the A spins and one resonance frequency of the B spins are shown.

A strong RF magnetic field $H_1 \cos\omega_o t$ modulates the precession of the A and B spins. In the frequency spectra of the dipolar magnetic fields new RF magnetic field induced peaks appear at the frequencies $\omega_A \pm \omega_o$ and $\omega_B \pm \omega_o$. This situation is shown in figure 2.b. These new peaks are weak as compared with the original ones since

$$\gamma_A H_1 \ , \ \gamma_B H_1 \ll \omega_o \qquad\qquad (1)$$

which is usually the case during a double resonance experiment.

The position of the RF magnetic field induced peaks can be changed by changing the RF field frequency ω_o. At some frequencies ω_o the RF magnetic field induced peaks in the frequency spectra of the dipolar magnetic fields produced by the A spins match the resonance frequencies of the B spins and vice versa. This happens when

$$\omega_o = \omega_A \pm \omega_B \qquad\qquad (2)$$

When such a situation occurs the A spins driven by the RF magnetic field induce transitions between energy levels of the B spins and at the same time the B spins driven by the RF magnetic field induce transitions between energy levels of the A spins.

Since a quantum of energy $\hbar\omega_o$ gained by the A spin system from the RF magnetic field is different from a quantum of energy $\hbar\omega_B$ needed to produce a transition in the B spin system,

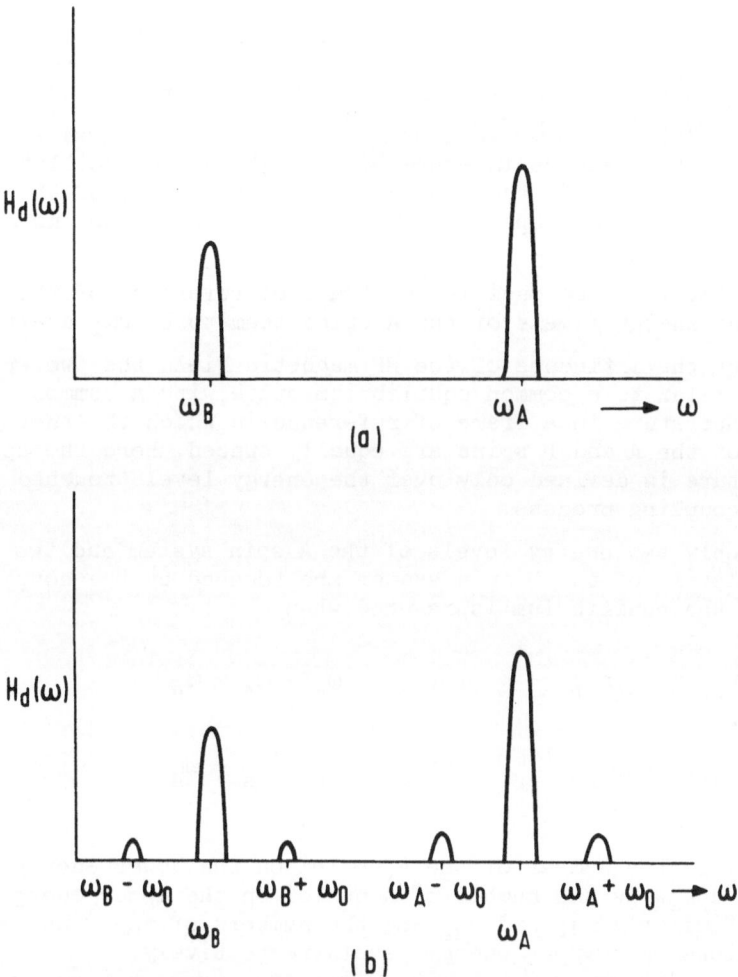

Fig. 2. Frequency spectra of the dipole magnetic fields produced by the A and B spins without (a) and with a RF magnetic field applied (b).

the energy difference $\pm \hbar\omega_A$ should be absorbed in or emitted
from the A spin system during such a process, depending on
whether $\omega_o = \omega_A + \omega_B$ or $\omega_o = \omega_A - \omega_B$. Every transition in the
B spin system due to this process is therefore accompanied
by a transition in the A spin system and similarly every transition in the A spin system produced by the B spins is accompanied
by a transition in the B spin system. These simultaneous
transitions in both spin systems are schematically represented
in figure 3 for the case of $\omega_o = \omega_A + \omega_B$ and for the case of
$\omega_o = \omega_A - \omega_B$.

This effect may be interpreted as that in the oscillating
frame of reference of the A spins the energy levels of the B
spins seem to be $\hbar\omega_A$ apart and thus resonant with the energy
levels of the A spins.

Similarly in the oscillating frame of reference of the B
spins the energy levels of the A spins seem to be $\hbar\omega_B$ apart.

Under the influence of the RF magnetic field the two spin
systems relax to a common equilibrium state with a common
spin temperature in a frame of reference in which the energy
levels of the A and B spins are equally spaced. Here the spin
temperature is defined only over the energy levels touched
by the coupling process.

If only two energy levels of the A spin system and two
energy levels of the B spin system are touched by the coupling
process the equilibrium is reached when

$$n_{1A}n_{2B} = n_{2A}n_{1B} \qquad \text{for} \qquad \omega_o = \omega_A - \omega_B$$

$$n_{1A}n_{1B} = n_{2A}n_{2B} \qquad \text{for} \qquad \omega_o = \omega_A + \omega_B$$

$$(3)$$

Here n_{1A} is the number of the A nuclei on the lower energy
level and n_{2A} is the number of A nuclei on the upper energy
level. Similarly n_{1B} and n_{2B} are the numbers of the B nuclei
on the lower and upper energy levels respectively.

For two purely magnetic spin systems with arbitrary spins
in a strong magnetic field the equilibrium is reached when

$$T_A/T_B = -\omega_A/\omega_B \qquad \text{for} \qquad \omega_o = \omega_A + \omega_B$$

$$(4)$$

$$T_A/T_B = \omega_A/\omega_B \qquad \text{for} \qquad \omega_o = \omega_A - \omega_B$$

Here T_A and T_B are the spin temperatures, and ω_A and ω_B are

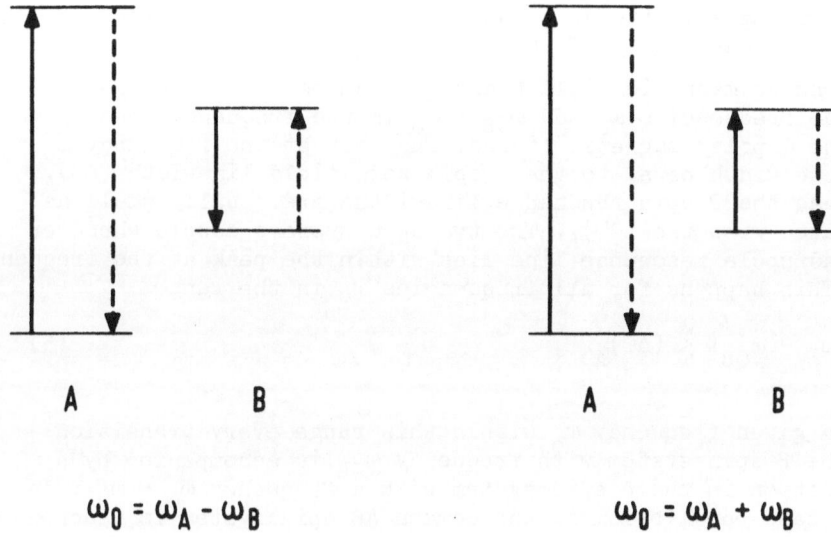

$$\omega_0 = \omega_A - \omega_B \qquad\qquad \omega_0 = \omega_A + \omega_B$$

Fig. 3. Mutual spin flips in the A and B spin systems for $\omega_o = \omega_A + \omega_B$ and $\omega_o = \omega_A - \omega_B$.

the Larmor frequencies of the A and B spin systems respectively.

When a system of purely magnetic A spins is put in zero magnetic field, the only motion the spins perform is due to magnetic dipolar coupling between the A spins themselves and between the A spins and any other spins which are not decoupled from the A spins by spin quenching. The frequency spectra of the dipolar magnetic fields produced by the A spins consist of a single peak at zero frequency with the width equal to the A spin zero field linewidth $(\Delta\omega)_A$.

The B spins for which we assume nonzero quadrupole coupling precess at their quadrupole resonance frequencies ω_{QB} in the local electric field gradients. The frequency spectra of the dipolar magnetic fields produced by the B spins consist of discrete peaks at the frequencies ω_{QB}.

The RF magnetic field $H_1\cos\omega_o t$ induces some extra peaks at the frequencies ω_o and $\omega_{QB} \pm \omega_o$ in the frequency spectra of the dipolar magnetic fields. The peak at the frequency ω_o has the width equal to the A spin zero field linewidth $(\Delta\omega)_A$, whereas the B spin resonance linewidths are usually small as compared with the $(\Delta\omega)_A$. The two spin systems couple whenever a quadrupole resonance line lies within the peak at the frequency ω_o. This happens for all frequencies ω_o in the range

$$|\omega_o - \omega_{QB}| \lesssim (\Delta\omega)_A \tag{5}$$

For a given frequency ω_o within this range every transition in the B spin system with frequency ω_{QB} is accompanied by a transition in the A spin system with a frequency $\omega_o - \omega_{QB}$. In this case equilibrium in the common AB spin system is reached when

$$n_1/n_2 = 1 - \hbar\,(\omega_{QB} - \omega_o)/kT_A \tag{6}$$

Here n_1 and n_2 are the populations of the two corresponding quadrupole energy levels of the B nuclei with energies $E_1 - E_2 = \hbar\omega_{QB}$, and T_A is the final spin temperature of the A spin system.

THEORY OF THE RF MAGNETIC FIELD INDUCED COUPLING BETWEEN SPIN SYSTEMS

In this section we give a theoretical description of the RF magnetic field induced coupling between spin systems for the case of two purely magnetic spin systems in a strong static magnetic field. A similar approach can be used for all cases described in this article.

The Hamiltonian of our system is

$$H = H_{ZA} + H_{ZB} + H_{dAA} + H_{dBB} + H_{dAB} + H_{RF\ A} + H_{RF\ B} \qquad (7)$$

Here H_{ZA} and H_{ZB} are the Zeeman Hamiltonians of the two spin systems:

$$H_{ZA} = \hbar\omega_A \sum_i I_{Az}^i$$

$$H_{ZB} = \hbar\omega_B \sum_k I_{Bz}^k \qquad (8)$$

H_{dAA}, H_{dBB} and H_{dAB} are the dipole-dipole interactions between the A, B and A and B spins respectively.

$H_{RF\ A}$ and $H_{RF\ B}$ describe the coupling of the A and B spins to the RF magnetic field $H_1\cos\omega t$.

$$H_{RF\ A} = \hbar\omega_{1A} \sum_i I_{Ax}^i \cos\omega t$$

$$H_{RF\ B} = \hbar\omega_{1B} \sum_k I_{Bx}^k \cos\omega t \qquad (9)$$

The A and B spins oscillate under the influence of the RF magnetic field. From now on we are going to observe the A spin system in the frame of reference which oscillates with the A spins and the B spin system in the frame of reference which oscillates together with the B spins.

The laboratory frame of reference and the oscillating one are connected by the transformation which excludes the term $H_{RF\ A} + H_{RF\ B}$ from the Hamiltonian.

Since

$$[H_{RF\ A}(t) + H_{RF\ B}(t),\ H_{RF\ A}(t') + H_{RF\ B}(t')] = 0 \qquad (10)$$

the transformation is

$$T = \exp\left[(i/h)\int (H_{RF\ A} + H_{RF\ B})dt\right] \qquad (11)$$

Performing the transformation one gets the Hamiltonian

$$H^{\bigstar} = H_{ZA} + H_{ZB} + H_{dAA} + H_{dBB} + H_{dAB} \tag{12}$$

plus some terms linear and quadratic in ω_{1A}/ω and ω_{1B}/ω which oscillate with frequencies ω and 2ω.

Since ω_{1A}/ω and ω_{1B}/ω are usually much smaller than unity we shall neglect all these terms except the ones coming from H_{dAB}. These terms couple the two spin systems.

The oscillating terms, coming from H_{dAB} — which does not commute neither with H_{ZA}, nor with H_{ZB}, and which can thus induce simultaneous transitions in both spin systems – are equal to

$$H_{dAB}^{(1)} = 3(\omega_{1A}/\omega) \sum_{i,k} C_{ik} Z_{ik} I_{Ay}^{i}(Y_{ik}I_{By}^{k}+X_{ik}I_{Bx}^{k})\sin\omega t \,+$$

$$3(\omega_{1B}/\omega) \sum_{i,k} C_{ik} Z_{ik} I_{By}^{k}(Y_{ik}I_{Ay}^{i}+X_{ik}I_{Ax}^{i}) \sin\omega t \tag{13}$$

$$H_{dAB}^{(2)} = \frac{1}{2} (\omega_{1A}\omega_{1B}/\omega^{2}) \sum_{i,k} C_{ik}I_{Ay}^{i}I_{By}^{k} \cos2\omega t \tag{14}$$

Here X_{ik}, Y_{ik} and Z_{ik} are the direction cosines of the distance vector r_{ik} between the i-th and k-th site and $C_{ik} = \gamma_A\gamma_B\hbar/r_{ik}^{3}$.

Assuming $\omega_A > \omega_B$ one can divide the term H_{ZA} into two terms

$$H_{ZA} = \pm \hbar\omega_B \sum_i I_{Az}^{i} + \hbar(\omega_A \mp \omega_B) \sum_i I_{Az}^{i} \tag{15}$$

The energy levels of the first term in expression 15 are equally spaced as the energy levels of H_{ZB}. For the (+) sign their sequence is the same as in the laboratory frame, whereas for the (−) sign it is inverted.

Let us now transform into a frame of reference in which the energy levels of the A and B spin systems are equally spaced. This is done by the transformation

$$T = \exp [i(\omega_A \mp \omega_B) \sum_i I_{Az}^{i}t] \tag{16}$$

The terms in H_{dAB} which can be made static in this frame of reference by proper choice of the RF field frequency are equal to:

$$H_{dAB}^{*(1)} = \frac{3}{2}(\omega_{1A}/\omega) \sum_{i,k} C_{ik}Z_{ik}(I_{By}^{k}Y_{ik}+I_{Bx}^{k}X_{ik})I_{Ax}^{i}\cos(\omega_A \mp \omega_B - \omega)t +$$

$$\frac{3}{2}(\omega_{1B}/\omega) \sum_{i,k} C_{ik}Z_{ik}(I_{Ax}^{i}Y_{ik}-I_{Ay}^{i}Y_{ik})I_{By}^{k}\cos(\omega_A \mp \omega_B - \omega)t \quad (17)$$

$$H_{dAB}^{*(2)} = \frac{1}{4}(\omega_{1A}\omega_{1B}/\omega^2) \sum_{i,k} C_{ik}I_{By}^{k}I_{Ay}^{i}\cos(\omega_A \mp \omega_B - 2\omega)t \quad (18)$$

The two spin systems couple when $H_{dAB}^{*(1)}$ or $H_{dAB}^{*(2)}$ is static. This happens

(1) when $\omega_A \mp \omega_B = \omega$. In this case a RF magnetic field induced peak and an original peak in the frequency spectra of the dipolar magnetic field match.

(2) when $\omega_A \mp \omega_B = 2\omega$. In this case two RF magnetic field induced peaks match.

The cross relaxation rate W_{AB} is in the case (1) approximately equal to [8]

$$W_{AB} \stackrel{\sim}{\sim} (\Delta\omega)_{AB} [(\omega_{1A}/\omega)^2 + (\omega_{1B}/\omega)^2] \quad (19)$$

whereas in the case (2) it is approximately equal to

$$W_{AB} \stackrel{\sim}{\sim} (\Delta\omega)_{AB} (\omega_{1A}^2\omega_{1B}^2/\omega^4) \quad (20)$$

Here $(\Delta\omega)_{AB}$ is the broadening of the A spin resonance line due to the B spins.

For a typical case of ^{14}N and ^{1}H spin systems in a static magnetic field of 4 kilogauss with a typical value $(\Delta\omega)_{NH}$ = 6 kHz and for a RF magnetic field amplitude 100 gauss the cross relaxation rate W_{NH} is in the case (1) approximately equal to 0.3 sec, whereas it is in the case (2) of the order of 10^5 sec and is thus negligibly long. For that reason we are considering only case (1) in this article.

The cross relaxation times in the present case are approximately a factor

$$1/(\omega_{1A}^2/\omega^2 + \omega_{1B}^2/\omega^2)$$

longer than in the case of rotating frame nuclear double resonance.
This is the main disadvantage of the measuring technique using
RF induced coupling between spin systems. However in the case
of zero magnetic field nuclear double resonance detection of
integer spin nuclei the presently discussed coupling mechanism
is the strongest one because of spin quenching. The measuring
technique using this coupling is in this case the only possible
double resonance detection scheme. In other cases the main
advantage of the new measuring technique is its simplicity and
convenience.

EXPERIMENTAL

The Case of Well Resolved Spectra of the A and B Spins

The technique described in this section can be used in all
cases, whenever the Zeeman, quadrupole or mixed resonance
frequencies of the A and B nuclei are significantly larger than
the resonance linewidths.

The experimental procedure is shown in Fig. 4.a. The A and
B spins are left until thermal equilibrium of both spin systems
with the lattice is reached. At this moment a strong RF magnetic
field pulse with frequency ω is applied to the sample. After
the end of this pulse a 90° pulse is applied to the A spin
system at some A spin resonance frequency ω_A and the A spin free
induction decay amplitude is measured.

Now the whole procedure is repeated without the strong RF
magnetic field pulse applied and the A spin free induction decay
amplitude at frequency ω_A is again measured. The difference
between the two free induction decays, which is called the
double resonance signal, is recorded as a function of frequency
ω.

When $\omega = \omega_A + \omega_B$, ω_B being a resonance frequency of the B
spins, a new state in the A spin system is reached during the
strong RF magnetic field pulse due to the RF magnetic field
induced coupling between spin systems. This new state results
in a changed A spin free induction decay amplitude and therefore
in a nonzero double resonance signal. In order to get an optimum
double resonance signal the strong RF field pulse should last
several cross relaxation times.

The sensitivity and the resolution of the technique can be
increased by applying a train of strong RF field coupling pulses

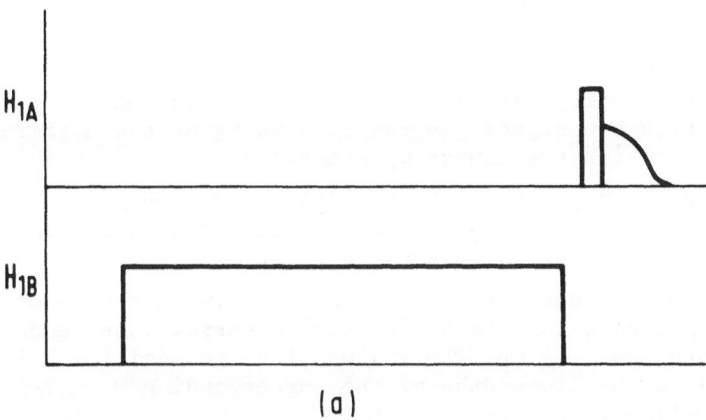

(a)

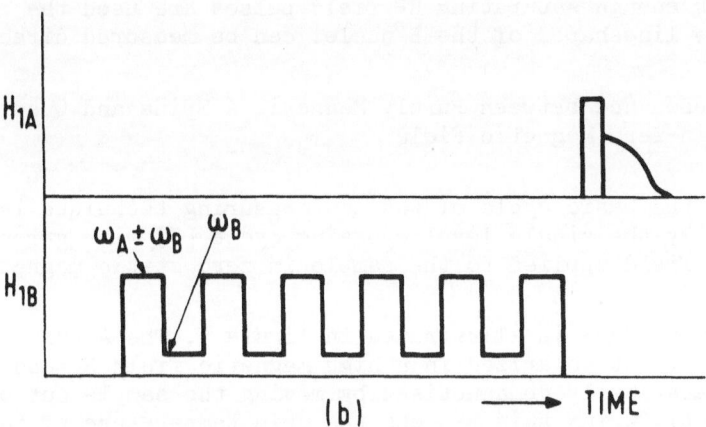

(b)   ⟶ TIME

Fig. 4. Schematic representation of the nuclear double resonance
process in the case of well resolved spectra of the A
and B nuclei
a) single coupling process
b) multiple coupling process.

at the frequency $\omega_A \pm \omega_B$ with weak RF field pulses at the
frequency ω_B inbetween (Fig. 4.b) to saturate the particular
transition between energy levels of the B spins. The weak RF
field pulses at the frequency ω_B can be for example obtained
by simultaneous attenuation and amplitude or frequency or phase
modulation of the carrying signal at the frequency $\omega_A \pm \omega_B$ with
the known frequency ω_A. No RF at the frequency ω_A should come
to the irradiation coil in order to prevent direct saturation
of the A spin energy levels. The same RF field sequence can be
obtained with two signal generators working at two different
frequencies which are always ω_A apart.

During the experiment the modulation frequency or the
frequency difference is set to the fixed value ω_A and the
frequency of the carrier signal is swept.

The double resonance spectra consist in this case of
broad single-coupling lines with strong narrow lines added at
the frequencies $\omega_A \pm \omega_B$. The widths of these additional lines
are equal to the linewidths of the corresponding B nuclei
resonance lines.

In a multiple coupling process the thermal equilibrium
between the B spins and lattice does not need to be reached.
When weak enough saturating RF field pulses are used the
resonance lineshapes of the B nuclei can be measured directly.

Double Resonance between Purely Magnetic A Spins and Quadrupole
B Spins in Zero Magnetic Field

Here the basic cycle of the new measuring technique is
essentially the single level crossing cycle [5] with strong RF
magnetic field applied to the sample in zero static magnetic
field.

The procedure is illustrated in figure 5. The A spin
system is first polarized in a high magnetic field H_O and
then adiabatically demagnetised by moving the sample out of
the magnet. During this process the spin temperature of the
A spin system decreases from the lattice temperature T_L to a
rather low value $T_A = T_L(H_L/H_O)$. Here H_L is the local magnetic
field at the A spin sites. The sample is now irradiated with
a strong RF magnetic field at a frequency ω for a time τ.
After the irradiation the sample is moved back into the magnet
and the remaining A spin magnetisation is determined by measuring
the A spin free induction decay amplitude following a $90°$ pulse.
Then the whole procedure is repeated without the strong RF
magnetic field applied. The difference between the A spin free
induction decay amplitudes with and without the strong RF

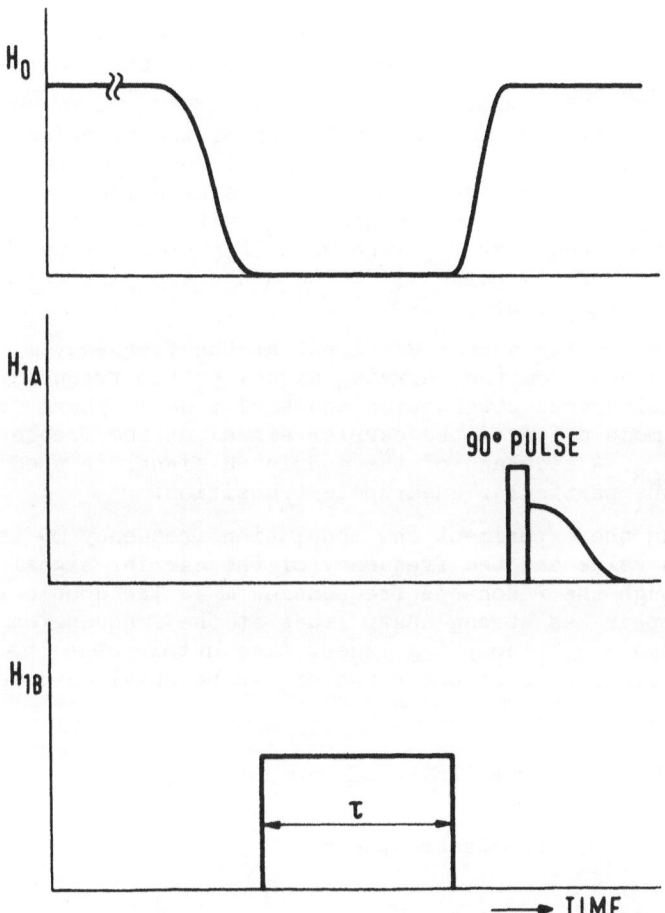

Fig. 5. Field sequence in a zero magnetic field nuclear double
resonance. Single coupling process.

magnetic field applied is recorded as a function of the RF
field frequency ω. It is different from zero when the RF
magnetic field couples the two spin systems, i.e. when $\omega \sim \omega_{QB}$.
Here ω_{QB} is any of the quadrupole transition frequencies of
the B nuclei.

If the B spin quadrupole resonance frequencies are lower
than the A spin Larmor frequency in high magnetic field H_O,
the two spin systems couple also when their energy levels
cross, which has some additional influence on the double
resonance spectra.

The sensitivity and the resolution of the technique is
increased when one uses instead of the strong RF magnetic field
coupling pulse a train of coupling pulses at a frequency ω near
to a quadrupole resonance frequency ω_{QB} with the weak RF field
pulses at the frequency ω_{QB} inbetween (Fig. 6). The purpose
of the weak RF field pulses is to saturate the particular
quadrupole transition.

Experimentally a weak RF signal at the frequency ω_{QB}
can be obtained from the carrying signal at the frequency ω
by its simultaneous attenuation and amplitude or phase or
frequency modulation of the carrier signal at the frequency
$\omega_M = \omega - \omega_{QB}$. A sideband of the modulated signal is used to
saturate the particular quadrupole transition.

During the experiment the modulation frequency ω_M is set
to a fixed value and the frequency of the carrier signal is
swept through the resonance frequencies ω_{QB}. The double resonance
spectra appear as strong sharp lines at the frequencies $\omega_{QB} \pm \omega_M$
on the broad single-coupling lines. Also in this case the
resonance lineshapes of the B nuclei can be obtained directly.

ANALYSIS OF THE DOUBLE RESONANCE PROCESS

The ^{1}H-^{14}N Double Resonance Spectra

The single coupling process

In this section we present the results of a calculation
of the double resonance spectra for a ^{1}H-^{14}N system in a single
level crossing cycle for the case of an applied strong RF magnetic
field. The analogous treatment for the weak RF magnetic field
case can be found in reference [4]. For sake of simplicity
spin lattice relaxation is omitted from the calculation.

The following conditions usually met in a double resonance
experiment have been considered to be valid throughout the

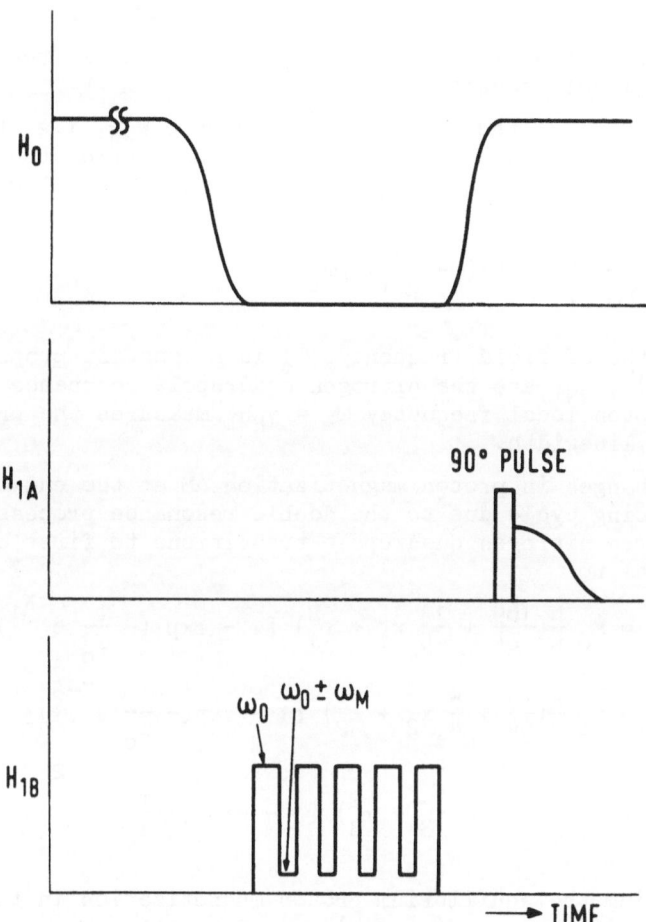

Fig. 6. Field sequence in a zero magnetic field nuclear double
      resonance. Multiple coupling process.

calculation.

(i) The number 3N of nitrogen nuclei per unit volume is much smaller than the number 2n of protons per the unit volume.

(ii) The nitrogen quadrupole transition frequencies are much lower than the Larmor frequency of protons in the strong static magnetic field H_o in which the protons are polarized.

(iii) An equilibrium in the common NH spin system is reached during every level crossing.

In addition, the cross relaxation rate W_{NH}, i.e. the rate in which the two spin systems are relaxing towards the common equilibrium state under the influence of a strong RF magnetic field was assumed to have the Gaussian form

$$W_{NH} = \frac{1}{T_o} \sum_i \exp[-(\omega - \omega_{QNi})^2/\omega_L^2] \tag{21}$$

Here ω is the RF field frequency, T_o is a constant proportional to $(\gamma_H H_1/\omega)^2$, ω_{QNi} are the nitrogen quadrupole resonance frequencies and the proton local frequency $\omega_L = \gamma_H H_L$ measures the proton zero field linewidth.

The changes in proton magnetization ΔM at the end of the level crossing cycle due to the double resonance process are for the three nitrogen quadrupole transitions to first order in N/n equal to

$$(\Delta M)_1 = M_o \frac{N}{n}(\frac{169}{64} + \frac{13}{4} x_1 + x_1^2) \, [1 - \exp(-\frac{\tau}{T_o} e^{-x_1^2})]$$

$$(\Delta M)_2 = M_o \frac{N}{n}(\frac{25}{64} + \frac{5}{4} x_2 + x_2^2) \, [1 - \exp(-\frac{\tau}{T_o} e^{-x_2^2})] \tag{22}$$

$$(\Delta M)_3 = M_o \frac{N}{n}(1 + 2 x_3 + x_3^2) \, [1 - \exp(-\frac{\tau}{T_o} e^{-x_3^2})]$$

with M_o being the equilibrium proton magnetization in the static magnetic field H_o, $x_i = (\omega - \omega_{QNi})/\omega_L$ and $\omega_{QN1} > \omega_{QN2} > \omega_{QN3}$. The calculated double resonance lines are plotted in figure 7 for two different ratios τ/T_o.

The double resonance signals are at $x_i = 0$ for a high enough ratio τ/T_o equal to the corresponding level crossing signals.

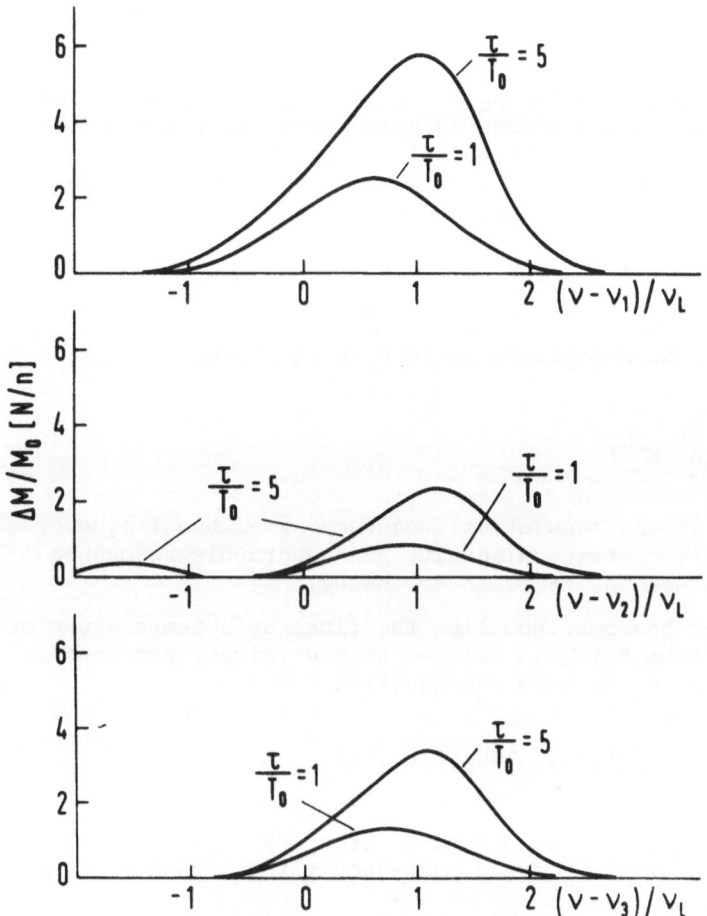

Fig. 7. Calculated ^{1}H–^{14}N double resonance lineshapes obtained by zero magnetic field nuclear double resonance in a single coupling process.

The multiple coupling process

In the calculation of the double resonance sensitivity we shall in this case completely neglect the level crossing process, but we can no longer neglect the proton spin lattice relaxation in zero magnetic field. Spin lattice relaxation of nitrogens is still assumed to be slow as compared with the double resonance process.

Assuming that during a single coupling pulse an instantaneous equilibrium in the common NH spin system is reached, the proton spin temperature at the end of the coupling pulse θ_f is related to the proton spin temperature at the beginning of the coupling pulse θ_i through the equation

$$\theta_f = \theta_i(1+\varepsilon) \tag{23}$$

Here the heat capacities ratio ε is equal to

$$\varepsilon = \frac{N}{n}(\omega_M/\omega_L)^2 \tag{24}$$

where ω_M is the modulation frequency, i.e. the frequency difference between the corresponding nitrogen quadrupole resonance frequency and the RF field frequency.

After the n-th coupling the final spin temperature of the proton system $\theta_f(n)$ is related to the initial proton spin temperature θ_i through the equation

$$\theta_f(n) = \theta_i(1+\varepsilon)^n \approx \theta_i \exp(n\varepsilon) \tag{25}$$

Let τ_1 be the length of a strong RF field pulse, τ_2 the length of a weak RF field pulse and τ the duration of the whole pulse train.

In order to reach an equilibrium in the common NH spin system during each coupling pulse the time τ_1 should be equal to several cross relaxation times W_{NH}^{-1}. In the following we shall assume $\tau_1 + \tau_2$ to be equal to $3W_{NH}^{-1}$. The number of couplings is thus equal to

$$n = \tau/(\tau_1 + \tau_2) = \tau W_{NH}/3 \tag{26}$$

In addition to the double resonance process the proton spin
temperature rises also because of the spin lattice relaxation.
Both processes are independent. The inverse proton spin
temperature which is proportional to the proton magnetization
in high magnetic field is at the end of the irradation equal to

$$\theta_f^{-1} = \theta_i^{-1} \exp(-\varepsilon W_{NH}\tau/3 - \tau/T_{1H}) \tag{27}$$

Here T_{1H} is the proton spin lattice relaxation time in zero
magnetic field.

The difference $\Delta\theta_f^{-1}$ between the final inverse proton spin
temperature with and without the strong RF magnetic field
irradation is maximum for [11]

$$\tau = T_{1H} \ln(1 + x)/x \tag{28}$$

where

$$x = \varepsilon W_{NH}T_{1H}/3 \tag{29}$$

In this case the difference $\Delta\theta_f^{-1}$, which is proportional to
the double resonance signal is equal to

$$\Delta\theta_f^{-1} = \theta_i^{-1}x/(1 + x)^{(1+1/x)} \tag{30}$$

The magnitude of the double resonance signal increases with x.
For the assumed Gaussian form of the cross relaxation rate W_{NH}
the maximum double resonance signals are obtained when $\omega_M = \omega_L$.

Since the cross relaxation rates W_{NH} are clearly for all
three nitrogen quadrupole transitions approximately equal, the
sensitivities of the detection of different nitrogen quadrupole
resonance frequencies by the multiple coupling technique are
also approximately equal.

Sensitivity of the Nuclear Quadrupole Double Resonance

The single coupling process

In this section we shall calculate the change in the A
spin free induction decay amplitude following a 90° pulse at
frequency ω_{QA} after the sample is irradiated with a strong RF

magnetic field pulse at frequency $\omega_{QA} \pm \omega_{QB}$. Here ω_{QB} is one of the quadrupole resonance frequencies of the B nuclei. We assume the cross relaxation between spin systems to be much faster than the spin lattice relaxation of both spin systems. We also assume that the RF magnetic field pulse lasts several cross relaxation times W_{BA}^{-1} so that an instantaneous equilibrium is reached in the common AB spin system during the pulse.

If the two spin systems are in thermal equilibrium with the lattice before the RF pulse is applied the A spin free induction decay signal S at the end of the RF field pulse will be equal to:

$$S = S_o \ [1 - (n_B/n_A + n_B)(\omega/\omega_{QA})] \tag{31}$$

Here n_A and n_B are the numbers of the A and B nuclei touched by the double resonance process, $\omega = \omega_{QA} \pm \omega_{QB}$, and S_o is the free induction decay signal of the A spin system at the frequency ω_{QA} after thermal equilibrium with the lattice is reached.

It can be easily seen that for $\omega_{QA} \gg \omega_{QB}$ all double resonance lines have the same intensity, whereas in a general case the double resonance line at the frequency $\omega_{QA} - \omega_{QB}$ is weaker than the one at the frequency $\omega_{QA} + \omega_{QB}$.

The multiple coupling process

In the analysis of this process we can no longer neglect the spin lattice relaxation of the A spin system, but we shall still assume that it is slower than the cross relaxation between the two spin systems. We shall again neglect the spin lattice relaxation of the B nuclei.

Let a strong RF field pulse in the train (Fig. 4.b) last several cross relaxation times W_{BA}^{-1} and let τ be the duration of a strong plus a weak field pulse.

Then a train of strong and weak RF field pulses at the frequencies $\omega_{QA} \pm \omega_{QB}$ and ω_{QB} respectively lasting for a time t which is longer than the spin lattice relaxation time T_{1A} of the A nuclei produces a new equilibrium state in the A spin system with the free induction decay signal at the frequency ω_{QA} equal to

$$S = S_o/(1 + \varepsilon T_{1A}/\tau) \tag{32}$$

Here

$$\varepsilon = n_B/(n_A + n_B) \tag{33}$$

The double resonance signals $\Delta S = S_0 - S$ at the frequencies $\omega_{QA} + \omega_{QB}$ and $\omega_{QA} - \omega_{QB}$ have equal intensities. The double resonance sensitivity increases with increasing T_{1A} and also increasing t as long t < T_{1A}; when t > T_{1A} the sensitivity reaches its maximum value and is independent on further lengthening of t.

Nuclear Double Resonance between Two Purely Magnetic Spin Systems

The single coupling process

In the analysis of the single coupling process we shall again neglect the spin lattice relaxation of both spin systems. Before the strong RF field pulse is applied the spin temperatures of both spin systems are equal to the lattice temperature T_L and the A spin magnetization is equal to the equilibrium A spin magnetization M_{AO} at the temperature T_L. After the coupling pulse at the frequency $\omega_A \pm \omega_B$ is applied, which lasts several cross relaxation times, the magnetization of the A spin system changes to

$$M_A = M_{AO}(1-\varepsilon) \tag{34}$$

where

$$\varepsilon = n_B I_B(I_B+1)[1 \pm (\omega_B/\omega_A)]/[n_A I_A (I_A+1)+n_B I_B(I_B+1)] \tag{35}$$

Here n_A and n_B are the numbers, ω_A and ω_B are the Larmor frequencies and I_A and I_B are the spins of the A and B nuclei respectively. Again the double resonance signal $\Delta M = M_{AO} - M_A$ at the frequency $\omega_A + \omega_B$ is stronger than the one at the frequency $\omega_A - \omega_B$.

The multiple coupling process

Similarly as in the case of $^1H-^{14}N$ double resonance and in the case of nuclear quadrupole double resonance we shall in the analysis of this process neglect the spin lattice relaxation of the B nuclei, but we shall no longer neglect the spin lattice relaxation of the A nuclei.

Before every coupling pulse the spin temperature of the B spin system is infinite due to the direct saturation of the B spin energy levels by the weak RF pulse.

After a train of strong and weak RF field pulses at the frequencies $\omega_A \pm \omega_B$ and ω_B respectively, which lasts longer than

the spin lattice relaxation time T_{1A} of the A system, the
magnetization of the A spins changes from M_{AO} to M_A

$$M_A = M_{AO} \, / \, (1 + \varepsilon T_{1A} / \, \tau) \qquad\qquad (36)$$

Here

$$\varepsilon = n_B I_B (I_B + 1) / [n_A I_A (I_A + 1) + n_B I_B (I_B + 1)] \qquad (37)$$

and τ is the duration of a weak plus a strong RF field pulse.
Also in this case the double resonance sensitivity increases
with T_{1A}.

EXPERIMENTAL RESULTS

The new technique has been experimentally tested.

a) In the case of zero magnetic field nuclear double resonance
between ^{1}H and ^{14}N nuclei in thymine. The double resonance
lineshape of the highest ^{14}N quadrupole transition
was measured at two different RF magnetic field amplitudes.
The single coupling technique was used. The results are shown
in figure 8. The measured lineshape qualitatively agrees
with the calculated one. When the RF magnetic field intensity
30 gauss was used the cross relaxation time was found to be
aprroximately 10 ms, which is well below our shortest
experimentally obtainable time for which the sample stays in
zero magnetic field.

b) In the case of nuclear double resonance between purely magnetic
^{23}Na and ^{19}F spins in NaF in a strong static magnetic field.
The double resonance lineshape at the frequency $\omega_F - \omega_{Na}$
in a single coupling process was measured. The result
is shown in figure 9.a. The RF field amplitude was approximately 80 gauss and the irradiation time was 1 s.The Larmor
frequency of ^{19}F was 18 MHz. In this case the cross relaxation
time was found to be 150 ms.

c) In the case of ^{1}H–^{14}N double resonance in $NH_4H_2PO_4$ powder
sample. The double resonance lineshape at the frequency
$\omega_H - \omega_N$ was measured. The proton Larmor frequency was 18 MHz,
the irradiation time was 1 s and the RF magnetic field amplitude
was 80 gauss. The results are shown in figure 9. b. The
unusual double resonance lineshape is due to the nuclear
quadrupole broadening of the ^{14}N resonance lines.

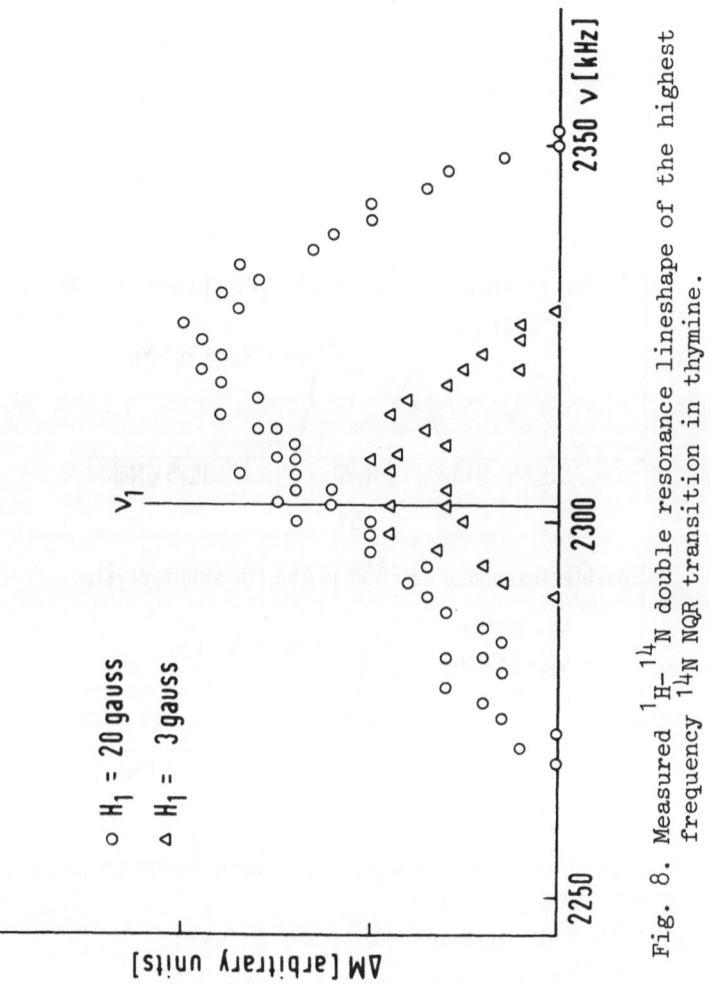

Fig. 8. Measured $^1H-^{14}N$ double resonance lineshape of the highest frequency ^{14}N NQR transition in thymine.

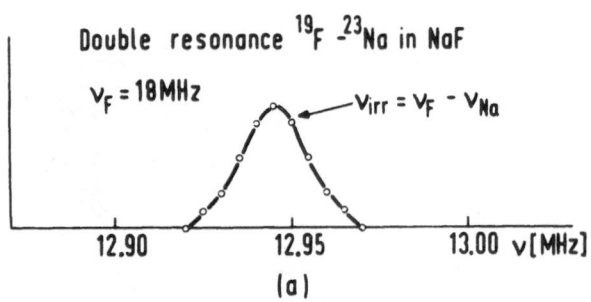

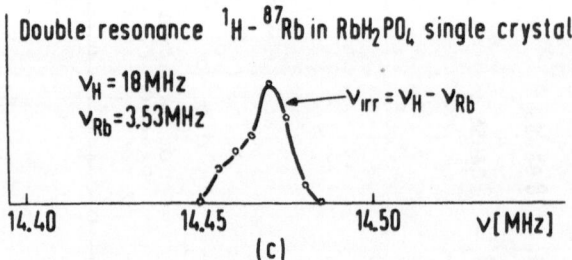

Fig. 9. Measured nuclear double resonance spectra in a strong
static magnetic field

a) ^{19}F-^{23}Na double resonance in NaF;

b) ^{1}H-^{14}N double resonance in NH$_4$H$_2$PO$_4$;

c) ^{1}H-^{87}Rb double resonance in RbH$_2$PO$_4$.

d) In the case of $^1H-^{87}Rb$ double resonance in RbH_2PO_4 single crystal. The experimental conditions were the same as in c). The ^{87}Rb resonance line at the frequency 3.53 MHz was chosen. The results are shown in figure 9.c.

All these results present various applications of the new double resonance technique in the single coupling case.

DISCUSSION

A new nuclear double resonance technique based on strong RF magnetic field induced coupling between spin systems has been presented. It has been analysed

a) in the case of zero field nuclear double resonance detection of ^{14}N NQR spectra through the proton signal,

b) in the case of nuclear quadrupole double resonance, and

c) in the case of nuclear double resonance between two purely magnetic spin systems in a strong static magnetic field.

In the case a) the sensitivity of the new technique is higher than the sensitivity of the level crossing technique which can be used for the same purpose. The new technique is of particular importance in some cases when the level crossing signals can not be obtained. This is the case when

(i) the spin lattice relaxation time of the B spin system is much shorter than the time for which the sample stays in zero static magnetic field

(ii) when the level crossing times are too short for sufficient energy exchange between the two spin systems

(iii) when the quadrupole resonance frequencies of the B nuclei are too high for the level crossing to occur.

In all these cases the new technique is a useful complementary tool to the level crossing technique.

In the cases b) and c) the cross relaxation times are longer than the cross relaxation times in the case of rotating frame nuclear double resonance resulting in a lower sensitivity. The main advantage of the new technique is in these cases the very simple experimental arrangement needed to obtain the double resonance signals. It can be also used to detect pure NQR spectra of integer spin nuclei by nuclear quadrupole double resonance, when the spin quenching makes the rotating frame nuclear double resonance impossible.

The double resonance linewidths are in the single coupling case mostly determined by the A spin resonance linewidths which is usually broad. Therefore the single coupling technique can be used for fast searching of unknown resonance frequencies of the B nuclei. The toll is a lower resolution.

This last disadvantage can be removed by the application of the multiple coupling technique where the double resonance spectra include also lines of the same linewidth as the corresponding resonance lines of the B nuclei. The multiple coupling technique is even suited for a direct lineshape measurement of the B nuclei resonance lines.

REFERENCES

1. S.R. HARTMANN and E.L. HAHN, Phys.Rev. 128, 2042 (1962)

2. F.M. LURIE and C.P. SLICHTER, Phys.Rev. 133, A1108 (1964)

3. R.E. SLUSHER and E.L. HAHN, Phys.Rev. 166, 332 (1968)

4. G.W. LEPPELMEIER and E.L. HAHN, Phys.Rev. 141, 724 (1966)

5. R. BLINC, M. MALI, R. OSREDKAR, A. PRELESNIK, J. SELIGER, I. ZUPANCIC and L. EHRENBERG, J.Chem.Phys. 57, 5087 (1972)

6. J. KOO, Ph.D.Thesis, University of California, Berkeley (unpublished)

7. D.T. EDMONDS, M.J. HUNT and A.L. MACKAY, J.Magn.Resonance 9, 66 (1973)

8. M. GOLDMAN, "Spin Temperature and Nuclear Magnetic Resonance in Solids" 7, Oxford U.P., London (1971)

9. A. ABRAGAM and W.G. PROCTOR, Compt.Rend. 246, 2253 (1958)

10. A. LANDESMAN, Journ.Phys.Chem.Solids 18, 210 (1961)

11. R. BLINC, "Proceedings of the Ampere International Summer School II, Basko Polje," Ed. R. Blinc, 51, J.Stefan Institute, Ljubljana, Yugoslavia

NMR STUDIES OF STRUCTURE AND CONFORMATION IN PEPTIDES AND PROTEINS

K. Wüthrich

Eidgenössische Technische Hochschule

Zürich, Switzerland

Investigations of the relations between covalent structure, molecular conformations and functional properties of biological macromolecules play a prominent role in modern biochemistry, molecular biology, and biophysics. While chemical methods have so far served to unravel the covalent structures of a large number of biopolymers [1], and X-ray crystallography in single crystals has set the standards for the description of the three-dimensional conformations in these molecules [2], NMR has become an attractive technique for investigations of the molecular conformations in solution. To a large extent the high sensitivity and high resolution required for biological applications of NMR have become available by recent advances in apparatus design and biochemical methods, i.e. on the one hand the introduction of superconducting magnets and Fourier transform techniques into the modern NMR spectrometers, on the other hand the preparation of isotope-labelled biopolymers, by chemical and biosynthetic techniques. In the present lecture, some basic notions on structure and conformation of one class of biopolymers, i.e. peptides and proteins, will be introduced, and the manifestations in the NMR parameters of different molecular structures and conformations will be discussed and illustrated with some selected examples.

STRUCTURE AND CONFORMATION IN PEPTIDES AND PROTEINS

The monometric units in a polypeptide chain are the amino acids. There are 20 common amino acids with different side chains R_i on the α-carbon atom (Fig. 1), which are either aliphatic hydrocarbons, or carry various functional groups including different aromatic rings. The primary structure of peptides and

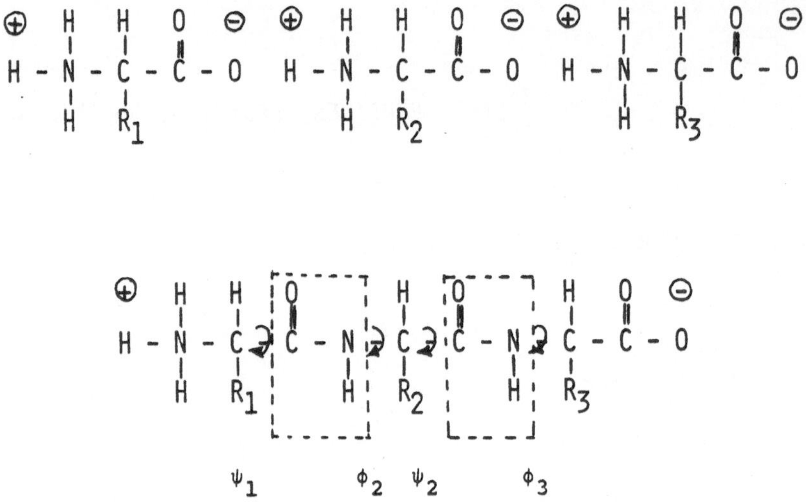

Fig. 1. The top row shows three individual amino acid residues which
        differ in the side chains R. on the α-carbon atom. Below,
        the three amino acids have been joined in a tripeptide.
        For considerations of the peptide conformations, the peptide
        units (dashed squares) are generally assumed to be planar.
        Different conformations ofthe peptide backbone can then
        be uniquely characterized by the torsion angles φ and
        ψ about the single bonds linking the peptide units with
        the α-carbon atoms [2, 3].

proteins consists of a linear array of amino acid residues which
are linked by peptide bonds (Fig. 1). As a consequence of the partial
double bond character of the N-C' bond, the peptide groups are
preferentially in a near-planar form. A widely used model for the
description of polypeptide chain conformations, which has also been
the starting point for numerous NMR investigations, is therefore
based on the assumption that all the peptide groups are rigid
planes with standard dimensions [2, 3] . While the trans form
(Fig. 2) is common for unsubstituted amide groups, cis peptide
groups (Fig. 2) are frequently encountered in N-substituted amino
acids, e.g. in proline. With the assumption of standard trans and

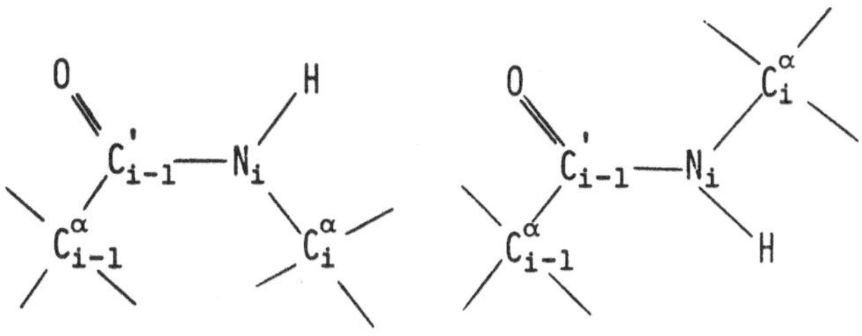

CIS   TRANS

Fig. 2. Cis and trans form of the peptide group.

cis peptide groups, the backbone conformations can unambiguously be described by the torsion angles ϕ_1 and ψ_1 about the $N-C^\alpha$ and $C^\alpha-C'$ bonds (Fig. 1). In this "rigid model", only certain combinations of ϕ and ψ are sterically allowed for each individual dipeptide fragment, the extent of the allowed regions in the $\phi-\psi$ plane depending on the amino acid side chains, and on the dimensions chosen for the peptide groups [2, 3].

PRIMARY STRUCTURE AND NMR PARAMETERS

The NMR spectra of individual amino acids are primarily determined by their covalent structures. They are readily interpreted by the empirical rules on the relations between structure and NMR parameters in diamagnetic organic molecules [4]. In addition to the protons bound to carbon atoms, which are non-exchanging in protic solvents, the amino acids contain several potentially exchangeable protons, of which the exact number depends on the solvent environment. In aqueous solution it varies with pH according to

$$\overset{\oplus}{H_3N} - \underset{\underset{H}{|}}{\overset{\overset{R}{|}}{C}} - COOH \underset{\overset{\oplus}{H}}{\rightleftharpoons} \overset{\oplus}{H_3N} - \underset{\underset{H}{|}}{\overset{\overset{R}{|}}{C}} - COO^{\ominus} \underset{\overset{\ominus}{OH}}{\rightleftharpoons} H_2N - \underset{\underset{H}{|}}{\overset{\overset{R}{|}}{C}} - COO^{\ominus}$$

(1)

Δδ (Hz at 60MHz)

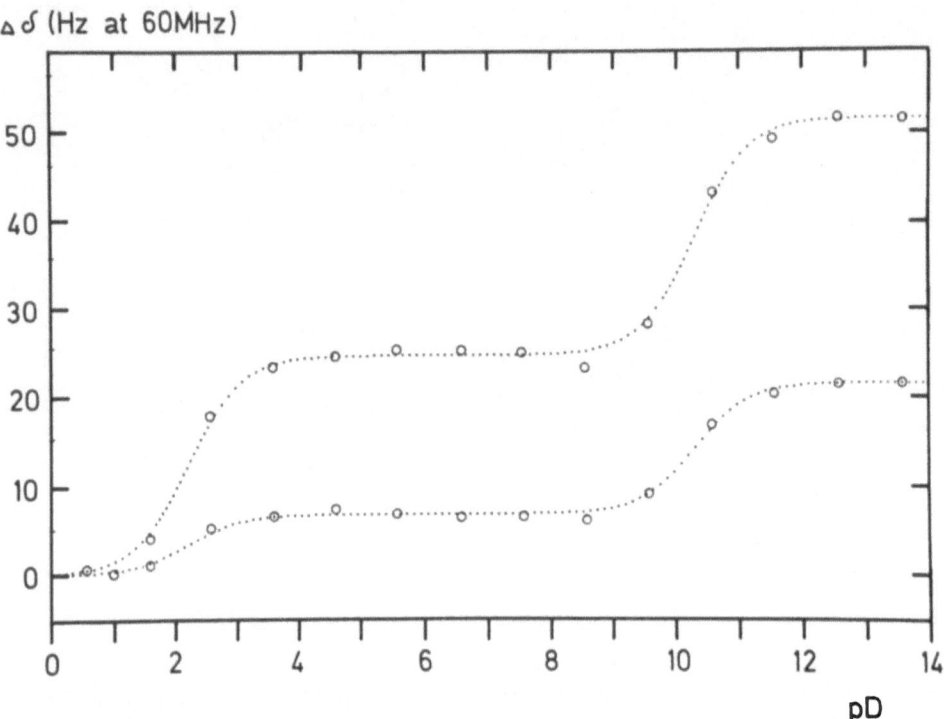

Fig. 3. Variation with pD of the chemical shifts of the C^{α}- and
        the methyl protons in L-alanine[5].

Because of the reactions (1), essentially all the resonances in the
individual amino acids vary with pH . This is illustrated in figure
3 for L-alanine, where R = - CH_3. In polypeptide chains (Fig. 1),
only the terminal amino and carboxylic acid groups of the backbone
can be titrated [5] . Hence, the pH dependence of the NMR spectra
can in favorable cases be employed for the identification of the
chain terminal amino acids [6]. The NMR of the non-terminal amino
acid residues (residue 2 in figure 1) can be quite different
from those of the free amino acids. Typical values for the chemical
shifts in the amino acid residues were obtained from observations
in linear oligopeptides, where the amino acid side chains extend
predominantly into the solvent [7, 8].

     In the context of the present discussion, a protein molecule
is taken to be in a "random coil form" when all the amino acid
side chains extend freely into the solvent. Since the chemical
shifts are primarily determined by the covalent structure of the
peptide and the magnetic susceptibility of the solvent environment,
one would expect that the NMR of the amino acid side chains in

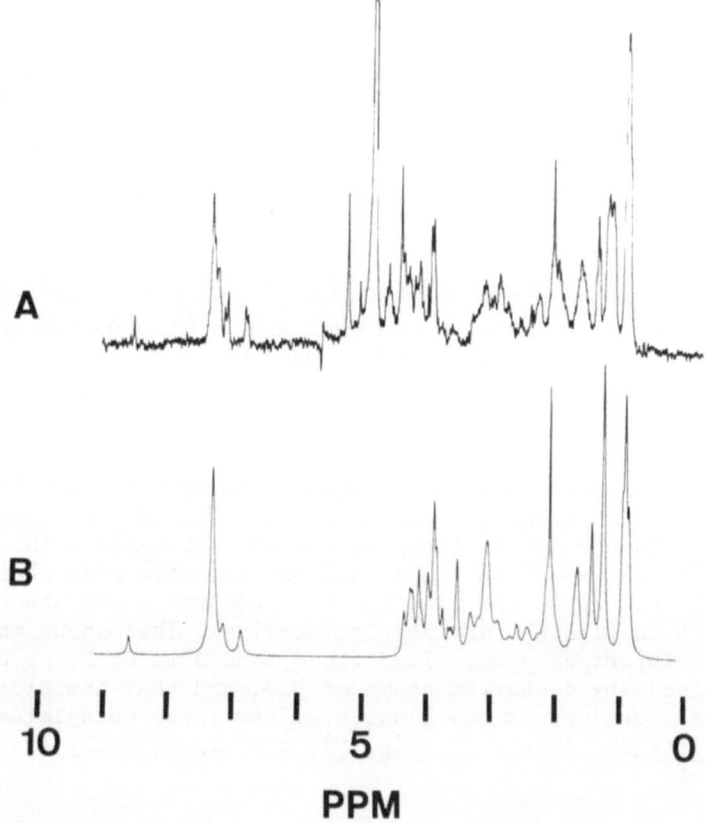

Fig. 4. ^{1}H-NMR spectrum at 220 MHz of calcitonin M in a D_2O-
        solution, pD = 5.0. A. Experimental spectrum B.
        Hypothetical spectrum computed on the basis of the amino
        acid composition (Fig. 5) as the sum of the resonances
        of the individual amino acid residues [7]. The strong
        narrow lines between 4 and 6 ppm in the spectrum A
        correspond to the resonance of the residual protons of
        the solvent and its spinning side bands.

such a random coil polypeptide chain correspond essentially to the
sum of the resonances of the constituent amino acid residues. This
is illustrated in figure 4 with the proton NMR spectrum of a
human polypeptide hormone, calcitonin M. This molecule, which
consists of one polypeptide chain with 32 amino acid residues
(Fig. 5), plays an important role in the regulation of the blood

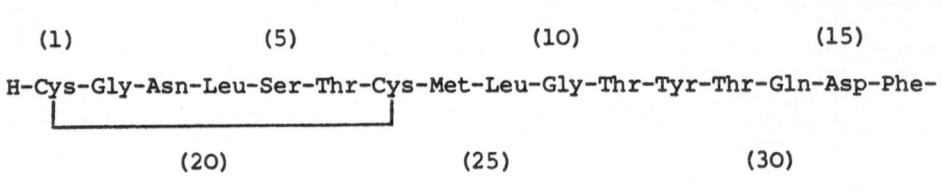

Fig. 5. Amino acid sequence in calcitonin M (R. Neher et al.,
        Helv. Chim. Acta 51, 1900 (1968)).

calcium level. NMR studies have shown that calcitonin M in
aqueous solution is predominantly in an extended random coil
form [9] . In the spectral regions of the aliphatic amino acid
side chains between 0 en 3 ppm, and the aromatic side chains
between 6 and 9 ppm, the experimental spectrum A and the computed
spectrum B in figure 4 are nearly identical. This shows that all
the labile hydrogen atoms, i.e. those bonded to O, N, or S, had
been replaced by deuterium atoms of D_2O, and that the different
amino acid residues of one kind, e.g. the three phenylalanines
in positions 16, 19 and 22 (Fig. 5), are magnetically essentially
equivalent.

    The similarities between the two spectra in figure 4 shows
that the ^{1}H-NMR of the amino acid side chains are essentially
independent of the amino acid sequence. Therefore, with the
exception of the above mentionned pH dependence of the chain
terminus, the ^{1}H-NMR spectra will not in general yield information
on the primary peptide structures. On the other hand, the amino
acid composition of a polypeptide chain determines the principal
features of its random coil NMR spectrum, and is thus in
principle accessible for investigation. Similar conclusions
can be drawn on the ^{13}C-NMR spectra of random coil peptides and
proteins.

MANIFESTATIONS OF DIFFERENT MOLECULAR CONFORMATIONS IN THE NMR
PARAMETERS

Taking into account the Karplus-type angular dependence of
vicinal spin-spin coupling constants, it is readily apparant from
figure 1 that a complete rigid model description of the peptide
backbone conformations in terms of the torsion angles ϕ_1 and
ψ_i could in principle be derived from NMR studies. The
homonuclear spin-spin coupling constant $J_{HN\alpha}$ between the amide
proton i and the C_i^α-proton (Fig. 1) is related to the torsion
angle ϕ_i. In peptides with ^{13}C-enriched carbonyl carbon positions,
the torsion angles ϕ_i can also be studied via the heteronuclear
vicinal spin-spin coupling $J_{^{13}C'\alpha}$ between the carbonyl carbon atom
(i-1) and the C_i^α-proton.
In ^{15}N-enriched peptides, the torsion angles ψ_i are accessible to
NMR investigation through the heteronuclear spin-spin coupling
constant $J_{^{15}N\alpha}$ between the C_i^α-proton and $^{15}N_{(i+1)}$. Additional
vicinal couplings occur between protons of neighbouring
carbon atoms of the amino acid side chains. Measurements of the
vicinal spin-spin couplings are thus in principle the most direct
way to investigate polypeptide conformations, which has been
employed with remarkable success for investigations of a variety
of cyclic oligopeptides [10, 11] . Even though there is a good
chance for the near future that the vicinal coupling constants will
become an interesting source of information also in other classes
of peptide molecules, several practical difficulties still have
to be overcome e.g. there are some uncertainties on the exact
form of the relations between the vicinal coupling constants
and the torsion angles [12] , the spectral resolution obtained
in larger molecules may not allow accurate measurements of the
spin-spin coupling constants, in flexible peptide chains the
observed vicinal coupling constants may correspond to the average
over a wide range of torsion angles rather than to a distinct value
of the latter, etc. Therefore, studies of a variety of additional
NMR phenomena relating to the molecular conformations are usually
also included, and in the investigations of non-cyclic polypeptide
chains these have actually yielded the bulk of the structural
information.

In many instances the molecular conformations of peptides
and proteins include the formation of intramolecular hydrogen
bonds, in particular between the amide protons and the carbonyl
oxygen atoms of different peptide groups (Fig. 1) [2, 3]. NMR
observation of the hydrogen-bonded amide protons can then yield
information on these weak intramolecular bonds which characterize
in part the spatial arrangement of the polypeptide chain. Such
experiments are mainly based on the following observations. When
the amide protons in a polypeptide chain are freely accessible to
the solvent molecules, they have typical chemical shifts for a

given solvent, and the temperature dependence of the chemical
shifts is also a characteristic quantity for a given solvent
system. Furthermore, the solvent-accessible amide protons are
rapidly exchanged with hydrogen atoms (or deuterium atoms) of
protic solvents. In molecules where some of the amide protons
are involved in intramolecular hydrogen bonds, or otherwise
shielded from the solvent, the chemical shifts and the
temperature coefficients for these protons will often be markedly
different from those of the solvent-exposed amide protons, and the
rate of the exchange reactions with the solvent will in general
be considerably slower, depending on the specific environment.

The NMR of numerous nuclei in a polypeptide chain can also
be affected by non-bonding interactions with their environment
in the native conformation of the molecule [13 - 15]. In the
interior of a globular protein, many segments of the polypeptide
chain are no longer freely accessible to the solvent, and are
instead in close contact with other amino acids. Across the protein
molecule, the different amino acid residues of a given kind are thus
in different micro-environments characterized by different local
magnetic susceptibilities, and hence their chemical shifts may
be different. As a consequence of this dispersion of the chemical
shifts of structurally identical protons, the ^{1}H-NMR spectra of
native globular proteins contain rather broad lines, which are
poorly resolved in the bigger molecules, and which consist of the
mutually overlapping resonances of the many magnetically slightly
non-equivalent protons. In figure 6 this is illustrated with the
^{1}H-NMR spectrum of the basic pancreatic trypsin inhibitor (BPTI),
a globular "miniprotein" consisting of one polypeptide chain
with 58 amino acid residues [16]. The overall ^{1}H-NMR spectral
features of a polypeptide chain can thus be used to distinguish
between extended random coil forms and globular conformations,
e.g. for a rapid qualitative characterization of the solution
conformation of a new protein or for studies of the denaturation
of a globular protein to a random coil form [13-15]. In favorable
cases the non-bonding intramolecular interactions can also result
in rather outstanding chemical shifts of individual resonances,
which then appear as well resolved lines in spectral regions
where no resonances of the individual amino acid residues occur
(Fig. 6) [13-15]. This will be further illustrated in the
subsequent lecture on hemoproteins. It is a fortunate coincidence
that some of the amino acids which have repeatedly been found to
be subject to outstanding chemical shifts, and hence to be accessible to NMR observation even in rather big proteins, have also
been reported to form an integral part of the active sites in
numerous proteins. Thus, even with the analysis of only a small
fraction of the ^{1}H-NMR lines in a protein, it is often possible
to obtain information on the active centers in these molecules.

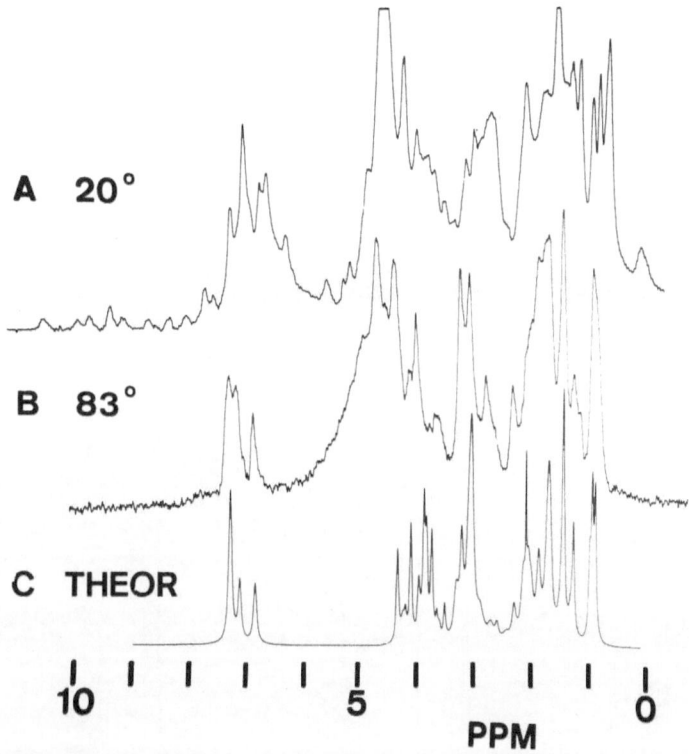

Fig. 6. ^{1}H-NMR spectra of a D_2O solution containing 0.01 M of
the basic pancreatic trypsin inhibitor (BPTI) and 6-M
guanidium chloride, pD = 7.1. A. Native globular conformation of BPTI at 20°. B. Denatured "random coil" conformation
with intact disulfide bonds. C. Hypothetical spectrum
for the random coil polypeptide chain computed on the
basis of the amino acid composition as the sum of the
resonances of the individual amino acid residues.

In the ^{13}C-NMR spectra, similar manifestations of different
protein conformations to those described above for the ^{1}H-NMR can
be observed. In the native globular form of the basic pancreatic
trypsin inhibitor, the dispersion of the chemical shifts among
different amino acids of a given type is again quite readily
apparent (Fig. 7). Here, a rather outstanding conformation-

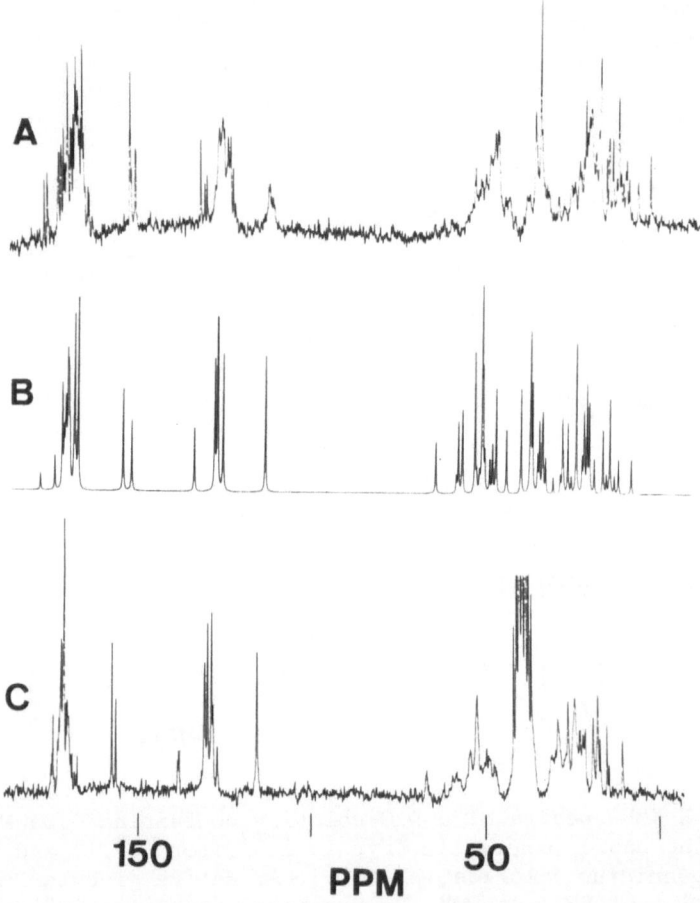

Fig. 7. Proton noise-decoupled natural abundance ^{13}C Fourier
transform NMR spectra at 25.14 MHz of the basic pancreatic
trypsin inhibitor. A. Native protein in D_2O solution.
B. Hypothetical spectrum for the random coil polypeptide
chain computed as the sum of the resonances of the
individual amino acid residues [8]. C. Denatured protein
in DMSO solution, representing a random coil conformation.

dependent ^{13}C-NMR phenomenon which bears on the cis-trans isomerism
(Fig. 2) of X-Pro bonds, where X stands for one of the common amino
acids, shall be discussed in more detail.

For the equilibrium between the cis- and trans-forms of X-
Pro, the free energy $\Delta G°$ is of the order -2.0 to +2.0 kcal Mol^{-1}
in a variety of linear oligopeptides [17, 18] , so that the two

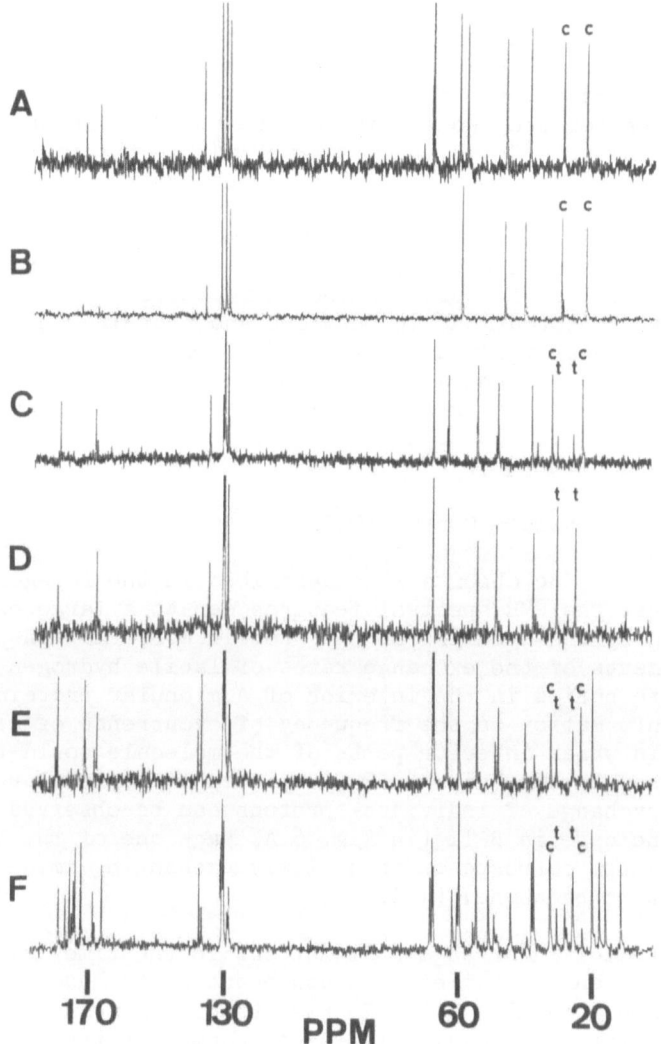

Fig. 8. Proton noise-decoupled ^{13}C-NMR spectra at 25.14 MHz of six prolinecontaining peptides. Approx. 0.1-molar solution in D_2O, pH = 5.0, has been studied at 26°.

A. cyclo-[-L-Phe-L-Pro-] (contains only cis-Pro).
B. cyclo-[-D-Phe-L-Pro-] (contains only cis-Pro).
C. H-L-Phe-L-Pro-OH.
D. H-D-Phe-L-Pro-OH (contains only trans-Pro).
E. H-L-Thr-L-Phe-L-Pro-OH.
F. H-L-Thr-L-Phe-L-Pro-L-Gln-L-Thr-L-Ala-L-Ile-Gly-OH.

The lines marked c and t correspond to the β- and γ-carbon atoms of proline in the cis and trans form of the Phe-Pro amide bond, respectively.

forms are simultaneously present at ambient temperature. Because
at the same time the barrier for interchange between the two
conformations is of the order $\Delta G^+ \approx 20$ kcal Mol^{-1} [18], separate
resonances for the two species can be observed in the ^{1}H-NMR[19] as
well as in the ^{13}C-NMR spectra [20-23]. In the latter, the C$^\beta$
and C$^\gamma$ resonances of cis- and trans-proline are particularly
nicely separated (Fig. 8), and therefore suitable for measurements
of the equilibrium between the two forms. This is of considerable
interest because the conformations of the X-Pro peptide bonds
are an important feature for the characterization of the overall
molecular conformations in peptides and proteins. Moreover, in
flexible linear peptides where the NMR spectra indicate the
presence of an extended "random coil conformation"(Fig. 4),
evidence for the occurrence of energetically preferred flexible
conformations can in principle be derived from studies of the
X-Pro isomerism [24].

DYNAMICS OF PROTEIN CONFORMATIONS

     Many of the observations described in the foregoing section
imply that the NMR spectral features can to a large extent be
affected by the dynamics of the protein conformations. For example,
measurements of the exchange rates of labile hydrogen atoms
which are buried in the interior of a globular protein, can
yield information on the frequency of occurrence of certain
events in which interior parts of the molecule would temporarily
be exposed to the solvent [25]. This is of particular interest
if the exchange of individual protons can be observed separately,
as is the case in BPTI (in Fig. 6 A, each one of the lines between
8 and 11 ppm corresponds to a slowly exchanging amide proton of
the peptide backbone) [16].

     Intramolecular segmental motions in the time domain 10^{-8} to
10^{-11} sec can be studied by measurement of the nuclear spin
relaxation times T_1. Using Fourier transform methods, T_1 can be
obtained for the individual ^{13}C-resonances in the spectrum of a
protein [26]. For proteins, the relevant correlation time for
T_1-relaxation of ^{13}C is determined by the combination of the
overall rotational tumbling of the macromolecule, and the intra-
molecular segmental motions. ^{13}C relaxation times can therefore
be employed to compare the segmental motions of individual amino
acid side chains in the globular and the random coil forms of the
polypeptide chain [26]. Experiments of interest can for example
include comparisons of the segmental mobility of the amino acids
in the active site of an enzyme in the presence and absence of
substrates, inhibitors, or other effector molecules.

ACKNOWLEDGMENT

The author acknowledges financial support by the Schweizerischer Nationalfonds for the research projects Nr. 3.423.70 and 3.4231.70 in the field of NMR studies of biopolymers.

REFERENCES

1. M.O. DAYHOFF ed., "Atlas of Protein Sequence and Structure", National Biomedical Research Foundation, Silver Spring, Md. (1969)

2. R.E. DICKERSON and I. GEIS, "The Structure and Action of Proteins", Harper and Row, New York (1969)

3. G.N. RAMACHANDRAN and V. SASISEKHARAN, Adv.Protein Chem. 23, 283 (1968)

4. F.A. BOVEY, "Nuclear Magnetic Resonance Spectroscopy", Academic Press, New York (1969)

5. G.C.K. ROBERTS and O. JARDETZKY, Adv.Protein Chem. 24, 447 (1970)

6. M. SCHEINBLATT, J.Amer.Chem.Soc. 88, 2845 (1966)

7. C.C. McDONALD and W.D. PHILLIPS, J.Amer.Chem.Soc. 91, 1513 (1969)

8. Ch. GRATHWOHL and K. WÜTHRICH, J.Magn.Res. 13, 217 (1974)

9. A. MASSON, Ph.D.Thesis: "Konformationsstudien an Calcitonin M und am Trypsin-Inhibitor BPTI", ETH Zürich (1974)

10. F.A. BOVEY, A.I. BREWQTER, D.J. PATEL, A.E. TONELLI and D.A. TORCHIA, Acc.Chem.Res. 5, 193 (1972)

11. R. SCHWYZER, Ch. GRATHWOHL, J.P. MERALDI, A. TUN-KYI, R. VOGEL and K. WÜTHRICH, Helv.Chim.Acta 55, 2545 (1972)

12. V.F. BYSTROV, V.T. IVANOV, S.L. PORTNOVA, T.A. BALASHOVA and Y.A. OVCHINNIKOV, Tetrahedron 29, 873 (1973)

13. C.C. McDONALD and W.D. PHILLIPS in "Fine Structure of Proteins and Nucleic Acids" eds. G.D. Fasman and S.N. Timasheff, Dekker, New York (1970)

14. K. WÜTHRICH, Chimia 24, 409 (1970)

15. K. WÜTHRICH, Experientia 30, 577 (1974)

16. A. MASSON and K. WÜTHRICH, FEBS Lett. 31, 114 (1973)

17. L.A. LA PLANCHE and M.T. ROGERS, J.Amer.Chem.Soc. 86, 337 (1964)

18. H.L. MAIA, K.G. ORRELL and H.N. RYDON, Chem.Comm. 1971, 1209

19. C.M. DEBER, F.A. BOVEY, J.P. CARVER and E.R. BLOUT, J.Amer.Chem. Soc. 92, 6191 (1970)

20. K. WÜTHRICH, A. TUN-KYI and R. SCHWYZER, FEBS Lett. 25, 104 (1972)

21. W.A. THOMAS and M.K. WILLIAMS, Chem.Comm. 1972, 994

22. F.A. BOVEY, in "Chemistry and Biology of Peptides", ed. J. Mei-
    enhofer, 3, Ann Arbor Sci.Publ., Ann Arbor (1972)

23. I.C.P. SMITH, R. DESLAURIERS and R. WALTER, in "Chemistry and
    Biology of Peptides", ed. J. Meienhofer, 3, Ann Arbor Sci.Publ.,
    Ann Arbor (1972)

24. K. WÜTHRICH and Ch. GRATHWOHL, FEBS Lett. 43, 337 (1974)

25. A. HVIDT and S.C. NIELSEN, Adv. Protein Chem. 21, 287 (1966)

26. A. ALLERHAND, D. DODDRELL, V. GLUSHKO, D.W. COCHRAN, E. WENKERT,
    P.J. LAWSON and F.R.N. GURD, J.Amer.Chem.Soc. 93, 544 (1971)

NMR STUDIES OF HEMOPROTEINS

K. Wüthrich

Eidgenössische Technische Hochschule

Zürich, Switzerland

In the foregoing lecture on "NMR Studies of Structure and
Conformation in Peptides and Proteins" some NMR spectral
phenomena relating to the spatial arrangement of the polypeptide
chains had been introduced. In this context it was mentionned
that certain non-bonding interactions between different segments
of a polypeptide chain in the interior of a globular protein
molecule can result in rather outstanding chemical shifts of
the NMR of individual groups of protons, which can then be
used as natural NMR probes for studies of the protein conformations.
This will now be illustrated by some experiments with hemoproteins.
Molecules of this class of conjugated proteins possess some
particularly interesting NMR features because they contain one
or several iron porphyrin complexes per molecule, which can
greatly affect the NMR spectra both in the diamagnetic and
paramagnetic electronic states of the heme iron.

STRUCTURE AND BIOLOGICAL ROLES OF HEMOPROTEINS

Hemoproteins are involved in many vital processes in living
organisms. Prominent among their biological functions are
those of hemoglobin, which is the oxygen transporting protein
in the blood, myoglobin, which binds and stores oxygen in the
muscles, various cytochromes, which act as electron transferring
oxidation-reduction carriers, and of various enzymatically
active hemoproteins which control diverse biochemical reactions.
A hemoprotein molecule consists of one or several polypeptide
chains, each of which is typically made up of 100 to 200 amino
acid residues, and one or several heme groups. The latter are

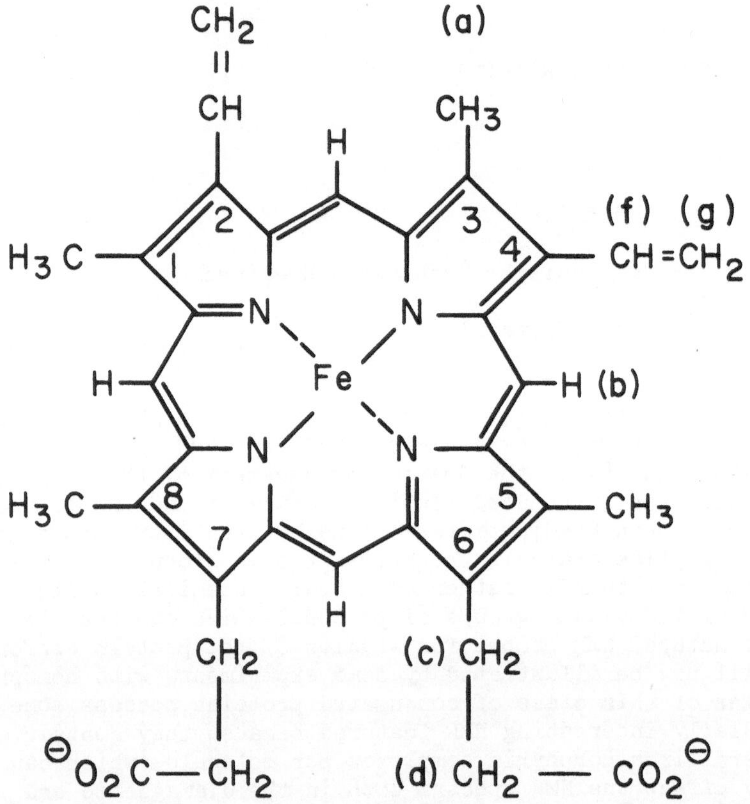

Fig. 1. The iron protoporphyrin IX complex, "protoheme IX", is the prosthetic group of cytochrome b_5, and a variety of other hemoproteins. In cytochrome b_5 the two axial coordination sites of the heme iron are occupied by two histidyl residues of the polypeptide chain. The numbers and letters given in the figure will be used for resonance identifications. The carbon atoms 1 to 8 are usually referred to as "β-carbons", the atoms b as "meso-carbons".

iron porphyrin complexes (Fig. 1) which can be combined with the polypeptide moiety of the molecule through one or several covalent bonds, and through a multitude of weaker interactions. The heme groups appear to be an integral part of the active centers in all the hemoproteins known to-date.

Four different electronic configurations of the heme iron are commonly encountered. Three of these are paramagnetic, i.e. the high spin ferric (Fe(III), S = 5/2), low spin ferric (Fe(III), S = 1/2), and high spin ferrous (Fe(II), S = 2) forms, whereas low spin ferrous iron (Fe(II), S = 0) is diamagnetic. Many biochemical reactions of hemoproteins involve changes of the oxidation and spin states of the heme iron, e.g. the oxygenation of myoglobin and hemoglobin

$$Mb^{II}(Fe(II),\ S = 2) + O_2 \rightleftharpoons Mb^{II}O_2(Fe^{2+},\ S = 0) \qquad (1)$$

and the electron transfer in cytochromes, e.g.

$$Cyt\ b_5^{III}(Fe^{3+},\ S = 1/2) + e^{\ominus} \rightleftharpoons Cyt\ b_5^{II}(Fe^{2+},\ S = 0) \quad (2)$$

In this paper we shall be mainly concerned with low spin ferrous and low spin ferric hemes and hemoproteins, in which the proton NMR have been extensively investigated [1].

^{1}H-NMR SPECTRA OF HEMOPROTEINS

From the viewpoint of NMR it is of particular relevance that hemoproteins are rather large molecules. This may be illustrated by the molecular weights of 12,000 for cytochrome b_5, 18,000 for myoglobin, and 64,000 for mammalian hemoglobins. Correspondingly, the polypeptide chains contain several hundred to several thousand carbon atoms and protons per molecule. The heme group in figure 1, on the other hand, contains 34 carbon atoms, and 30 protons. Yet because of the paramagnetism of the heme iron, the resonances of the relatively few nuclei of the heme group can be quite prominent in the NMR spectra of hemoproteins. The ^{1}H-NMR spectrum of the oxidized paramagnetic form of cytochrome b_5 (Fig. 2) is quite typical for a low spin ferric hemoprotein. Outside the spectral region from 0 to 10 ppm, which contains the bulk of the resonances of the ca. 500 protons in this molecule, there are numerous well resolved resonance lines in quite outstanding high and low field positions. These large chemical shifts are a consequence of the electronic paramagnetism of the heme group, as evidenced by their dependence on temperature [1, 2], and some of the largely shifted lines

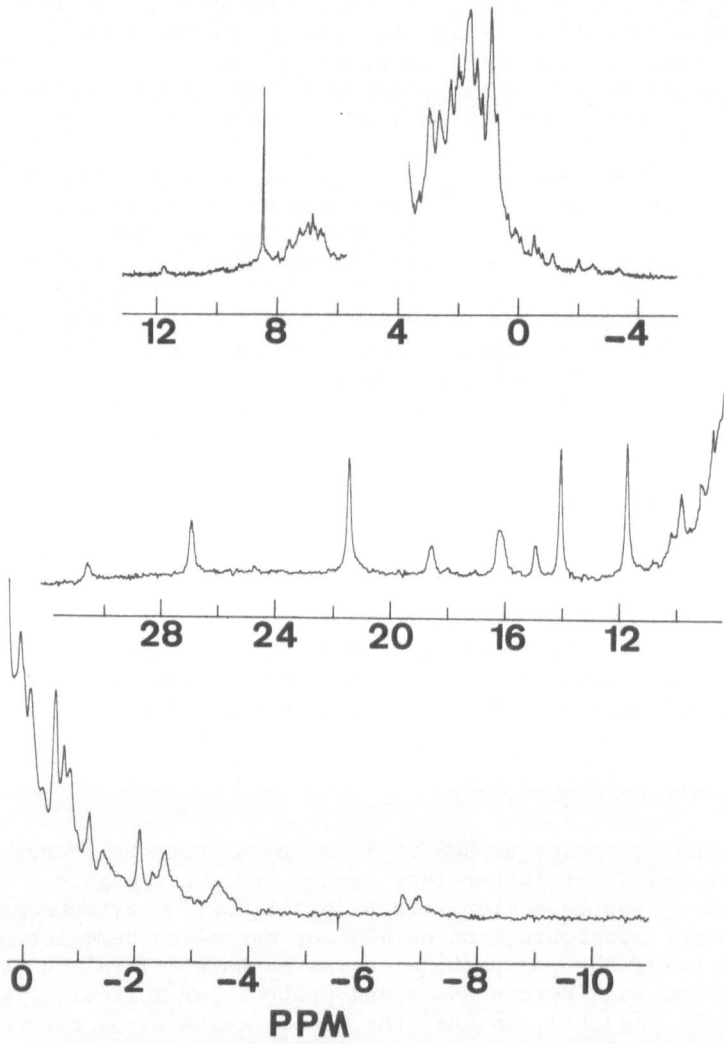

Fig. 2. 220 MHz proton NMR spectrum of 0.008-M ferricytochrome
b$_5$ in 0.2 M deuterated phosphate buffer, pD 6.3, at
29 C. The three spectral regions are represented with
different horizontal and vertical scales. (The resonance
intensities of the resolved lines in the regions 10 to
30 and 0 to -10 ppm correspond to 1 to 3 protons, whereas
the spectral region from 0 to 10 ppm contains the
resonances of approx. 500 protons). The spectral region
from 4 to 6 ppm, which contains the resonance of HDO
and its spinning side bands, has been omitted. (From
Keller and Wüthrich [2]).

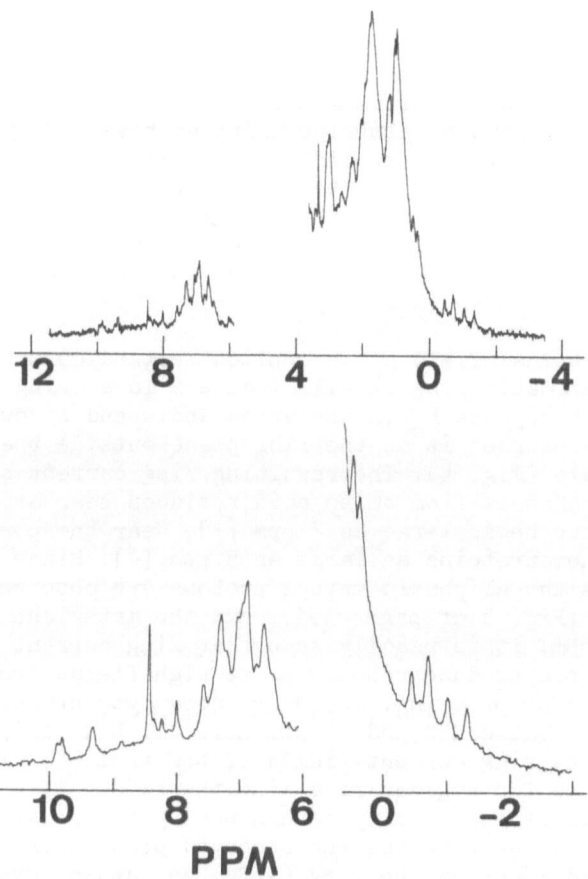

Fig. 3. Proton NMR spectrum at 220 MHz of a ca. 0.008-M
solution of ferrocytochrome b_5 in deuterated 0.2-M
phosphate buffer, pD = 7.0 at 29 C. (From Keller and
Wüthrich [2]).

correspond to protons of the heme group. In the diamagnetic
reduced state of cytochrome b_5 (Fig. 3) [2], the only really
outstanding chemical shifts are those of the five methyl
resonances between 0 and −2 ppm.

LOCAL MAGNETIC FIELDS IN HEMOPROTEINS

    Because of the limited spectral resolution even at the
highest presently available magnetic fields, detailed analyses
of the ^{1}H-NMR spectra of globular proteins have so far
generally concentrated on a relatively small number of resolved
resonances which had been shifted to outstanding positions
by interactions with their environment in the native form of
the molecule. Typical examples are the chemical shifts caused
by the ring current fields of aromatic rings [3], and by the
electron paramagnetism in low spin ferric hemes [1].

    If an external field H_o is applied perpendicular to the
plane of an aromatic ring it will induce a local ring current
field H_R which opposes H_o in the areas above and below the ring
plane, and reinforces it in the ring plane outside the contours
of the molecule (Fig. 4). The resulting ring current shifts
of proton resonances from amino acid residues near aromatic
amino acids can be as large as 2 ppm [3], near the porphyrin ring
(Fig. 1) in hemoproteins as large as 5 ppm [1]. Since in random
coil peptides the aliphatic methyl protons are observed at
around 1 ppm (Fig. 4 of page 351), and the methylene protons
at 1.5 to 3 ppm, it is readily seen that ring current fields
can in principle produce resonances at high fields from 0 ppm
in the spectra of proteins. In the reduced cytochrome b_5 (Fig. 3),
the resonances between 0 and −2 ppm have all been shifted to
higher field by ring current fields of the aromatic rings in the
molecule. Since for a given ring size the extend of the ring
current shifts is essentially determined by the relative
orientations of the ring and the observed protons (Fig. 4),
these resolved lines can be very useful as natural NMR probes,
e.g. to detect conformational changes in the molecule when the
latter is subject to variable external conditions, or to
interactions with other molecules.

    In the paramagnetic states of a hemoprotein, the resonance
positions in ppm for the individual protons can be expressed
as a sum of three terms

$$\Delta\nu = \Delta\nu_d + \Delta\nu_c + \Delta\nu_{pc} \tag{3}$$

$\Delta\nu_d$ is the position which the resonance would have in the absence

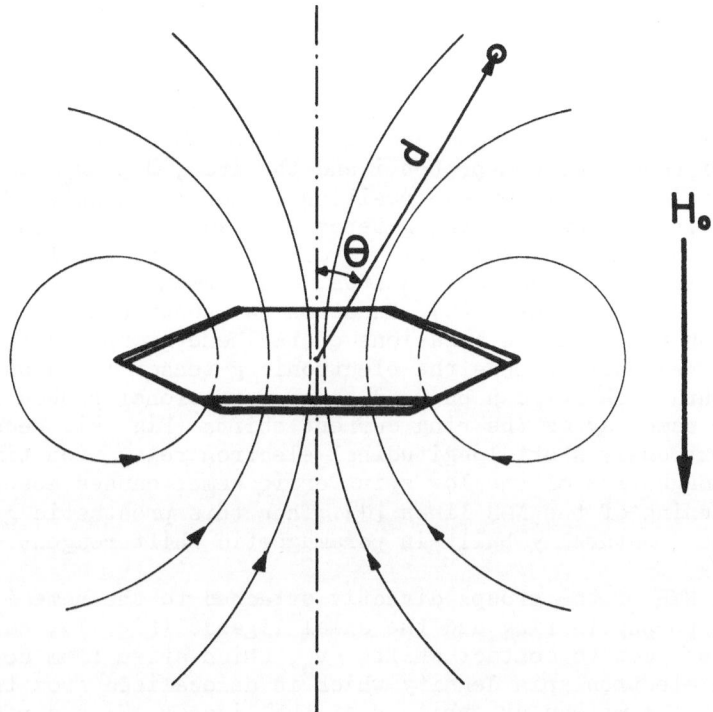

Fig. 4. The local magnetic ring current field of an aromatic
        molecule. H_o is the external polarizing field. For a
        given ring the field strength at any point in space
        is determined by the position (d, θ) relative to the
        aromatic molecule.

of the unpaired electron, and $\Delta\nu_c$ and $\Delta\nu_{pc}$ are the contact shift
and pseudocontact shift which arise from the interactions with
the unpaired electron. Quite generally, all the protons in the
molecule are to some extent influenced by pseudocontact coupling
[4]. The pseudocontact shift for the i-th proton in an ortho-
rhombic system is given by equation (4) [2, 5]

$$\Delta\nu_{pci} = F_i \frac{\beta^2 S(S+1)}{9kTr_i^3} [g_x^2(1 - 3\cos^2\omega_{xi}) + g_y^2(1 - 3\cos^2\omega_{yi})$$

$$+ g_z^2(1 - 3\cos^2\omega_{zi})] \tag{4}$$

β is the Bohr magneton, S the effective spin (S = 1/2 for low spin Fe^{3+}), k the Boltzmann constant, T the absolute temperature, r_i the distance between proton i and the iron, ω_{xi}, ω_{yi} and ω_{zi} the direction cosines of the position vector r_i relative to the principal axes of the g-tensor. F_i accounts for the influence of spin delocalization from the iron to the ligands, thermal mixing of electronic states, and Zeeman mixing with low lying excited states [5]. Equation (4) shows that $\Delta\nu_{pc}$ depends on the relative locations of the heme group and the observed nuclei. Once the electronic g-tensor is known, pseudocontact shifts can be used as conformational probes in much the same way as the ring current shifts (Fig. 4). Because of the extremely short longitudinal electron relaxation time, the paramagnetism of the low spin ferric hemes causes essentially no broadening of the NMR lines [1]. Thus this prosthetic group acts like a naturally built-in paramagnetic shift reagent.

The NMR of the groups directly attached to the heme iron, i.e. the porphyrin ring and the axial ligands (Fig. 1), can also be subject to contact shifts $\Delta\nu_c$, which arise from coupling with the electron spin density which is delocalized from the iron into the molecular orbitals of the ligands. With certain simplifying assumptions, $\Delta\nu_c$ is given by [6]

$$\Delta\nu_{ci} = - A_i|\gamma e|S(S+1) / 3|\gamma_I|kT \tag{5}$$

where γ_e and γ_I are the gyromagnetic ratios for the electron and the nuclear spin, respectively, and A_i is the contact interaction constant for the nucleus i.

PROTEIN CONFORMATIONS IN SINGLE CRYSTALS AND IN SOLUTION

When the single crystal atomic coordinates of a protein are available, and the local magnetic fields of the individual components are known, the conformation-dependent proton NMR chemical shifts which would arise if the molecular conformations in the crystal and in solution were identical, can in principle

be computed. Comparison of the calculated chemical shifts with
the observed spectra then provides a criterion for the
investigation of the relations between the single crystal and
solution conformations. When a close correspondence between the
protein conformations in the crystal and in solution is evidenced
by the above procedures, some resonance lines can in favorable
cases be assigned to specific amino acid residues. These can
then be used as probes for conformational studies of well defined
regions of the molecule.

The procedures outlined here had been followed in the
interpretation of the data on cytochrome b_5 (Figs. 2 and 3) [2].
Only the ring current fields of the aromatic rings and the
pseudocontact shifts in the ferric protein had been considered
in this analysis. As can be seen from figure 5, the ring current
shifts and the pseudocontact shifts have opposite sign for all
the resonances in the high field region of the spectrum.
From comparison with the single crystal X-ray data, the high
field lines in ferrocytochrome b_5 were assigned to five methyl
groups, where four methyls are part of amino acids near the
heme, and the remaining one is located near a tryptophanyl
residue at a distance of approximately 17 Å from the heme iron.
These resonance assignments could then also be confirmed by
the observations in the oxidized protein (Fig. 5).

Even though it is of course of much help if one can rely
on the X-ray data for the resonance assignments, NMR studies
are by no means restricted to proteins which had previously
been studied by X-ray methods, and individual lines could in
various proteins also be identified without reference to the
atomic coordinates. Typical examples include comparative
studies of proteins differing in their primary structures
by single amino acid substitutions, and combination of the
NMR data with a variety of other chemical and biochemical
evidence.

NMR STUDIES OF THE ELECTRONIC STATES IN THE HEME GROUPS

From the foregoing sections it is quite evident that a
detailed knowledge of the electronic states in the heme groups
is essential for investigations of the molecular conformations
in paramagnetic hemoproteins. In addition, studies of the
relations between the electronic structures of the heme
groups and the biological roles of the hemoproteins have for
many years attracted a lot of interest. Therefore much effort
has also gone into NMR studies of the electronic states in iron
porphyrin complexes [1, 7-10]. In the following, some combined

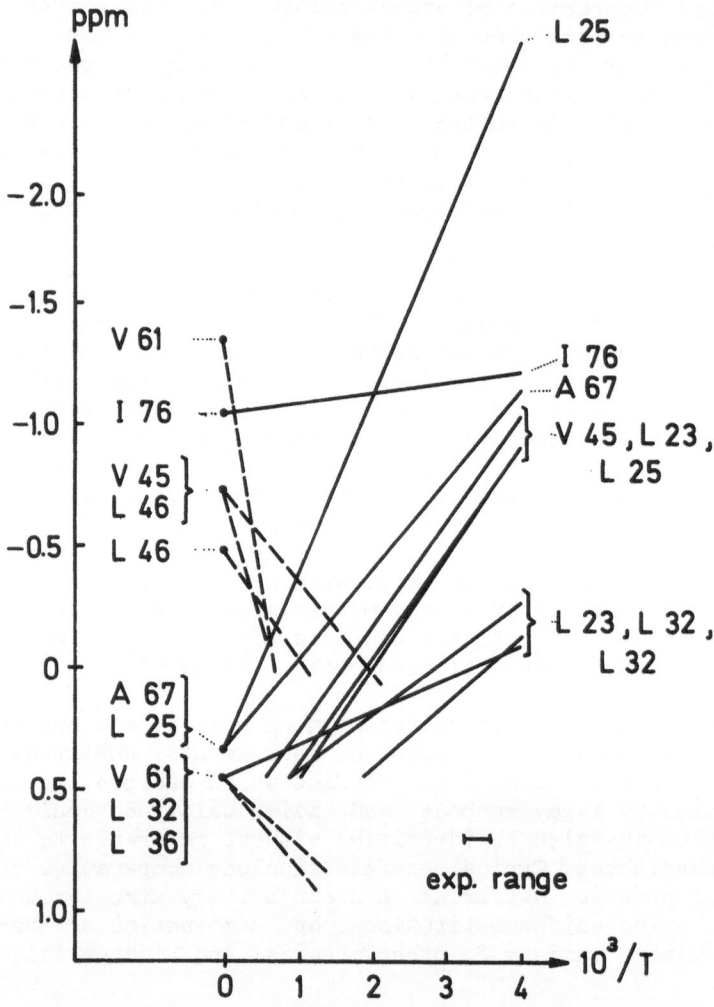

Fig. 5. Dependence on the reciprocal of temperature and
         assignments of the high field methyl resonances in
         ferric and ferrous cytochrome b_5 (V = valine, I =
         isoleucine, L = leucine, A = alanine). The circles
         on the left indicate the observed resonance positions
         in the reduced protein (Fig. 3). The broken lines indicate
         the dependence on the reciprocal of temperature for these
         resonances in the ferric protein, which were obtained
         from theoretical considerations. The solid lines show
         the extrapolation to 1/T = 0 of the measured temperature
         dependence of the high field methyl resonances in the
         oxidized protein.(From Keller and Wüthrich [2]).

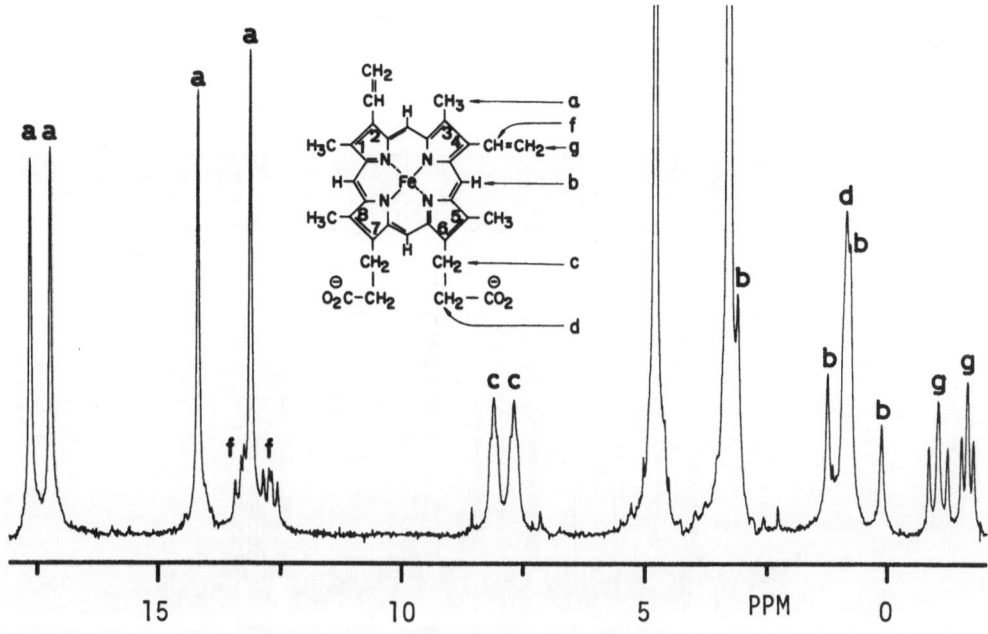

Fig. 6. Fourier transform proton NMR spectrum at 100 MHz of the
dicyanide complex of iron (III) protoporphyrin IX in CD_3OD.
T = 29°. The structure of the complex, where the
axial cyanide ligands have been omitted, and the resonance
assignments are also indicated. (From Wüthrich and
Baumann [11]).

^{1}H- and ^{13}C-NMR experiments will be discussed.

    The proton NMR spectrum of a paramagnetic low spin ferric
complex of protoporphyrin IX (Fig. 1) is shown in figure 6. Because
the proton resonances are well separated by the hyperfine shifts,
most of the carbon-13 resonances could be identified in a
heteronuclear proton-carbon-13 double resonance experiment with

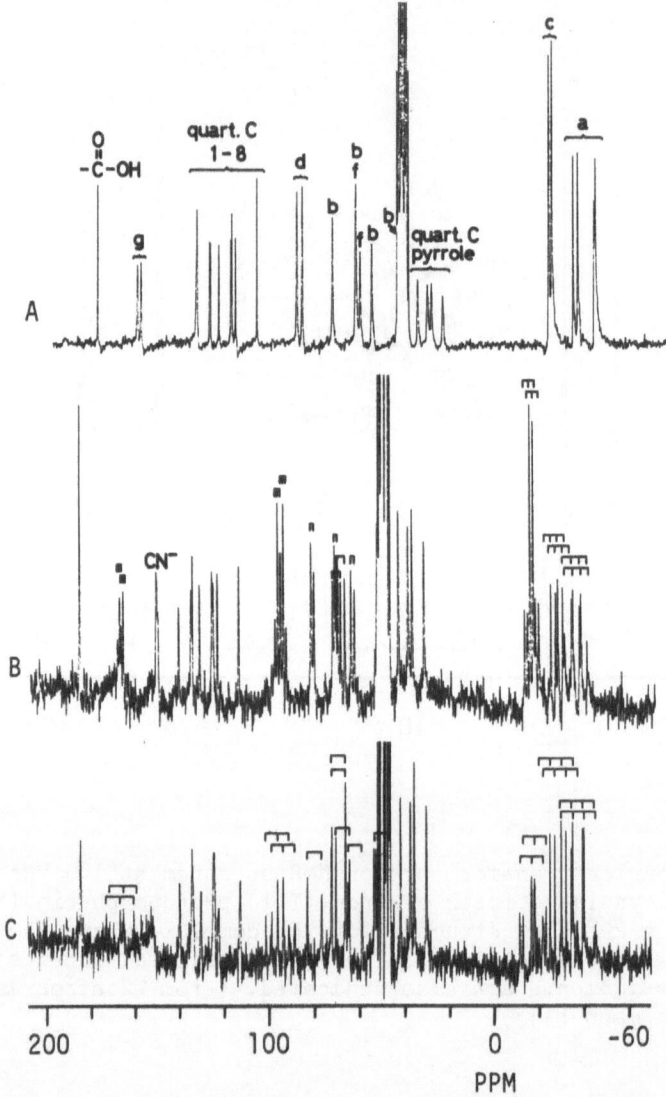

Fig. 7. Fourier transform ^{13}C-NMR spectra at 25.14 MHz of the dicyanide complex of iron (III) protoporphyrin IX in CD_3OD. T = 29°. The strong resonance at 49 ppm comes from the solvent. A. proton noise decoupled. B. off-resonance proton irradiation at -5 ppm (see Fig. 6). C. no proton irradiation. The resonance assignments (see Fig. 6 for the nomenclature) which resulted from these and additional experiments [11, 15], are also given. (From Wüthrich and Baumann [12]).

off-resonance proton irradiation at -5 ppm [11]. Three carbon-13 spectra are shown in figure 7. In the proton noise-decoupled spectrum 7 A, and in the spectrum 7 C, which was recorded without proton irradiation, all the resonances expected from the structure of protoporphyrin IX are observed. In the off-resonance experiment, figure 7 B, the different methylene and methine carbons are readily identified from the different residual spin-spin couplings. Similarly the ^{13}C resonances in the corresponding diamagnetic Zn^{2+} complex were identified [11], and the hyperfine shifts for the individual ^{1}H- and ^{13}C-resonances were obtained from the relative chemical shifts between the two spectra. With the formalism of equations (3) to (5), and using the Mc Connell [13] and Karplus-Frankel [14] relations between the contact coupling constant A and the spin densities localized on the aromatic ring carbon atoms, a quite detailed description of the spin density distribution in the heme groups can then be derived from the combined ^{1}H- and ^{13}C-NMR data [11, 15].

ACKNOWLEDGMENT

The author acknowledges financial support by the Schweizerischer Nationalfonds for the research projects Nr. 3.423.70 and 3.4231.70 in the field of NMR studies of biopolymers.

REFERENCES

1. K. WÜTHRICH, Structure and Bonding 8, 53 (1970)

2. R.M. KELLER and K. WÜTHRICH, Biochim.Biophys.Acta 285, 326 (1972)

3. C.C. McDONALD and W.D. PHILLIPS, Fine Structure of Proteins and Nucleic Acids, ed. by G.D. Fasman and S.N. Timasheff, 1, M. Dekker, New York (1970)

4. H.M. McCONNELL and R.E. ROBERTSON, J.Chem.Phys. 29, 1361 (1958)

5. R.J. KURLAND and B.R. McGARVEY, J.Magn.Res. 2, 286 (1970)

6. N. BLOEMBERGEN, J.Chem.Phys. 27, 595 (1957)

7. K. WÜTHRICH, R.G. SHULMAN, B.J. WYLUDA and W.S. CAUGHEY, Proc. Nat.Acad.Sci. US 62, 636 (1969)

8. R.G. SHULMAN, S.H. GLARUM and M. KARPLUS, J.Mol.Biol. 57, 93 (1971)

9. H.A.O. HILL and K.G. MORALLEE, J.Amer.Chem.Soc. 94, 731 (1972)

10. R.S. KURLAND, R.G. LITTLE, D.G. DAVIS and C. HO, Biochemistry 10, 2237 (1971)

11. K. WÜTHRICH and R. BAUMANN, Helv.Chim.Acta 57, 336 (1974)

12. K. WÜTHRICH and R. BAUMANN, Ann.New York Acad.Sci. 222, 709 (1973)

13. H.M. McCONNELL, J.Chem.Phys. 24, 764 (1956)

14. M. KARPLUS and G.K. FRAENKEL, J.Chem.Phys. 35, 1312 (1961)

15. K. WÜTHRICH and R. BAUMANN, Helv.Chim.Acta 56, 585 (1973)

HIGH RESOLUTION NMR INVESTIGATION OF NUCLEIC ACID STRUCTURES

D.R. Kearns

University of California

Riverside, California, 92502, U.S.A.

INTRODUCTION

During the past 10-15 years high resolution NMR has been widely used to investigate the structure of proteins in solution. Surprisingly, the use of this powerful technique in the investigation of polynucleotide structures is of much more recent origin [1 - 3]. Part of the delay is due to the fact that early attempts to use NMR in the study of high molecular weight (say 30,000) DNA and RNA samples were quite disappointing. The other problem arises from the fact that so many resonances are located in the same spectral regions that very high resolution spectrometers are required. The development of the 220 MHz and especially the 300 MHz spectrometers has helped to overcome some of these problems. The other breakthrough was the discovery by Kearns and Shulman of a set of especially downfield shifted resonances arising from the presence of Watson-Crick base pairs in double helices of RNA [4, 5]. These studies laid the ground work for using NMR to obtain structural information about RNA and DNA molecules in solution that cannot presently be obtained by any other technique. Before discussing the way in which NMR is used in these structural studies, I first want to briefly comment on some of the important roles which the various polynucleotides play in biochemical processes.

The central dogma of molecular biology was that DNA directs the synthesis of RNA which in turn directs the synthesis of proteins [6]. The code which is used for this process has been worked out and it is now known that a triplet of nucleic acid bases (G = guanine, A = adenine, C = cytosine, T = thymine and U = uridine in RNA, see structures in figure 1) in the DNA are

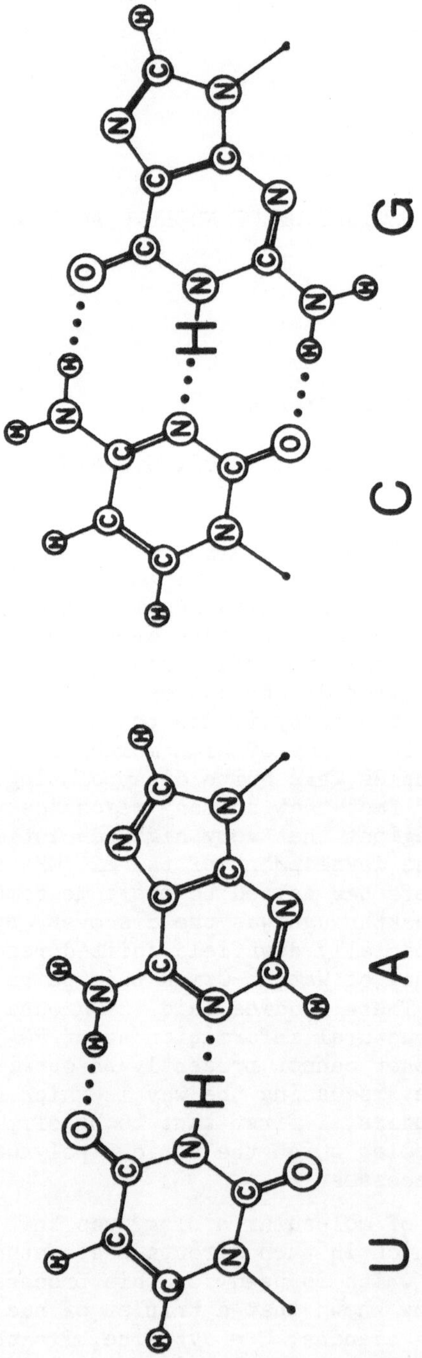

Fig. 1. Schematic drawing showing the four common bases in the typical Watson-Crick base paired structures. Uracil differs from thymine in that a hydrogen atom is replaced by a methyl group.

associated with every amino acid. DNA is the repository of genetic
information in a cell, and this is "read out" first by synthesis
of low molecular weight complimentary copies of appropriate portions
of the DNA in the form of "messenger RNA". These "messenger RNA"
molecules must now be translated from a nucleic acid language
into an amino acid language and it is the transfer RNA (tRNA)
molecules which are involved in this important process. The
tRNA molecules (molecular weights typically around 28,000) must
"read the code words" contained in the messenger RNA and introduce
the appropriate amino acids in the polypeptide sequence as
required. The "factories" where this translation takes place
are called ribosomes (large assemblies of proteins and polynucleotides) and there is considerable interest now in determining
how the components of the ribosome are actually arranged.

The work which I shall describe deals primarily with the use
of high resolution NMR in the determination of the structure
of tRNA molecules in solution. This is logical since most of our
information to date has been obtained with these molecules.
Furthermore, the techniques developed in the tRNA studies can,
and have been used in the study of other RNA molecules. Studies
of the conformation of natural and synthetic DNAs are only
just beginning and we briefly discuss the way in which we believe
NMR can be used to provide very precise structural data on
these important molecules.

THE ASSIGNMENT PROBLEM

To derive structural information from proton NMR spectra we
must first assign resonances to the various types of protons
in the molecule, and then be able to understand the way in
which certain structural features affect the positions of the
resonances. We first consider the gross assignment of tRNA
spectra and then show how resonances observed in the low field
region (11 to 15 ppm downfield* from DSS) can be used to provide
the desired structural information.

The complete NMR spectrum of tRNA$_{\text{yeast}}^{\text{Phe}}$ in H_2O from 6 to 15 ppm
is presented in figure 2. It is convenient to divide the spectrum
into the following different regions:

---

* All positions in parts per million (ppm) downfield from
  the usual standard DSS.

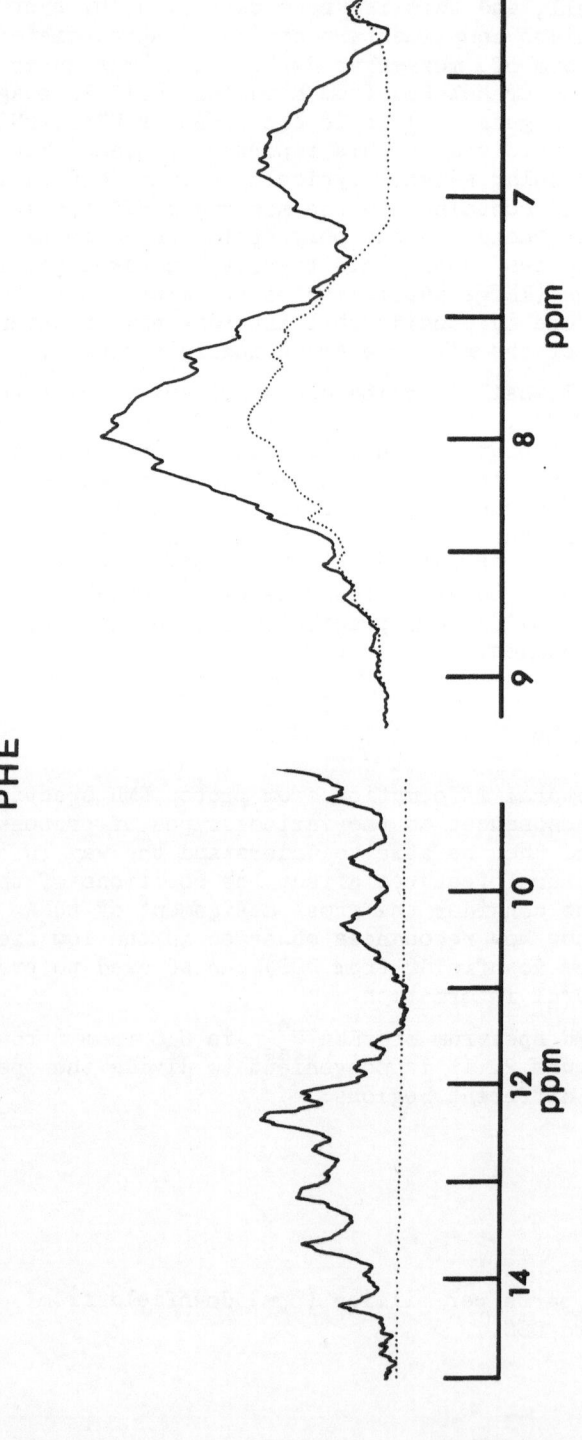

PHE

Fig. 2. Proton NMR spectrum at 220 MHz of a solution of tRNA$^{Phe}_{yeast}$. Solid lines are resonances obtained with H_2O solvent, dotted lines with D_2O. (Temp. 35 C, concentration 2 mM, NaCl 0.1 M, Mg^{2+} 5 mM, pH 5.0).

Region   I : 6 to 9 ppm
Region   II : 11 to 15 ppm
Region III : 9 to 11 ppm

1) Region I (6 to 9 ppm): NMR studies of mononucleotides suggest
   that at least three different types of resonances might be
   observed in this region: the non-exchangeable protons
   associated with the bases, i.e. adenine H-8 and H-2,
   guanine H-8, and pyrimidine H-6, the free amino protons
   associated with non-paired bases, and the hydrogen bonded
   amino protons in base pairs. The exchangeable protons are
   easily distinguished from the non-exchangeable protons by
   comparing the spectra obtained in D_2O with those obtained in
   H_2O; these results are also shown in figure 2. From a
   comparison of these spectra with an external standard the
   total number of non-exchangeable protons in this region was
   found to be 89 $\pm$ 4 and this agrees perfectly with the value
   of 89 calculated from the known base composition of
   $tRNA^{Phe}_{yeast}$ [5].

   The number of exchangeable proton resonances located in the
   6 to 9 ppm region can be obtained as the difference between
   the H_2O and D_2O spectra and this comparison (shown in figure
   2) indicates that there are two classes of exchangeable
   protons in the 6 to 9 ppm region; these are the class
   falling above and the class falling below 7.4 ppm, the point
   at which the spectra in H_2O and D_2O coincide.

   From their positions and pH dependence, resonances observed
   between 6.3 to 7.4 ppm are assigned to the free amino
   protons of non-paired adenine, guanine and cytosine residues.
   This assignment is based on the fact that free mononucleotides
   exhibit similar resonances.

   The exchangeable resonances in tRNA between 7.4 and 9.0 ppm
   are assigned to hydrogen bonded amino protons on the basis
   of the resonance positions in model systems. This assignment
   is consistent with our finding that their exchange rates are
   less pH sensitive since hydrogen bonding generally slows down
   the rate of exchange with H_2O.

2) Region II (11 to 15 ppm): Resonances in this region are ascribed
   to the hydrogen bonded ring NH protons and each proton detected
   corresponds to one base pair since each Watson-Crick base pair
   contains just one ring NH proton, i.e. U_3H or G_1H. This
   assignment is also supported by comparison with the resonance
   positions observed in model compounds. In the model compounds
   no other resonances are observed near the 11 to 15 ppm region
   [7]. In contrast to the free amino protons, ring NH protons
   which are not hydrogen bonded or otherwise "protected" are
   not observed in the model compounds presumably due to their
   very rapid exchange with water. The number of protons

contributing to the low field region has been determined in two
ways. First, several intensity measurements at 220 MHz were
made and compared with the lowest field methyl resonance
in metcyanomyoglobin and this gave 21 ± 3 protons in the
11.5 - 15 ppm region [5]. Internal intensity comparisons
based upon assigning the resolved resonance at 14.4 ppm, an
intensity corresponding to one proton, and/or the slightly
split resonances at 13.8 and 13.7 ppm, an intensity of three
protons, indicated that there are 18 ± 1 protons in the 11
to 15 ppm (including the line at 11.55 ppm).

3) Region III (9.3 to 11.0 ppm): The resonances in this region
(typically 4.5) are the most difficult to assign. It is
worth mentioning, however, that resonances in this region
broaden and disappear under conditions which barely affect
the resonances between 11.0 and 15.0 ppm. These and other
results suggest that the resonances between 9.3 to 11.0 ppm
are related to the three dimensional structure of tRNA. One
possibility is that these resonances are due to ring NH
protons which are associated with non-standard base pairs.
Another possibility is that these resonances are associated
with ring NH protons of U and G residues which, although not
base paired, are in some special "protected" environment where
the proton exchange rate is decreased to the extent that a
resonance can be detected. In addition to the sensitivity
of these resonances to both temperature and pH, this
assignment is consistent with the model system studies which
demonstrate that the U_3H proton resonance shifts from 8 ppm
in $CHCl_3$ to 11.5 ppm in DMSO, and all the way to 14.5 ppm in
a Watson-Crick base pair in aqueous solution.

DETAILED INTERPRETATION OF THE LOW FIELD NMR SPECTRA OF tRNA

     The high resolution low field NMR spectrum of tRNA clearly
demonstrates that all AU and all GC base pairs do not give
resonances at the same position in the spectrum. Obviously
there are important sequence effects on the positions of the
resonances from each base pair. After considering various
factors which might be responsible for the variety of chemical
shifts exhibited by the different AU and GC base pairs, we have
concluded that the major contributions come from nearest neighbour
ring current effects. Consequently, once the secondary structure
(complete specification of the base paired helical regions) of a
polynucleotide is known, all that is needed in order to predict
its low field NMR spectrum is:

1) the location of the low field resonances for standard
   (unshifted by nearest neighbour ring current effects) AU
   and GC base pairs, and

2) the magnitude of the ring current shifts which the four
   different bases exert on the protons on a neighbouring
   base pair.

The locations of the "standard base pairs" have been deduced
from studies of various model systems and different purified
tRNAs, and on the basis of this work we have concluded that
$(AU)^\circ = 14.5$ ppm, $(CG)^\circ = 13.5$ ppm, and $(A\Psi)^\circ = 13.5$ ppm [8-10].

The problem of determining the nearest neighbour ring current
effects was largely solved by using recent ring current calculations
of Giessner-Prettre and Pullman [11] in conjunction with the
RNA structural diagrams of Arnott [12] . Figure 3 shows the
relative positions of two adjacent base pairs in a standard
12-fold right handed RNA helix, as viewed down the helix axis.
In this particular diagram the hydrogen bonded ring nitrogen
proton of the Pu_1-Py_1* base pair is directly over a portion
of the purine base Pu_2, but it is rather far from the
pyrimidine Py_2. This same diagram also indicates that the
hydrogen bonded ring nitrogen proton from the Pu_2-Py_2 base
pair is directly under the Pu_1 but rather far removed from Py_1.
From these and similar diagrams it is an easy matter to determine
the location of a specific hydrogen bonded ring nitrogen proton
with respect to the bases of adjacent base pairs. The theoretical
maps of Giessner-Prettre and Pullman shown in figure 4 can then
be used to compute the magnitudes of the ring current shifts
exerted on each proton by neighbouring bases. A tabulation of
these results is presented in table 1 for all possible combinations
of base pairs and neighbours. As an example, consider the ring
current shifts exerted on the AU base pair sandwiched between the
two GC base pairs in the structure

```
          5' 3'
          GC
          AU
          GC
          3' 5'
```

From table 1 we see that the low field AU resonance would be
shifted 0.2 ppm by the C above it and 0.6 ppm by the G below it,

---

* Pu_1-Py_1: $purine_1$-$pyrimidine_1$

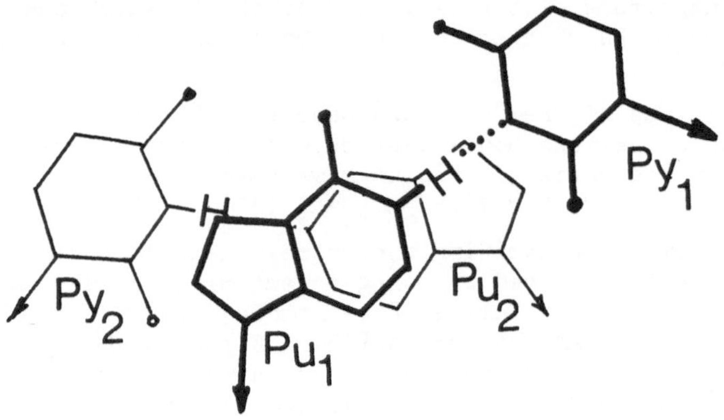

Fig. 3. Schematic diagram showing the overlap of bases when a
         pyrimidine follows a purine on the same chain (from S.
         Arnott [12]).

resulting in a displacement of the AU resonance from 14.5 ppm, its
unshifted position, to 13.7 ppm. In an entirely similar fashion,
it is possible to use these data to predict the low field NMR
spectrum for any polynucleotide, provided secondary structure
is known.

     Of the 60 tRNA which have been sequenced to date [13] all
of these can be put into universally accepted cloverleaf
structure shown in figure 5 for tRNA^{Val}* from yeast. This
structure has a set of four helical arms, a CCA terminal region
and four looped out regions. These are believed to be features
common to all tRNA molecules. By straightforward application of
the ring current shift rules to yeast tRNA^{Val} we obtain the
results shown in table 2 and figure 5. The agreement between the
calculated and observed spectrum is sufficiently good that we are

_____

* The specific amino acid for which a tRNA codes for is indicated
  by a superscript. Hence, tRNA^{Val} is responsible for reading
  the code word for valine (Val) whenever it appears in a message.

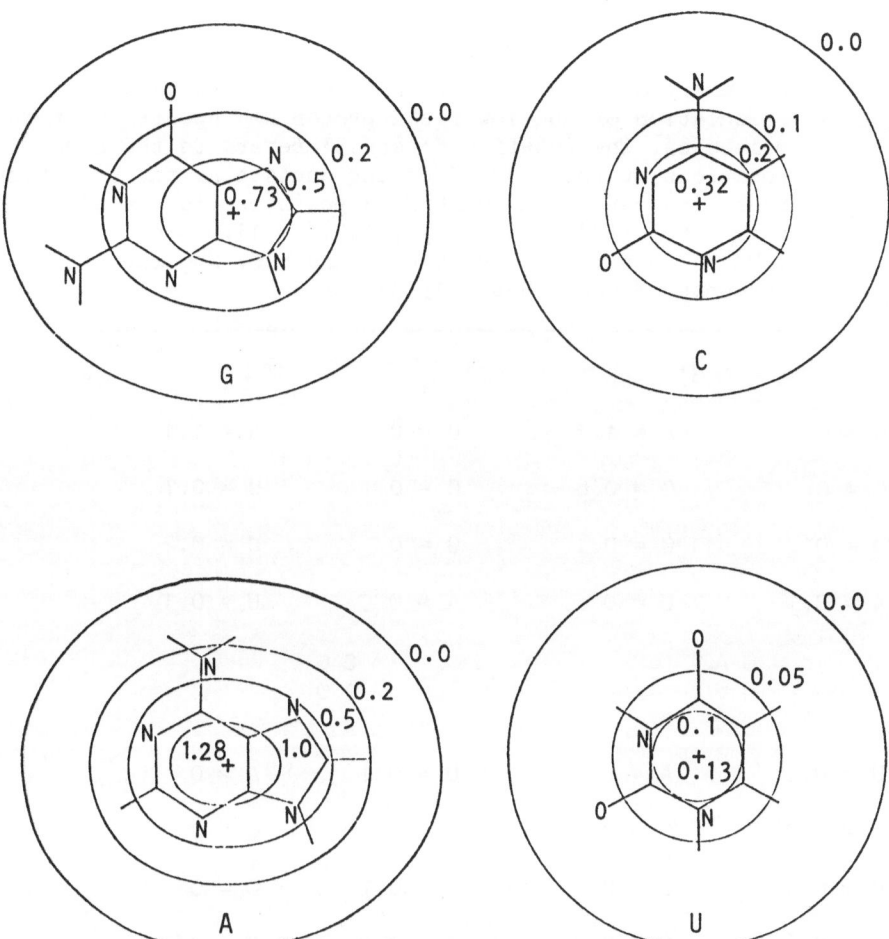

Fig. 4. Predicted ring current shifts (in part per million) due
to A, U, G or C in a plane 3.4 Å distant from the
molecular surface (from Giessner–Prettre and Pullman,
1970).

Table 1. A summary of ring current shift parameters used in the calculation of the low field proton NMR spectra of tRNA molecules. The notation 5' and 3' refers to the sugar positions at the ends of chains containing the neighbouring bases indicated. The unshifted positions for usual base pairs are: $(AU)^\circ$ = 14.5 ppm, $(GC)^\circ$ = 13.6 ppm and $(A\Psi)^\circ$ = 13.5 ppm. (D.R. Kearns and R.G. Shulman, Accounts of Chem. Res., 1973).

5'	3'	5'	3'
U = 0	A = 1.1	U = 0	A = 1.1
C = 0	G = 0.6	C = 0	G = 0.7
G = 0	C = 0.1	G = 0	C = 0.2
A = 0.1	U = 0	A = 0	U = 0.1
	U A		C G
	A U		G C
U = 0.1	A = 0	U = 0.1	A = 0
C = 0.2	G = 0	C = 0.25	G = 0
G = 0.6	C = 0	G = 0.7	C = 0
A = 0.6	U = 0	A = 1.0	U = 0
3'	5'	3'	5'

BAKERS' YEAST

$tRNA_1^{val}$

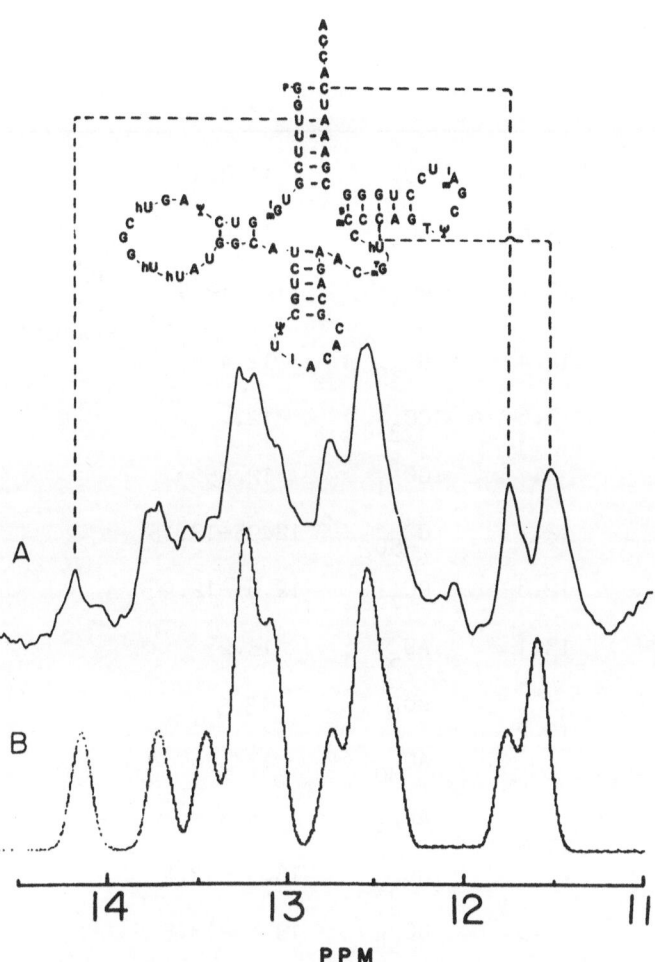

Fig. 5. Comparison of the observed and computed low field NMR
spectrum of yeast $tRNA_1^{Val}$.
A) 300 MHz proton NMR spectrum of $tRNA_1^{Val}$ at 25 C in
a buffered solution containing 0.1 $\underline{M}$ NaCl, but no
Mg^{++}.
B) Calculated spectrum based on ring current shift
predictions.

Table 2. Predicted and observed values of chemical shifts for $tRNA_I^{Val}$ base pairs.

Peak	Intensity	τ_{obs}	Assignment	τ_{calc}	Remarks
A	2	11.5	GC_6	11.8	
		11.6	GC_{52}	11.9	
A'	1	11.9	GC_1	11.9–12.9	A_{74} stacked on CCA helix. G_2 not looped out. Must be stacked in helix.
		12.4	GC_{29}	12.4	
		12.6	GC_{31}	12.4	
B	5		GC_{51}	12.65	
		12.7	GC_{50}	12.95–13.15	
		12.8	GC_7	12.8 –12.9	
		13.1	AU_5	13.2	
			AU_4	13.3	
		13.2			
C	5		AU_{30}	13.3	
			AU_{53}	13.3	
		13.3	GC_{32}	13.3 –13.4	
		13.5	GC_{54}	13.4 –13.5	
D	2	13.7	AU_{28}	13.2 –14.3	AU_{45} partially stacked.
E	<1	14.3	AU_3	14.4	

convinced that ring current effects are the major factor
responsible for shifting of the low field AU and GC resonances,

TEMPERATURE DEPENDENCE OF LOW FIELD NMR

1. Simple Hairpin Fragments - fMet

   The temperature dependence of the low field NMR spectrum
provides valuable information about the stability of the double
helical structure and the kinetics of the helix-to-coil transition.
As an example, we have studied the temperature dependence of the
anticodon hairpin fragment from E. Coli tRNA^{fMet} and these
results are shown in figure 6 [14]. The assignment of the spectrum
is given in table 3. As the temperature is raised from 17 to 42 C
the resonance assigned to AU_{28} (12.8 ppm) broadens and then
disappears. In contrast to the behaviour of AU_{28} the resonances
from the four GC base pairs are virtually unaffected by
temperature up to about 50 C where some broadening of the
resonances from GC_{29} (11.9 ppm) and GC_{32} (12.9 ppm) becomes
evident. At 58 C, GC_{29} is severely broadened, and GC_{32} is
further broadened but the resonances from GC_{31} (12.8 ppm) and
GC_{30} (12.3 ppm) are still largely unaffected. Finally, when
the temperature is raised to 71 C,which is within 7 C of the T_m
found optically, all resonances are lost.

   The fact that the resonances disappear by broadening rather
than by simply diminishing in intensity, suggests that melting
in these NMR experiments occurs when the lifetime of a proton
in a base pair hydrogen bond is reduced to the millisecond time
range. Theoretically, the linewidth and the lifetime are related
by the expression

$$\tau = 1/\pi \nu_{1/2}$$

so that a linewidth of 150 Hz (0.5 ppm at 300 MHz) would correspond
to a lifetime of a base pair with respect to opening of about
2 msec. Since the NMR melting appears to depend only on the rate
constant for the opening of base pair in the helix, whereas the
"true" melting point depends upon the ratio of the rate constants
for opening and closing the helix, the T_m measured by NMR will not
necessarily correspond to the T_m determined optically. From
the optical melting data it appears that T_m = 78 C, or about 7 C
higher than the NMR T_m.

   In light of the above considerations we suggest the following
interpretation for our observations. The early "melting" of AU_{28}
can undoubtedly be attributed to the fact that AU base pairs
are weaker than GC pairs, and that terminal base pairs are much

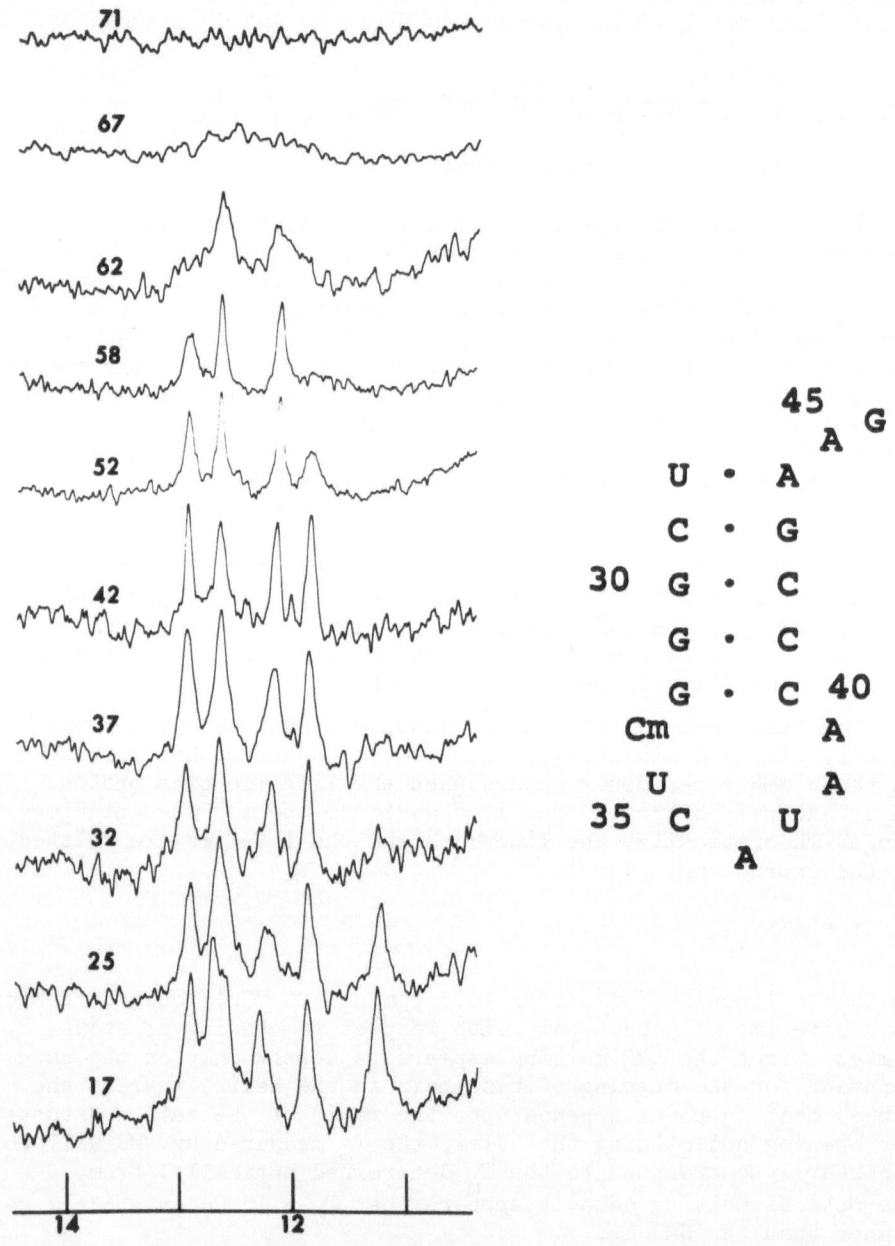

Fig. 6. 300 MHz proton NMR spectrum of the anticodon hairpin
        (structure shown in figure) of E. Coli tRNA^{fMet} at
        various temperatures.

Table 3. Comparison of the observed and calculated resonance
positions for the low field protons of the anticodon
hairpin from E. Coli tRNA^{fMet}. The NMR T_m obtained
experimentally for each resonance is also shown in this
table.

Base pair	Ring current predicted resonance position	Observed resonance position	Error	NMR T_m
ppm	ppm	ppm	ppm	C
AU$_{28}$	13.2	12.8	+ 0.40	$\sim$ 42
GC$_{29}$	11.8	11.9	− 0.10	52–58
GC$_{30}$	12.2	12.3	− 0.10	67
GC$_{31}$	12.65	12.75	− 0.10	67
GC$_{32}$	13.15	12.9	+ 0.25	62
U$_{34}$ or U$_{37}$	-	11.3	-	30

more susceptible to opening than are interior base pairs. Despite
the fact that AU$_{28}$ is no longer locked in a base paired state,
the NMR spectra clearly indicate that A$_{44}$ is still stacked on G$_{43}$
up to at least 52 C. The fact that the position of this resonance
GC$_{29}$ is unchanged from 17 to 52 C requires that A$_{44}$ remain stacked
on G$_{43}$ up to 52 C. At 52 C, the resonance from GC$_{29}$ (11.9 ppm)
is beginning to broaden and by 58 C it has broadened beyond
recognition . Since resonances from the other GC base pairs remain
sharp we conclude the stacking of A$_{44}$ and G$_{43}$ is lost between
52 and 58 C.

Although the resonance from the other terminal base pair,
GC$_{32}$ (12.9 ppm) is beginning to broaden at 52 C the resonances
from the interior GC base pairs (GC$_{30}$ and GC$_{31}$) remain sharp.
At 62 C the resonances from GC$_{32}$ and GC$_{29}$ have broadened beyond
recognition and now the resonances from GC$_{30}$ and GC$_{31}$ are

noticeably broadened. When the temperature is raised to 67 C even
these last two resonances have almost disappeared. We are thus
able to assign an NMR T_m to opening up of each individual base
pair in the fragment and these are also presented in table 3.
While these NMR T_m do not have the same significance as the
optical T_m they nevertheless provide an approximate measure
of the stability of different base pairs in a helix and should
provide a clear indication of the way in which the melting of
the hairpin helix melts.

The fact that the positions of the resonances from the
four GC base pairs are independent of temperature from 17
to almost 58 C permits us to draw the further conclusion that
the geometry of the double helix is independent of temperature
over this range. This has interesting implications with regard
to melting experiments.

So far we have not commented on the assignment of the
resonance at 11.3 ppm. The fact that this resonance disappears
upon heating to only 42 C suggests it arises from a slowly
exchanging ring nitrogen proton of a G or U residue. We believe
that the exposed G_{46} is not a likely candidate. However, U_{34} and
U_{37} which are located in the anticodon loop could be subjected
to protection by other bases in the loop and this suggests that
one of these U residues is responsible for the unassigned
resonance. If this latter possibility is correct, we should
expect to see similar resonances in NMR spectra of other tRNA
species, and we do.

2. NMR Melting of an Intact tRNA Molecule – Yeast tRNA^{Leu}

The temperature dependence of the NMR spectrum provides an
independent method for testing models of tRNA structure. This
follows from the fact that if the helical portions of the
molecule melts sequentially over a range of temperatures, rather
than cooperatively over a relatively limited temperature range,
then groups of resonances associated with a particular helix (or
perhaps a pair of helices) should simultaneously disappear from
the spectrum and the model should be able to account for their
loss.

The temperature dependence of the denatured conformer tRNA^{Leu}
spectrum is shown in figure 7 and it is evident from these data
that the loss of resonances occurs in a very non-cooperative
fashion [15].

Comparison of the 25 and 35 C spectra indicates some loss
of intensity at 14.2 ppm and at 13.5 ppm and by comparison of
the 25 and 45 C spectra indicates a slight loss of intensity
at 13.5 ppm and by 52 C there are rather substantial losses in
intensity. The comparison of the 35 and 55 C spectra shown in

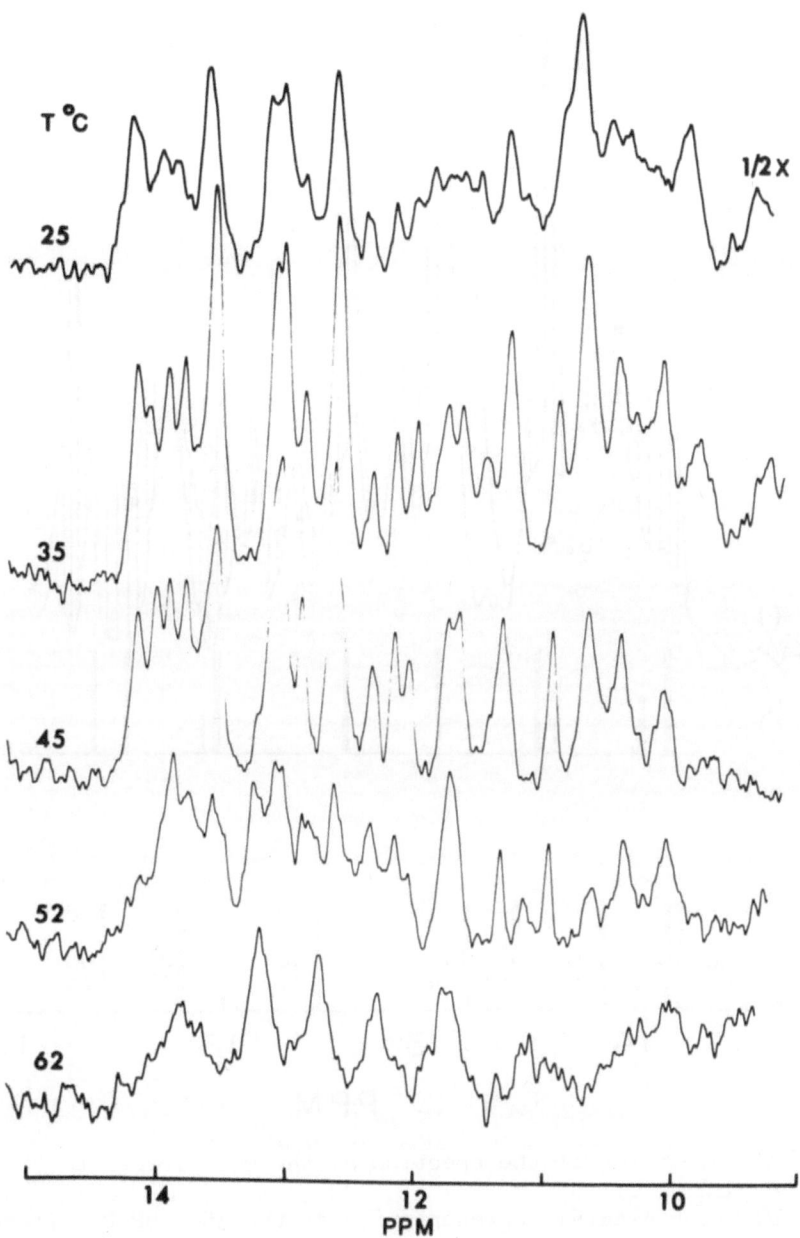

Fig. 7. Temperature dependence of the 300 MHz proton NMR spectrum of the denatured conformer of yeast tRNA$_3^{Leu}$.

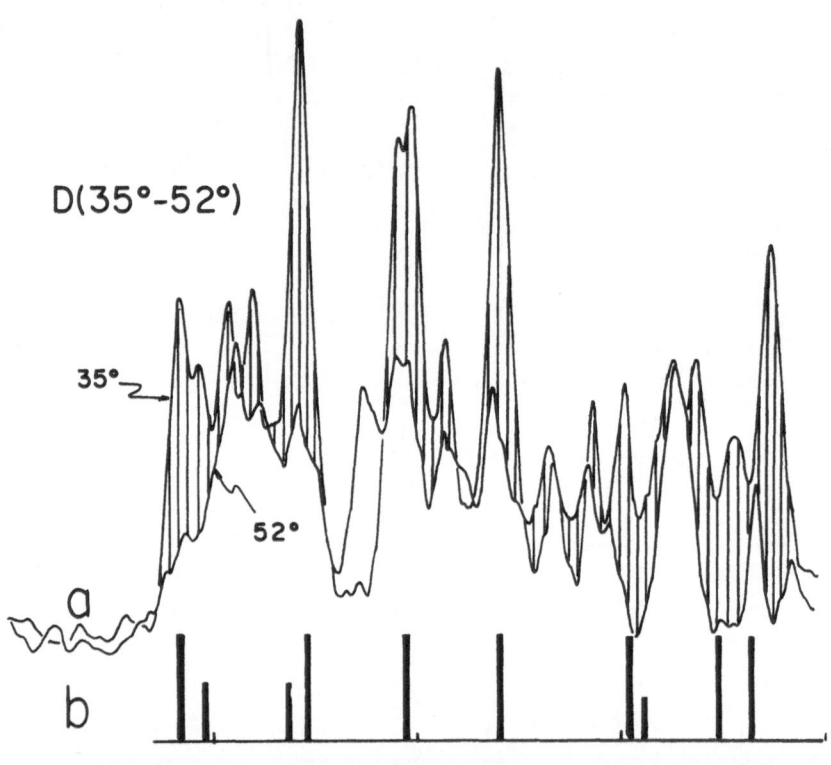

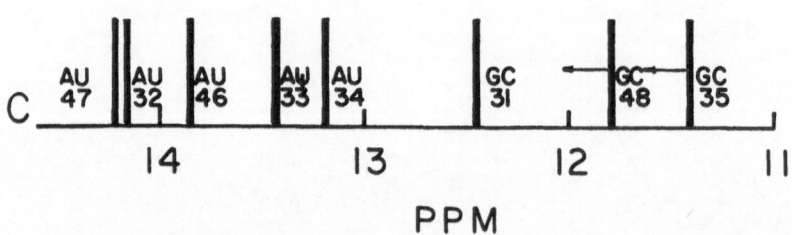

Fig. 8. a) Comparison of the spectrum of the D-conformer at 35
           and 52 C.
        b) Stick diagram representation of the 35 – 52 C difference
           spectrum.
        c) Predicted difference spectrum assuming the loss of the
           minor stem (H_{14}) and H_{27} from the model shown in fig. 10.

figure 8 graphically illustrates the changes   which occur over
this temperature range. An analogous comparison of the 52 and 62 C
spectra is shown in figure 9. Integration of the spectrum
indicates a loss of approximately 9 resonances at 52 C. In
figure 8 the observed intensity losses are compared with
those which would be predicted for the loss of the minor stem
(H_{14}) and the helix (H_{22}) which is unique to the denatured
conformer (see figure 10). According to our model these two
short helices are expected to be the least stable ones and we
find that their predicted NMR spectra agrees reasonably well
with the set of resonances lost when the sample is heated to
52 C. Heating the D-conformer from 52 to 62 C results in a
further reduction in intensity corresponding to a loss of ~ 4
more resonances. Comparison of the loss spectrum with the NMR
spectrum predicted for the TΨC stem (which contains only 4
base pairs in the model) shows that there is good agreement.
At 62 C temperature the total intensity remaining in the
NMR spectrum corresponds to approximately 5 - 6 protons per
molecule and 5 distinct peaks are observed. The amino acid
acceptor stem satisfactorily accounts for the observed resonances.
The sequential melting of the D-conformer involves then first
loss of the minor stem (H_{14}) and the new stem (H_{22}) followed
by the TΨC stem (H_{29}). The amino acid acceptor stem (H_{33}) remains
to the last.

The fact that we can give a reasonably good account of
the temperature dependence of the D-conformer provides an
additional piece of evidence supporting the proposed model [15, 16].

PARAMAGNETIC RARE EARTH PROBES OF tRNA STRUCTURE

The determination of the structure of tRNA molecules in
solution is an important problem because of the key role which
these molecules play in the translation of messenger RNA and
in the control of cell metabolism [17, 18]. We have discussed
the way in which high resolution NMR can be used to determine
the base pairing structure of tRNA molecules in solution. We
now want to discuss the way in which paramagnetic rare earth
ions can be used to probe the tertiary structure of tRNA
molecules in solution. In the course of this study we also
obtain information on stoichiometry of the metal binding, the
nature of the binding sites, and the order of metal addition
to these sites. The binding data introduce important constraints
on the relative orientation of the CCA stem and the DHU stem and
these new results, taken in conjunction with our other NMR data,
establish much of the three-dimensional structure of this molecule
in solution. The example which we shall discuss is yeast tRNA[Phe]
and the low field spectrum of this molecule is shown along with
the cloverleaf structure in figure 11 and the assignments are

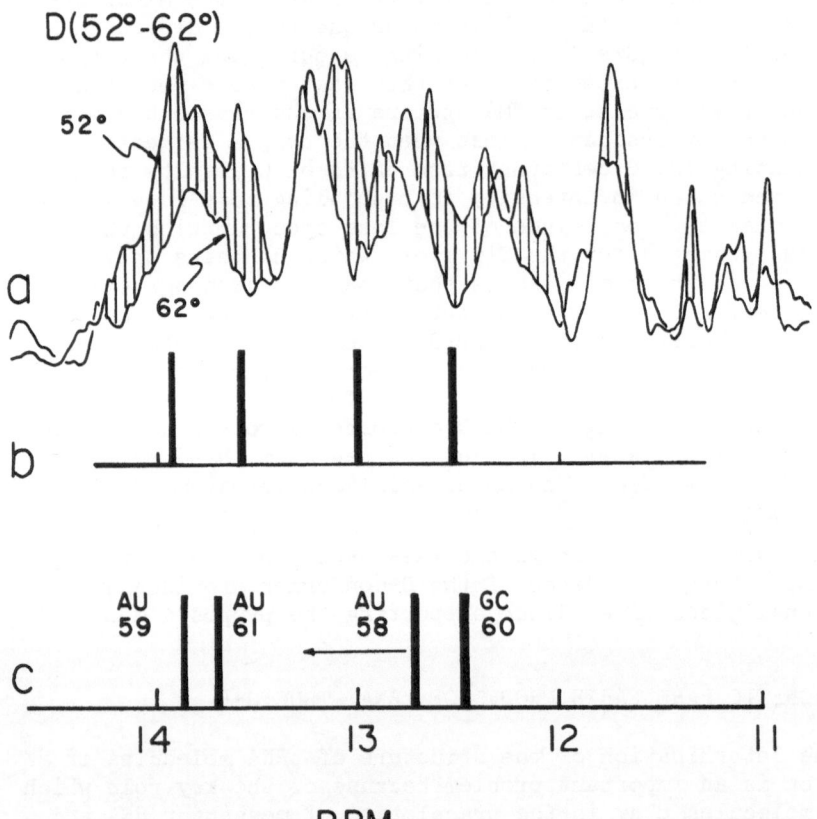

Fig. 9. a) Comparison of the spectrum of the D-conformer at 52 and
           62 C.
        b) Stick diagram representation of the 52 - 62 C
           difference spectrum
        c) Predicted difference spectrum assuming the loss of the
           TΨC stem. The position of AU_{58} is uncertain due to end
           effects.

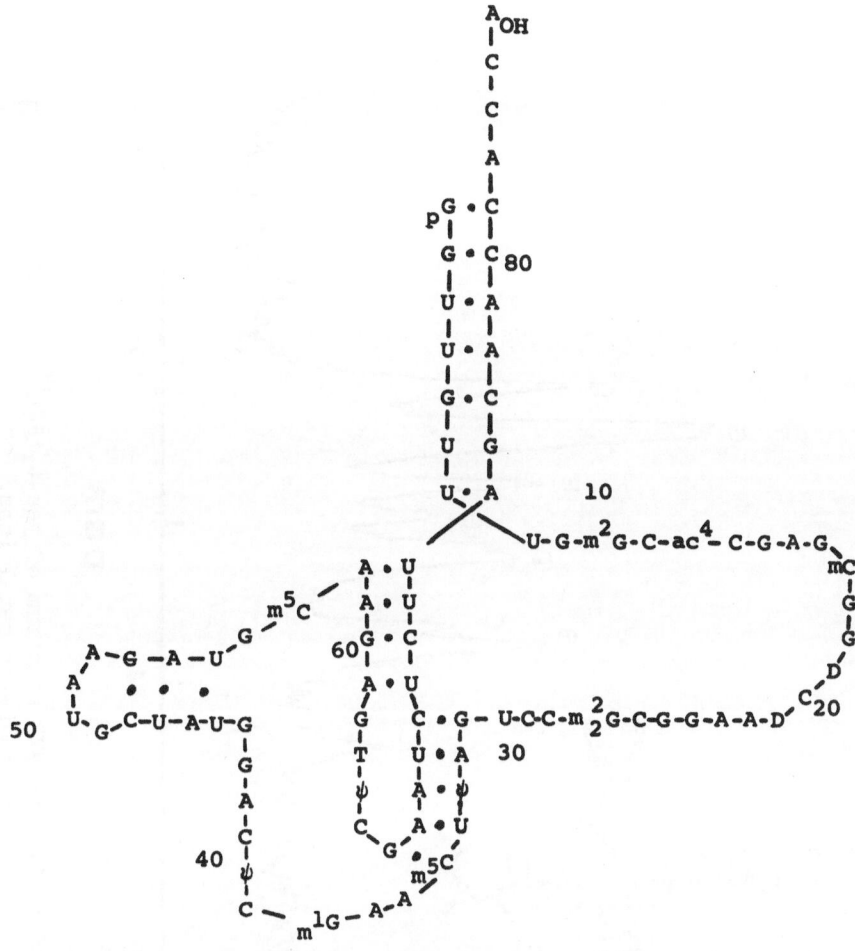

Fig. 10. Proposed model for the secondary structure of the denatured conformer of yeast tRNA$_3^{Leu}$.

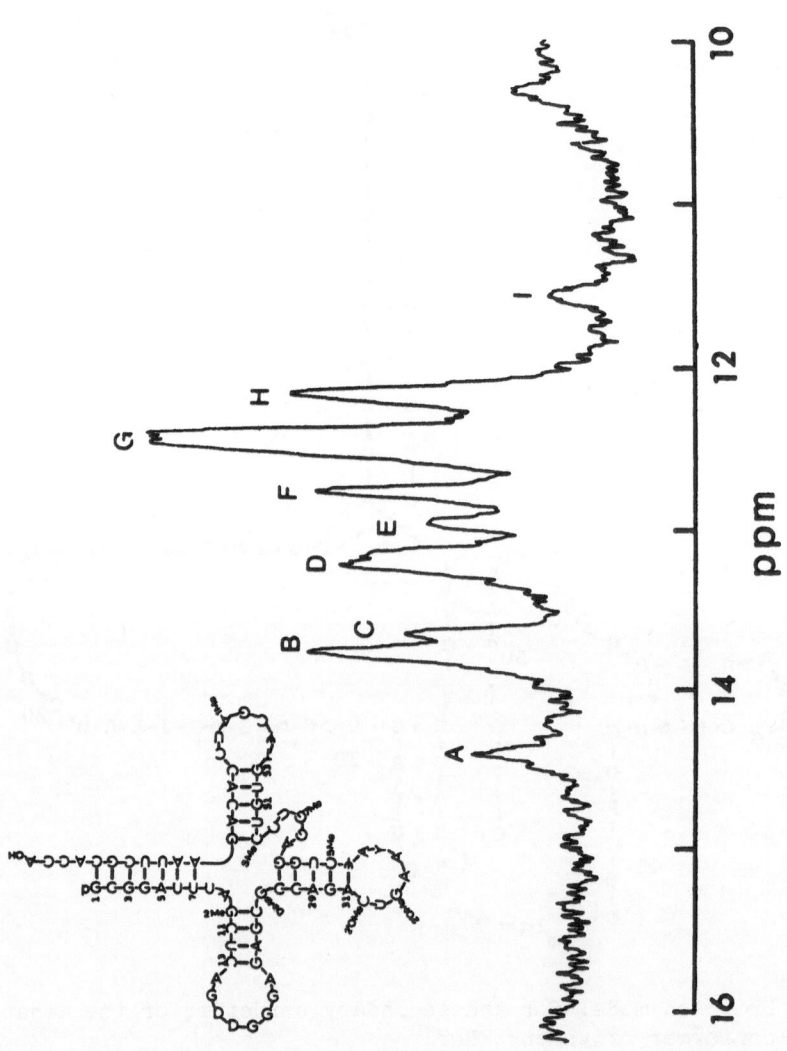

Fig. 11. The NMR spectrum of yeast tRNA^{Phe} in the absence of paramagnetic metal ions.

Table 4. The assignments of the low field resonances in yeast
   tRNA^{Phe} [3].

Resonance (Fig. 1)	Intensity	Position	Assignment [3, 5]
		ppm	
A	1.7	14.4	AU # 5, 6
B } C	3	13.8 } 13.7	AU # 7, 12, 52
D	3	13.3	AU # 29, 50
			AΨ # 31
E	1	13.0	GC # 11
F	2	12.8	GC # 2, 10
G	6	12.5 ± 2	GC # 1, 3, 30
			49, 51, 53
H	2	12.2	GC # 27, 28
I	1	11.6	GC # 13

given in table 4. When Eu^{3+} is added to tRNA^{Phe}, many resonances
in the low field spectrum are progressively shifted (Fig. 12)
[19]. (The fact that addition of Pr^{3+}, instead of Eu^{3+},
produced shifts of the resonances in the opposite direction
proves that the shifts are of paramagnetic origin). One of
the first peaks to be affected by the addition of the Eu^{3+} is
peak A (at 14.4 ppm) which on successive additions of Eu^{3+} is
broadened and upfield shifted until it merges with peak B, C
(at ∿ 2 Eu/tRNA). The fact that this resonance shifts gradually
with the addition of the Eu^{3+} indicates that the rate of
exchange of the Eu^{3+} among the strong binding sites is fast
on the NMR time scale ($\tau \leq 15$ msec in these particular
experiments) [20].

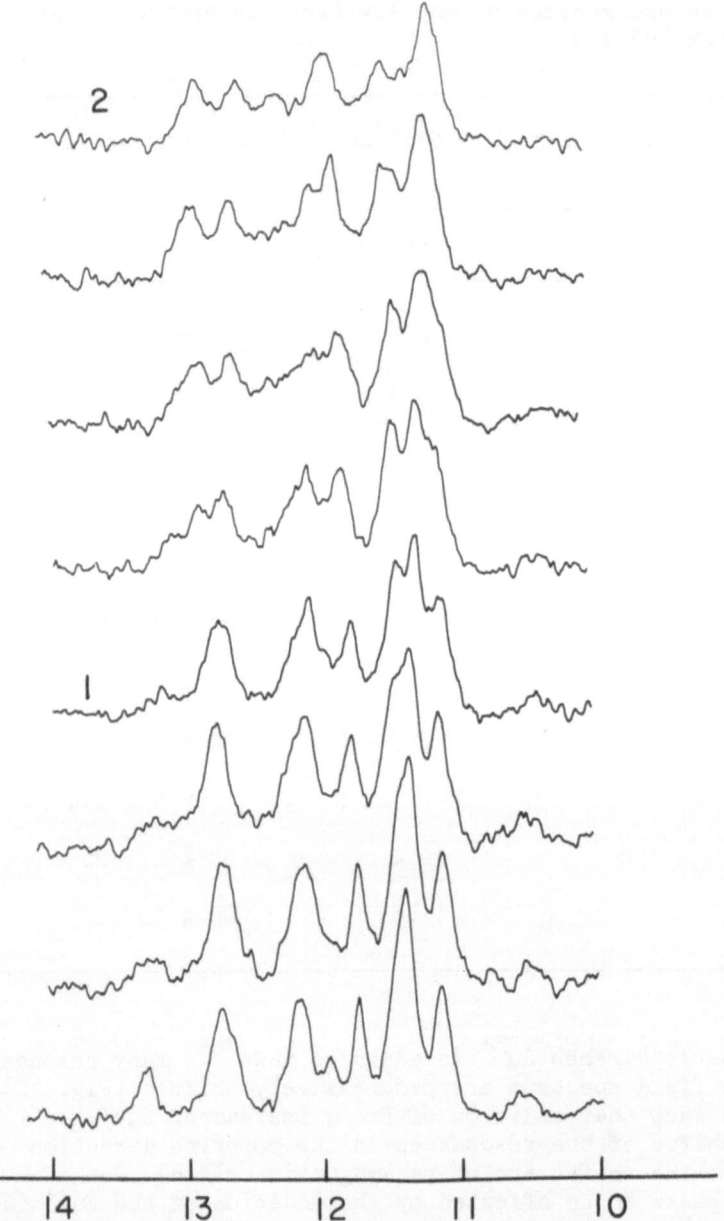

Fig. 12. The NMR spectra of tRNA^{Phe} at 40 C with various amounts
of added Eu^{3+} ion (Eu/tRNA ratio as indicated to the
left of the spectra).

If the exchange rates had been slow then peak A would have split
into two peaks and the fully shifted one would have grown in
intensity at the expense of the unshifted one as more Eu^{3+} was
added.

The spectra shown in figure 12 also provide information about
the number and the relative strengths of the various Eu^{3+} binding
sites. Since most of the shifting seems to be complete by the
time 4 - 5 Eu^{3+} ions per tRNA have been added, we conclude that
there are at least four strong binding sites for Eu^{3+}. However,
the fact that peak A shifts before some of the other peaks in
the spectrum shift, indicates that the four binding sites are
not of equal strength. Obviously the binding site which is
located near base pair AU_6 (peak A) is somewhat stronger than
some of the other sites and this is supported by the results
of optical studies which also show that the binding of Eu^{3+} is
sequential. These same optical studies also indicate that the
binding constant is on the order of $10^6 - 10^7$ $1/\underline{M}$ so that in the
NMR experiments all of the Eu^{+3} is bound to tRNA. Therefore, the
fact that peak A is still shifting after the Eu/tRNA ratio
reaches 1/1 could be taken as an indication that the second
Eu^{3+} binding site is located near the strongest Eu^{3+} binding
site.

In addition to the upfield shift of peak A, the other
pronounced change which occurs in the spectrum at low levels
of Eu^{3+}, is a downfield shift of peak E. The shift of this peak
is so large that with the first addition of Eu^{3+}, it appears as
a shoulder on the high field side of peak D. By the second
addition of Eu^{3+} this shoulder is gone, presumably because peak E
is now coincident with peak D. Since the binding of only one
Eu^{3+} ion shifts at least two resonances, the binding site
must simultaneously be close to at least two different base
pairs. Peak A, as we have noted, is assigned to the AU_6 base pair
in the CCA stem, whereas peak E has been assigned to the GC_{11}
base pair in the DHU stem. If the assignment of this latter
resonance is correct, then we conclude that the first Eu^{3+} ion
is simultaneously bound to phosphate groups adjacent to both
of these base pairs. A model of the three-dimensional structure
of $tRNA^{Phe}$ which employs a metal binding to phosphates adjacent
to U_6 and C_{11} is shown in figure 13. In this particular model,
the DHU and TΨC loops are brought into close proximity, and
this fits well with oligomer binding data and enzymatic
digestions studies which indicate that the TΨC loop is quite
protected by the folding of the molecule [21].

The Eu^{3+} binding studies described here are of interest in
relation to other biochemical and structural studies which have
been carried out. First of all, it has been shown that a number
of different rare earth ions, including Eu^{3+}, can replace Mg^{2+} in

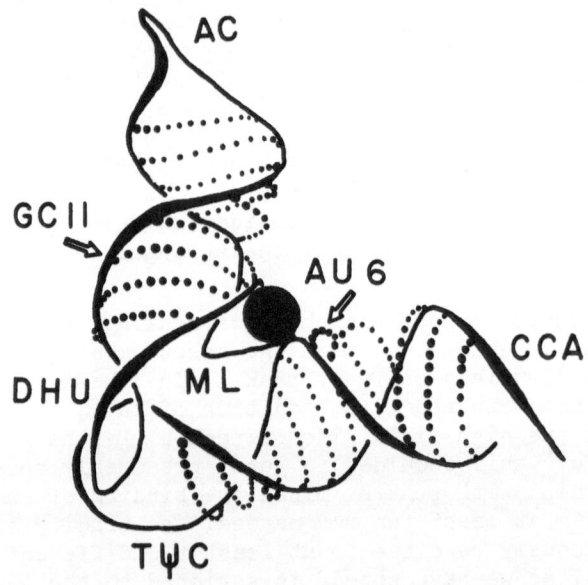

Fig. 13. A model showing one of four possible three dimensional
structures for tRNA^{Phe} that are consistent with the Eu^{3+}
induced shifts.

promoting the aminoacylation of tRNA [22] and this suggests that insofar as this particular biochemical reaction is concerned, the tRNA molecules are in their native state when they bind Eu^{3+} instead of Mg^{2+}. Secondly, earth ions have been used as optical probes of tRNA structure [23, 24]. Finally, and perhaps more importantly, rare earth ions are currently being used to obtain isomorphous derivatives in tRNA crystal structure work [25, 26]. When the results of these studies become available, it should be possible to determine whether or not the structure of the tRNA molecules are the same in the crystal as they are in solution.

CONFORMATION OF DNA: POLY d(A-T)

The conformation of DNA in solution depends both on nucleotide sequence and solvent conditions and this is believed to have considerable biological significance. The sensitivity of conformation to environmental effects is particularly well illustrated by the synthetic DNA copolymer dAT. Depending upon conditions used to prepare fibers, Davies and Baldwin [27] obtained X-ray diffraction patterns corresponding to the A, B, and C forms of DNA and a new, previously unreported form (D form). Although a variety of techniques have been used to study the structure of DNA in solution, none of these measurements have provided the quantitative structural data now required to understand the various physico-chemical properties of DNA in solution.

We now discuss the way high resolution proton NMR can be used to obtain data on the structure of a synthetic DNA in solution. In this preliminary work we have examined the spectrum of $(dA - dT)_n \cdot (dA - dT)_n$ in 0.1 $\underline{M}$ NaCl solution. The temperature dependence of the NMR spectra in several different spectral regions are presented in figures 14 and 15. The assignments of the resonances to individual protons of the T and A bases (see table 5) were obtained by comparing the high temperature spectra of the polynucleotide with the free mononucleotide spectra. As the spectra shown in the figures demonstrate, many of the resonances are upfield shifted in the low temperature spectra relative to their positions in the high temperature spectra and it is these shifts that are used to obtain the structural information. The procedure is as follows.

The work with the tRNA demonstrated that in the double helix the proton resonances of the individual bases are up-field shifted as a result of ring current field effects from neighbouring bases and the ring current fields computed by Giessner-Prettre and Pullman rather accurately account for the

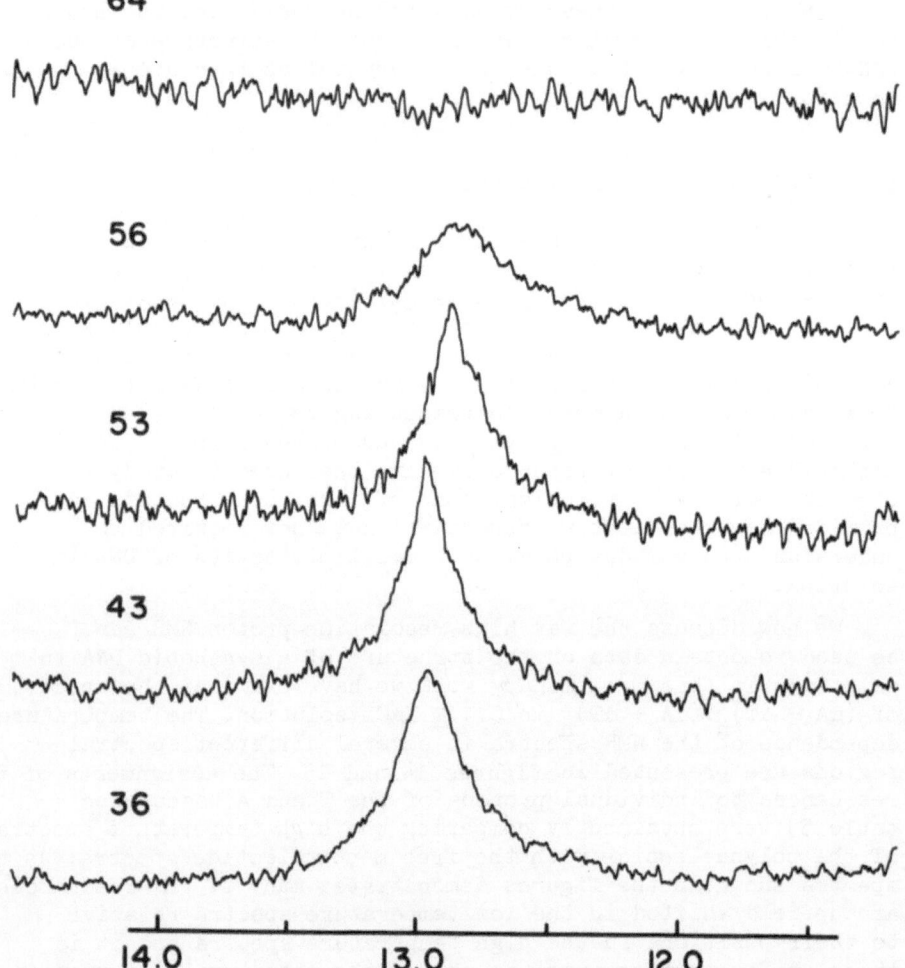

Fig. 14. Temperature dependence of the 300 MHz proton NMR spectrum of $(dA - dT)_n \cdot (dA - dT)_n$. First part.

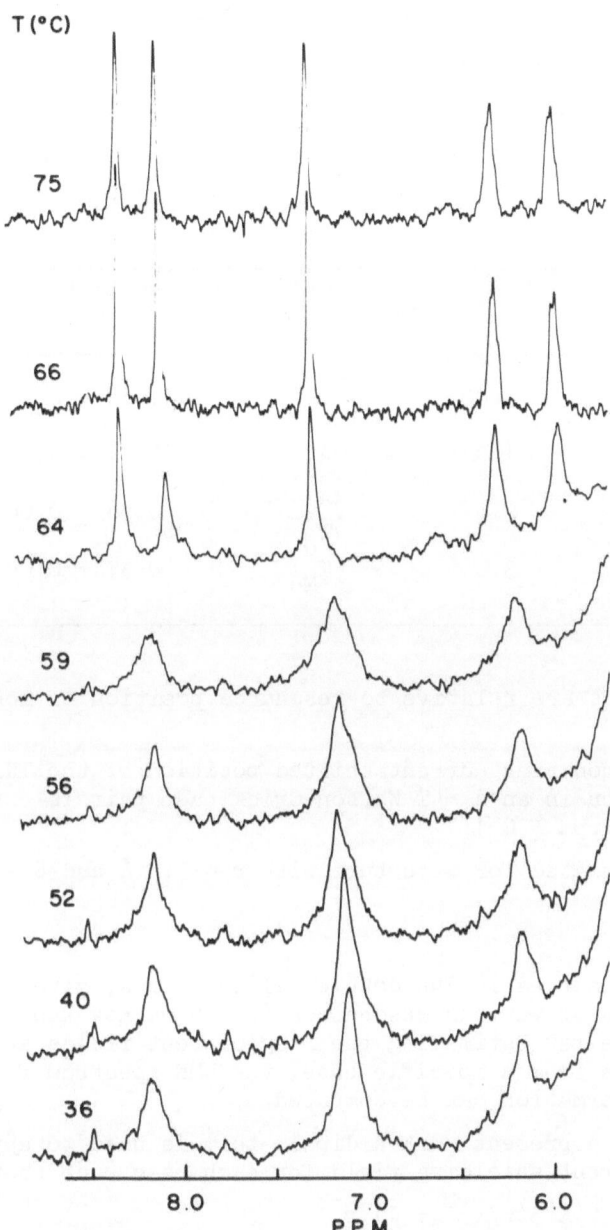

Fig. 15. Temperature dependence of the 300 MHz proton NMR spectrum of $(dA - dT)_n \cdot (dA - dT)_n$. Second part.

Table 5. A summary of the NMR spectral properties of $(dA - dT)_n$. $(dA - dT)_n$

Resonance position	Relative intensity	Assignment	Observed upfield shift in double helix (a)	Calculated upfield shift (c)
ppm				ppm
13.0	1.00	TN_3	1.6 $\pm$ 0.1 (b)	1.5
8.3	1.0	A_8	0.14 $\pm$ 0.05	0.02
7.2	2.0	$\begin{cases} A_2 \\ T_6 \end{cases}$	0.98 $\pm$ 0.06	0.97
1.3	3.0	T_{Me}	0.37 $\pm$ 0.09	0.49

(a) Shifts are relative to resonance position in mononucleotide ppm.

(b) The non-ring current shifted position of the TN_3 ring nitrogen proton in an A – T Watson-Crick base pair is estimated to be 14.6 ppm $\pm$ 0.1.

(c) Calculated for structure with r = 3.0 Å and θ = 35°.

shifts observed in RNA double helices. Thus, given the unshifted positions of various resonances (which we now know), and a procedure for estimating the ring current fields at different distances from a specific base, the NMR spectrum for a particular DNA conformation can be computed.

In the present work a dipole term is used to approximate the ring current shielding field for each base such that

$$\Delta H/H = K[3(z/r)^2 - 1]\, r^{-3}$$

where $r^2 = x^2 + y^2 + z^2$ for a pyrimidine base; and $r^2 = (x/1.27)^2 + y^2 + z^2$ for a purine base. K is adjusted so that the "hot spot" [(0, 0, 3.4) on the coordinate system] on the z = 3.4 Å plane is in close agreement with Pullman's results (i.e. 1.28 ppm for adenine and 0.30 ppm for thymine).

The shielding field for a T - A base pair is then obtained by adding the adenine term to the thymine term which is moved from adenine by the vector (-5.5, 2.7, 0.0).

The resonance which appears in the spectrum at 13.0 ppm can be assigned to the TN_3 proton in an A - T Watson-Crick base pair. As the temperature is raised from 36 to 56 C the 13.0 ppm resonance is considerably broadened and by 64 C it has disappeared entirely. On the basis of earlier studies we conclude that the broadening and subsequent disappearance of this low field resonance (which precedes complete disruption of the double helix) occurs when the rate of opening of the A - T hydrogen bond approach ~ 400 sec^{-1}. The resonances at 8.3 ppm (intensity one proton per base pair) and 7.2 ppm (intensity of two protons per base pair) are assigned to the A_8, A_2 and T_6 protons respectively. At 59 C the noticeable broadening of all these resonances is attributed to the onset of the helix-to-coil transition. At 64 C the helix has completely melted and the resonances are sharp and located at approximately the same positions as in the appropriate mononucleotides. The $\sim$ 60 C melting point inferred from the NMR spectra is in good agreement with the T_m determined optically.

The resonances which are located near 6.0 ppm are assigned to the A and T 1' ribose protons.

There is an additional resonance at 1.3 ppm (intensity of 3 protons per base pair) in the 36 C spectrum which can be assigned to the methyl protons of T. When the double helix is melted this resonance shifts $\sim$ 0.3 ppm to lower field to the position where it is observed in free TMP.

DETERMINATION OF DNA STRUCTURAL PARAMETERS

In order to actually determine the structure of $(dA - dT)_n \cdot (dA - dT)_n$ from the spectral data we have computed the ring current shifts of the five resonances listed in table 5 for a variety of different DNA conformations. The two geometrical parameters which we found most easy to vary were the helix radius and the number of base pairs per turn. In this initial calculation we assumed the bases were oriented with their planes perpendicular to the helix axis and with an interbase distance of 3.4 Å. Thus, in the coordinate system which we are using, the B form of DNA corresponds to a helix radius of 2.0 Å and θ = 36°.

The results of the ring current shift computations are shown in figure 16 where they are compared with the experimental shifts. It is evident from the curves shown in figure 16 that the best agreement between the observed and computed shifts is

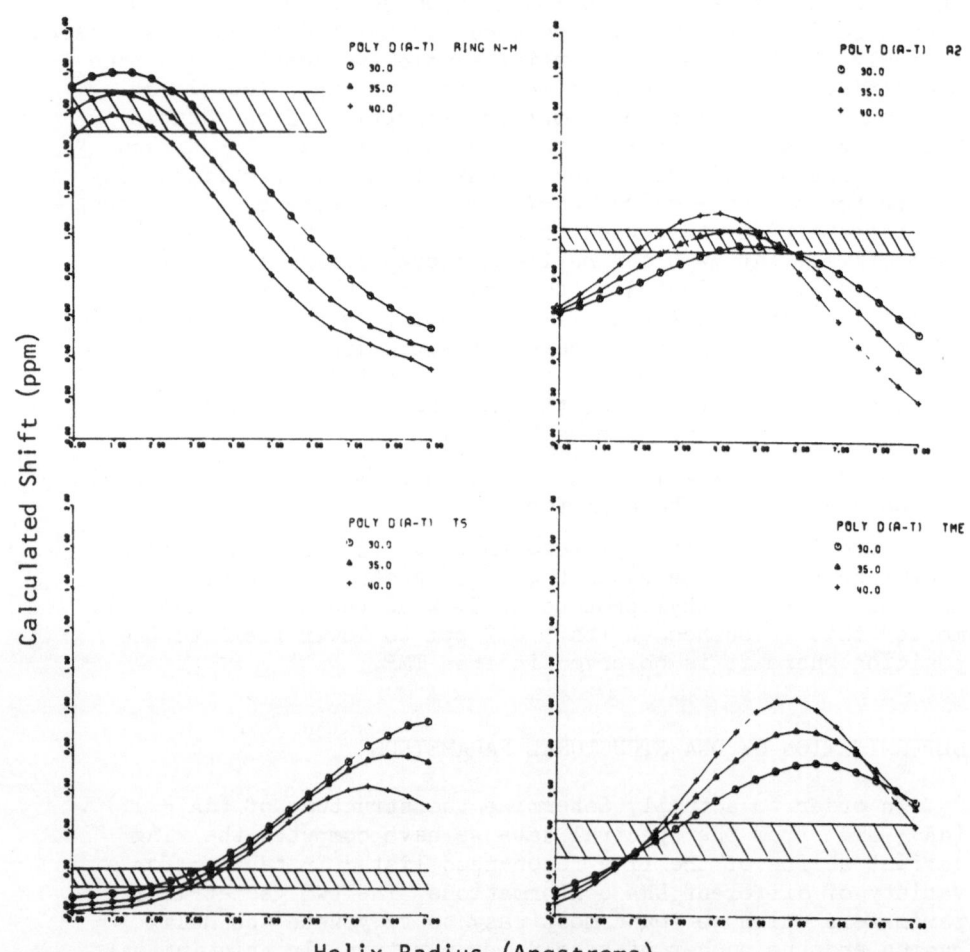

Helix Radius (Angstroms)

Fig. 16. A comparison of the upfield shifts of the
$(dA - dT)_n \cdot (dA - dT)_n$ resonances with those computed for
various possible double helix geometries.

obtained for a helix radius of r = 3.0 Å and a turn angle
of θ = 35°. A less good fit is obtained using r = 2.0 Å
and θ = 30°. The predicted shift for the A_8 proton (not shown)
is quite small (0.0 - 0.1 ppm) for all geometrics examined.
This is in agreement with the observed value of $\sim$ 0.1 $\pm$ 0.05 ppm.

FUTURE NMR STUDIES

Because success in the application of high resolution
NMR to the determination of the structure of polynucleotides in
solution is of recent origin, there are a wide range of problems
yet to be examined by this method. The dynamic aspects of polynucleotide structure are of special interest, and NMR may
provide the ideal method for studying such problems. The use
of partially relaxed Fourier transform NMR spectroscopy will add
an important dimension to such studies. In our discussion we
have emphasized proton NMR, but there have been recent reports
on the use of ^{13}C and ^{31}P NMR which look promising [28 - 30].
Although furhter refinements are required, it is possible that
the techniques which are being developed for carrying out
high resolution NMR investigations of solids will be directly
applicable to very high molecular weight samples such as large
proteins and DNA. There is the possibility that it may be
possible to construct a high resolution 500 - 600 MHz NMR
spectrometer and such an instrument would find ready use in the
studies which we have described here.

REFERENCES

1. C.C. McDONALD, W.D. PHILLIPS and J.R. PENSWICK, Biopolymers 3,
   609 (1965)

2. J.P. McTAGUE, V. ROSS and J.H. GIBBS, Biopolymers 3, 163 (1964)

3. I.C.P. SMITH, T. YAMANE and R.G. SHULMAN, Canad.J.Biochem. 47,
   480 (1969)

4. D.R. KEARNS, D. PATEL and R.G. SHULMAN, Nature 229, 338 (1971)

5. D.R. KEARNS, D. PATEL, R.G. SHULMAN and T. YAMANE, J.Mol.Biol.
   61, 265 (1971)

6. C.R. WOESE, The Genetic Code: The Molecular Basis for Genetic
   Expression, Harper & Row, New York (1967)

7. L. KATZ and S. PENMAN, J.Mol.Biol. 15, 220 (1966)

8. D.R. LIGHTFOOT, K.L. WONG,D.R. KEARNS, B.R. REID and R.G. SHUL-
   MAN, J.Mol.Biol. 78, 71 (1973)

9. R.G. SHULMAN, C.W. HILBERS, D.R. KEARNS, B.R. REID and Y.P. WONG,
   J.Mol.Biol. 78, 57 (1973)

10. D.R. KEARNS, D.R. LIGHTFOOT, K.L. WONG, Y.P. WONG, B.R. REID, L. CARY and R.G. SHULMAN, Annals of N.Y.Acad.Sci. 222, 324 (1973)

11. C. GEISSNER-PRETTRE and B. PULLMAN, J.Theoret.Biol. 27, 87 (1970)

12. S. ARNOTT, Progress in Biophys. and Mol.Biol. 22, 181 (1971)

13. B.G. BARRELL and B.F.C. CLARK, Handbook of Nucleic Acid Sequences, Joynson-Bruvvers Ltd., Oxford, England (1974)

14. K.L. WONG and D.R. KEARNS, Biolpolymers 10, 13 (1974)

15. D.R. KEARNS, Y.P. WONG, S.H. CHANG and E. HAWKINS, Biochem. (1974)

16. D.R. KEARNS, Y.P. WONG, E. HAWKINS and S.H. Chang, Nature 247, 541 (1974)

17. H.G. ZACHAU, Angew Chem.Internat.Edit. 8, 711 (1969)

18. R.H. LOFTFIELD, preprint of a chapter to be included in"Protein Synthesis"1, ed. E. McConkey, M. Dekker

19. C.R. JONES and David R. KEARNS, J.Amer.Chem.Soc. 96, 3651 (1974)

20. A. CARRINGTON and A.D. McLACHLAN, Introduction to Magnetic Resonance with Applications to Chemistry and Chemical Physics, ed. S.A. Rice, Harper & Row, New York (1967)

21. C.R. JONES and D.R. KEARNS, Proc.Nat.Acad.Sci. USA (1974)

22. M.S. KAYNE and M. COHN, Biochem.Biophys.Res.Commun. 46, 1285 (1972

23. J.M. WOLFSON and D.R. KEARNS, J.Amer.Chem.Soc. 96n,3653 (1974)

24. C. FORMOSO, Biochem.Biophys.Res.Commun. 53, 1084 (1973)

25. J.D. ROBERTUS, J.E. LADNER, J.T. FINCH, D. RHODES, R.S. BROWN, B.F.C. CLARK and A. KLUG, Nature 250, 546 (1974)

26. S.H. KIM, F.L. SUDDATH, G.J. QUIGLEY, A. McPHERSON, J.L. SUSSMAN, A.H.J. WANG, N.C. SEEMAN, A. RICH, Science 185, 435 (1974)

27. D.R. DAVIES and R.L. BALDWIN, J.Mol.Biol. 6, 251 (1963)

28. M. GUERON, FEBS Letters 19, 264 (1971)

29. L.M. WEINER, J.M. BACKER and A.I. REZVUKHIN, FEBS Letters 41, 40 (1974)

30. R.A. KOMOROSKI and A. ALLERHAND, Biochemistry 13, 369 (1974)

PARAFFINIC CHAINS: THE OBSERVATION OF STATIC DIPOLAR INTERACTIONS

IN THE PRESENCE OF ANISOTROPIC MOTIONAL NARROWING[*]

M. Guéron

Ecole Polytechnique

Paris, France

Potassium laurate $CH_3(CH_2)_{10}CO_2K$ can be mixed with water to form a variety of phases depending on water content and temperature (Fig. 1). In the various liquid crystal phases, the phase is composed of aqueous regions and laurate regions, the interface being formed by the contiguous polar heads of the laurate molecules. The laurate molecules are in effect linear chains: in some phases the chains are straight and aligned, in others they are twisted and disordered. Furthermore it is known from nmr, spin labels and fluorescence studies that there is considerable local motion. Nevertheless x-ray scattering shows that the liquid crystal phases exhibit long range order. The interface surfaces form a lattice which can have cubic, lamellar or hexagonal symmetry.

The cubic phase has the appearance of a solid: its macroscopic viscosity is high. Nevertheless the paraffinic protons give rise to a narrow resonance line, indicating a large microscopic mobility and complete averaging of the dipole-dipole interaction. The translational diffusion coefficient has been measured by the usual spin-echo and pulsed gradient method and is indeed found to be quite high: 2×10^{-6} cm^2/s, corresponding to a jump time of 10^{-9} s when computed with a mean free path equal to the intermolecular distance ~ 7 Å [1].

In contrast, the proton resonance of randomly oriented hexagonal or lamellar phases has a narrow peak with very intense

---

[*] This lecture is based on the work of J. Charvolin and P. Rigny described in [1].

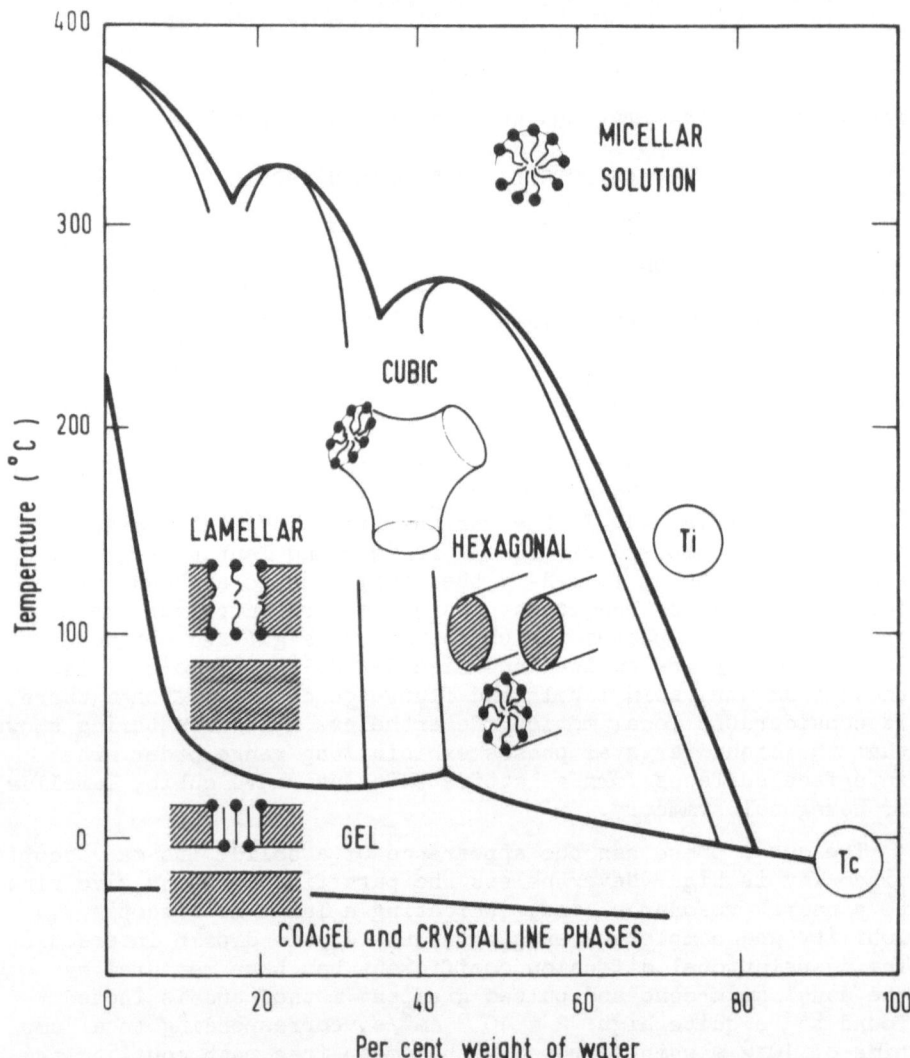

Fig. 1. A typical phase diagram for a potassium soap-water
       system. The structures are schematically drawn. The
       numbers are only intended to indicate typical conditions
       under which phase transitions occur [1].

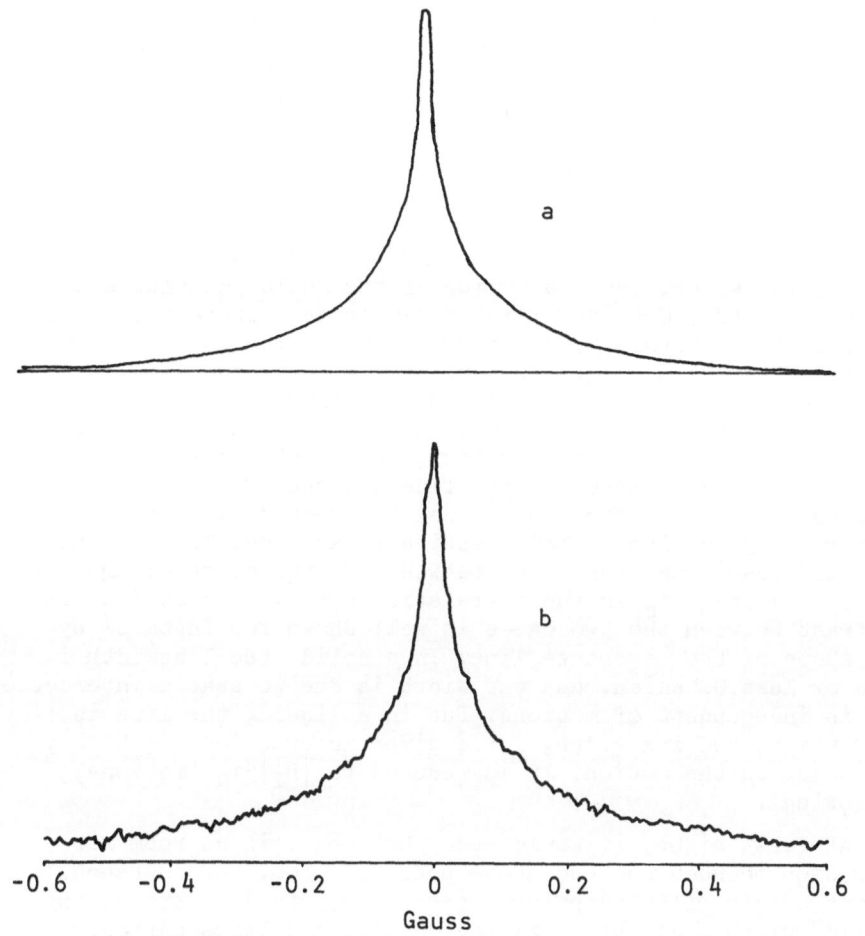

Fig. 2. a) Lineshape calculated from eq. (5) with $f(x) = \exp(-\pi x^2)$
The singularity has been removed by neglecting the
effect of the protons with $3\cos^2\theta - 1 = 0$ [2].
b) Lineshape observed by Lawson and Flautt.

and broad wings (Fig. 2). It has been thought at first that
this originated in the contribution of different protons which
could have different mobilities, thus giving broad and narrow
components to the line. On the other hand Wennerström [2] showed
that the lineshape could be explained if one assumes that
anisotropic motion of the chains results in a reduced, but non-
zero, average dipolar interaction between the paraffinic protons.

    The demonstration of a static dipolar interaction was given
by Charvolin and Rigny [2] who also studied the molecular motions

using proton relaxation. These experiments provide a good example
of the problems raised by systems which partake of both the
solid and liquid state.

DEMONSTRATION OF A STATIC DIPOLAR INTERACTION

Let us first recall the difference between a solid and a
liquid from the point of view of nmr. Consider a system of
interacting spins, where a motion of the nuclei modulates the
interaction $\hbar H_I$. Consider first a simple case where the motion
averages H_I to zero: $\langle H_I(t) \rangle = 0$. If the motion is very slow,
the fact that H_I averages to zero over a very long time is not
significant, and the main effect of the interaction will be
due to the instantaneous value of H_I. This is the "solid" case.
On the other hand, for fast motion, the first order effect of H_I
will be reduced to zero and the fluctuations of the H_I will
give second order effects which can be treated by perturbation
theory and give rise to spin-lattice relaxation. This is the
"liquid" case. The transition between the two cases occurs for
$|H_I| \tau_c = 1$ where τ_c is the correlation time of the motion. The
contrast between the two cases is well shown for instance by
the shape of the resonance line. In a solid, the linewidth is
more or less Gaussian, and the width is due to static interactions
and is independent of motions. But in a liquid, the line is
Lorentzian, and its width, for a given interaction, is highly
dependent on the motion. It is reduced to $|H_I|^2 \tau_c$ (motional
narrowing).

At first sight, it would seem that there is no room for
confusion between the two above possibilities. For instance,
if the origin of H_1 (dipolar interaction, etc.) is known, the
"solid" width Δ can be estimated easily, and lines narrower
than Δ will be evidence for motion: think, for example of nmr
in metallic lithium at room temperature. It is when we investigate
the lineshape and residual width δ of the motion-narrowed line
that care must be exercised. If we are dealing with an isotropic
motion is which $\langle H_I(t) \rangle = 0$, then indeed $\delta \sim |H|^2 \tau_c$. However, in
particular in the case of asymmetric molecules the motion may
be very anisotropic, so that, on the time scale t_0 of $|H_I|$:
$t_0 \sim (1/|H_I|)$, the average of $|H_I|$ is not zero. We then have an
ambivalent case, where $\{H_I\}$ gives rise to solid-like features,
while the fluctuating component $H_I - \{H_I\}$ can be treated on a
liquid-like basis: if the motion is very fast, the relaxation
time will be long, giving a small contribution $\delta_{Relax} \sim |H_I|^2 \tau_c$
to the linewidth. We have indicated by curly brackets the
average of H_1 on the time scale t_0. We see that if

$$\delta_{Relax} \ll \{H_I\} \ll |H_I|$$

the residual linewidth $\delta \sim \{H_I\}$ will be much less than the rigid linewidth Δ but nevertheless entirely due to the "solid" component $\{H_I\}$ and it will therefore be quite unrelated to the molecular motion.

Instead of the anisotropic motion discussed above, broad components to the nmr line could have other origins: they could be due to protons having an isotropic but slow motion, or also to field or sample inhomogeneity. We need a way to distinguish between these four possibilities. The lineshape cannot provide an answer, nor can the free precession which is simply its Fourier transform. On the other hand pulse manipulations of the magnetic system provide the appropriate tool.

We start with a 90° pulse and look at the free precession. Applying a 180° pulse (which is 90° out of phase although this is not necessary) during the free precession decay will give an echo if the decay is due to field or sample inhomogeneity. As shown in figure 3, none is observed; so this cause can be excluded. On the contrary, an echo is observed following a 90° pulse 90° out of phase. In the absence of inhomogeneity (as shown by the previous experiment), this can only be a dipolar echo, thereby demonstrating the existence of an average static dipolar interaction [3].

As to the fourth possibility, that is the existence of protons having a slow isotropic motion, we can try to evaluate the proportion of such protons by studying the amplitude of the dipolar echo. Indeed, only protons having anisotropic motion contribute to the echo, so that one should be able to say from the echo amplitude (as compared to the amplitude of the free precession) whether all protons have anisotropic motion, or only a certain proportion of them. This comparison has been made and shows that all, or most, protons have an anisotropic motion.

In summary, it has been shown that the motion of the paraffinic protons is anisotropic, and that the static dipolar interaction is entirely responsible for the observed linewidth.

INVESTIGATION OF MOLECULAR MOTIONS

The measurement of the relaxation times affords a means to probe the molecular motions. The rates for Zeeman relaxation (T_1^{-1}), relaxation in the rotating frame $(T_{1\rho}^{-1})$ and relaxation of the dipolar energy (T_D^{-1}) are proportional to the spectraldensity of the fluctuating Hamiltonian, respectively at the Larmor frequency ω_0, the Larmor frequency in the rotating field $\omega_1 = \gamma b_1$, and zero frequency. For an exponential correlation function we therefore expect relaxation rates of the form:

M. GUÉRON

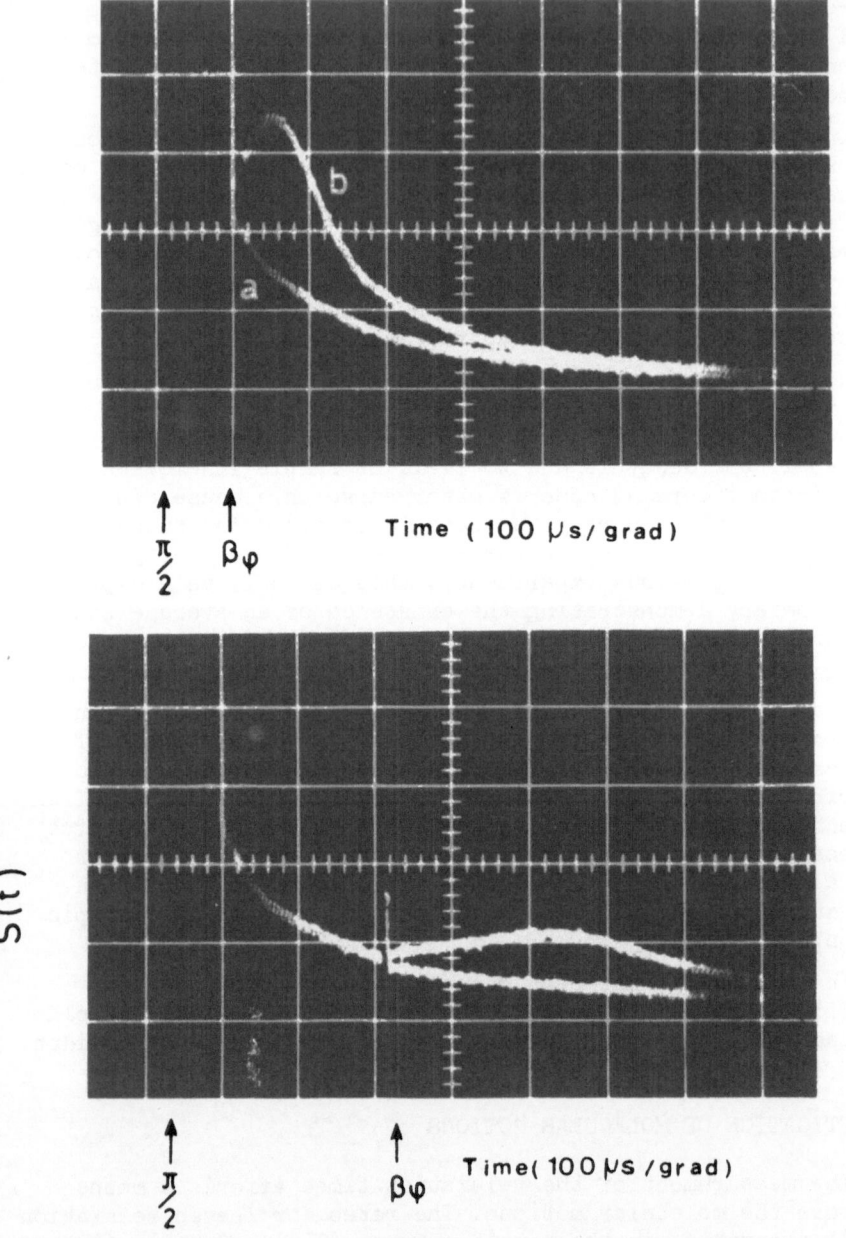

Fig. 3. Dipolar echoes observed, for short and long times, in a
lamellar $C_{12}K$-28 % D_2O sample at 90 C. The effects of two
pulse sequences are compared: (a) [II/2, (II)90], the second
pulse does not alter the FID. (b) [II/2, (II/2)90], the
second pulse is followed by a dipolar echo [1].

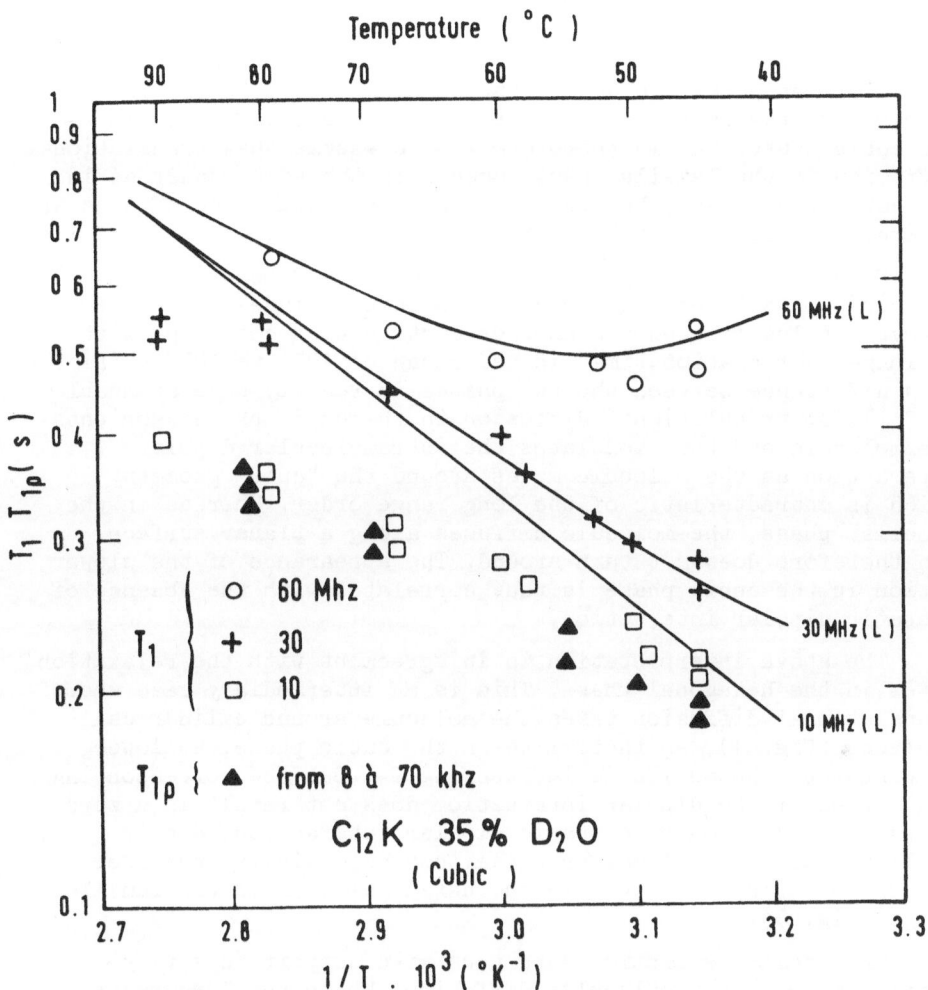

Fig. 4. Comparison of nuclear relaxation in the lamellar and cubic
phases. The drawn curves are based on T_1 measurements of the
Lamellar phase.   The points are the measured
values of T_1 and $T_{1\rho}$ for a cubic $C_{12}K$-35 % D_2O sample [1].

$$J(\omega) \underset{\sim}{\sim} 2|H^2|\tau_c/(1+\omega^2\tau_c^2) \qquad\qquad (1)$$

where ω is the relevant frequency in rad/s.

In the lamellar phase the variation of T_1 vs. frequency is in agreement with (1) and corresponds to a correlation line of $\sim 10^{-9}$ s at 90 C, and an activation energy of 0.25 eV. These numbers are the same as those for translational diffusion in the cubic phase. One is therefore led to assume that translational diffusion in the lamellar phase occurs in the same manner as in the cubic phase, and that the corresponding modulation of dipolar interactions is responsible for T_1.

Should one then expect the same relaxation behaviour in the cubic phase? For high frequencies this is indeed the case. However at low frequencies another contribution comes in, with a longer correlation time, in the range of 10^{-7} to 10^{-8} s (Fig. 4). This difference between the two phases is readily understandable as follows: translational diffusion in the cubic phase reorients the molecule and thus modulates the intramolecular dipolar interaction as the molecule moves around the "cube" geometry which is characteristic of the long range order. Whereas in the lamellar phase, the molecule diffuses along a planar surface, and therefore does not turn around. The appearance of the slower motion in the cubic phase is thus correlated with the absence of a static dipolar interaction.

The above interpretation is in agreement with the relaxation rates in the hexagonal phase. This is an intermediary case where translational diffusion takes the molecule around cylindrical surfaces (Fig. 1), so that, like in the cubic phase, a slower component to the motion is influencing T_1. But the corresponding modulation of the dipolar interaction does not result in a zero average, there remains a static dipolar interaction, albeit smaller than in the lamellar phase: correspondingly, the free precession decay is slower in the hexagonal than in the lamellar phase (Fig. 5).

The Zeeman relaxation data thus give support to a very simple image of the molecular diffusion. While the long-range order is stable, individual molecules travel along those long-range structures; in this way long-range order and local disorder are reconciled, as are the macroscopic rigidity and the rapid motion on the molecular scale. It remains for further work to analyze these data quantitatively, in particular to compare the longer correlation times to the diffusion time around a lattice "cube" in the cubic phase. In the lamellar phase, one would like to have an estimate of the relative contributions of inter- and intramolecular interactions to the Zeeman relaxation.

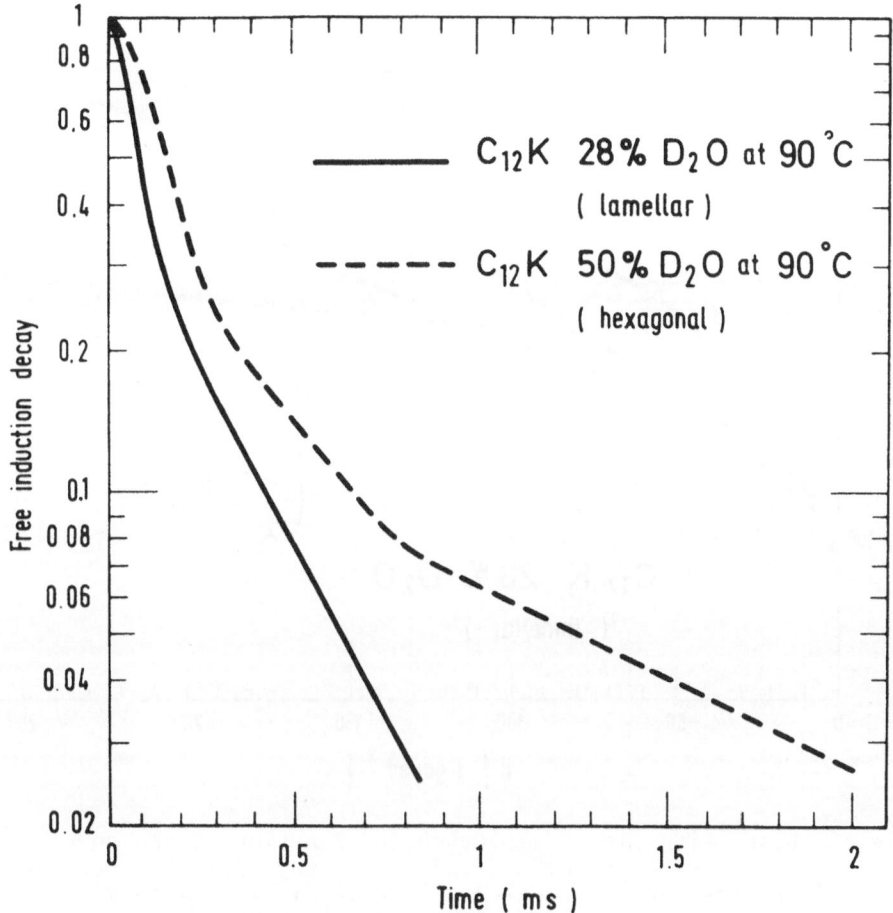

Fig. 5. Comparison of the FID of paraffinic protons in lamellar
        and hexagonal samples at 90 C [1].

        A difficulty of such interpretations resides in the use of
powder samples in which various orientations of the lamellar
give rise to a distribution of the different relaxation parameters.

        Turning now to $T_{1\rho}$ measurements in the lamellar phase, we
find evidence for further correlation times, corresponding to

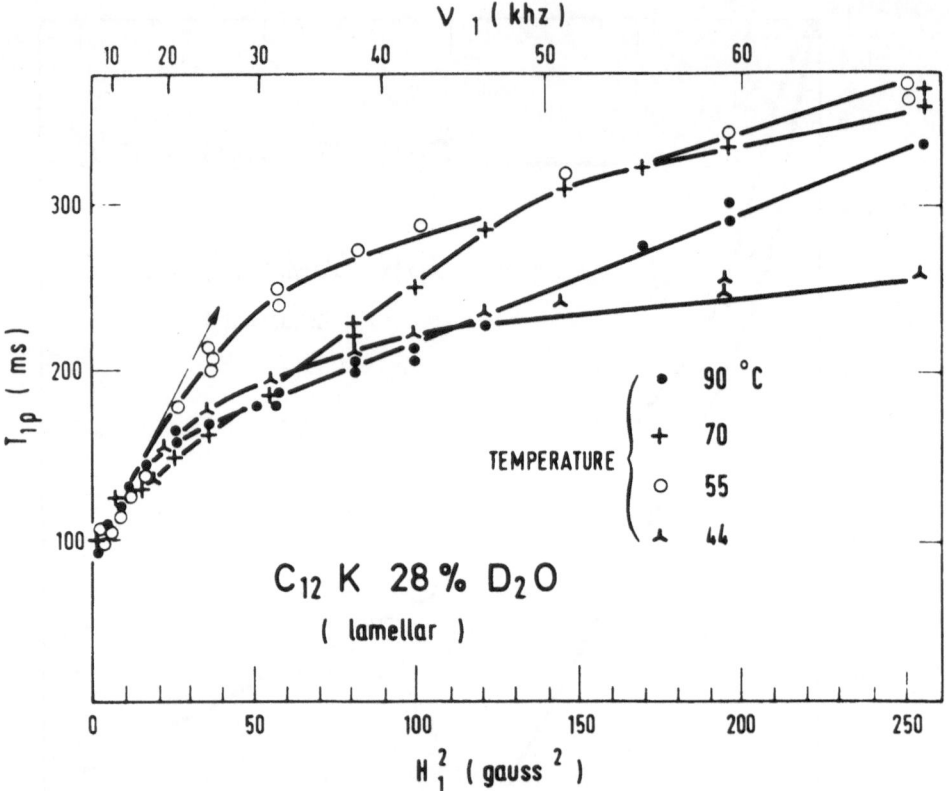

Fig. 6. Relaxation time $T_{1\rho}$ plotted, as function of the square
        of the strength of the radiofrequency field H_1, measured
        at different temperatures in a lamellar $C_{12}K$ 28 % D_2O
        sample (only points for H_1 larger than 1 G are shown) [1].

the modulation of other interactions.

   According to expression (1), a plot of $T_{1\rho}$ vs. ω_1^2 (the
square of the r.f. field) would be a straight line if there
were only one interaction modulated with one correlation time.
Instead the plot is curved, showing a distribution of correlation
times the largest being in the range of 10^{-5} seconds (Fig. 6).

   Lastly the measurement of the dipolar signal, following
the method of Jeener et al. [4], gives the value of $|H^2|\tau_c$. It
is interesting to note that this value cannot be obtained from a

$T_{1\rho}$ measurement with $\omega_1 = \gamma b_1 \to 0$, because this measurement, for b_1 smaller than static dipolar interactions, reflects mixing between the Zeeman and dipolar energies, and not dynamic relaxation phenomena. Indeed the result of the $T_{1\rho}$ measurement for $\omega_1 \to 0$ gives an apparant $T_{1\rho}(0)$ shorter than T_D, showing that the measured decay in the "$T_{1\rho}$" experiment cannot be due to relaxation, since one must have $T_{D} < T_{1\rho}$ for the real $T_{1\rho}$. In fact one expects

$$T_D = \frac{1}{2}T_{1\rho}(0)$$

We must now try to understand the scheme of these different relaxation times and the interactions to which they relate. We shall consider the lamellar phase. The only type of interaction present is the dipolar interaction between protons.

First, there is the intermolecular interaction. This is modulated completely by translational diffusion and is therefore only related to the fastest correlation time.

Next, and larger, are the intramolecular interactions. A proton H_m linked to carbon number m along the chain is close to the other proton H'_m, and also to both protons on each of carbons (m-1) and (m+1).

These five protons are the ones with which it will mainly interact. These interactions will be modulated by rigid rotation of the molecule, and also by the rotational isomerizations of the chain around the C-C bonds. These complex movements can be analyzed in normal modes for which a distribution of correlation times τ_c is expected. Each normal mode would modulate in a specific way the dipolar interactions.

From the point of view of relaxation, one should then analyze the various modes, separating them in two classes depending on whether $|H_I|\tau_c$ is smaller or larger than one. Here τ_c is the correlation time relevant to the normal mode being considered and H_1 is the portion of the dipolar interaction that is modulated by the motion due to this normal mode.

In the general case, one would find parts of the dipolar interaction not modulated by any of the normal modes, other parts being modulated only by slow modes ($|H_I|\tau_c>1$) and others being rapidly modulated ($|H_I|\tau_c<1$). According to the proposed model of the motion however, no slow modes (correlation time 10^{-5} s) are expected. The fluctuations due to the motions should then appear in the $T_{1\rho}$ experiments, and not contribute to T_D.

The comparison of the preceding scheme with the relaxation measurements is obviously difficult, and is rendered even more complex by the fact that the motions differ according to which proton is considered along the chain. But the demonstration of a range of correlation times extending form 10^{-9} s (T_1 measurements) to 10^{-5} s($T_{1\rho}$) is entirely compatible with the predictions of the theoretical model which we have only very briefly outlined.

ACKNOWLEDGMENT

We thank Dr. P. Rigny for helpful discussions.

REFERENCES

1. J. CHARVOLIN and P. RIGNY, J.Chem.Phys. 58, 3999 (1973) and references quoted therein

2. H. WENNERSTROM, Chem.Phys.Lett. 18, 41 (1973)

3. The average anisotropy in the lamellar phase has been studied by the quadrupolar splitting of deuteron nmr in oriented lamellae of perdeuterated laurate: J. CHARVOLIN, P. MANNEVILLE and B. DELOCHE, Chem.Phys.Lett. 23, 345 (1973)

4. J. JEENER and P. BROEKAERT, Phys.Rev. 157, 232 (1967)

LIST OF STUDENTS

B ADRIAENSSENS Guy, Universiteit Leuven, Celestijnenlaan 200 D, Leuven
  ALEWAETERS Gerrit, Universiteit Brussel, A. Buyllaan 105, Brussel
  BROEKAERT Paul, Université de Bruxelles,av.F.Roosevelt 50, Bruxelles
  DARVILLE Jacques, Institut de Physique, Sart-Tilman, Liège
  HOUGARDY Jacques, Sciences de la Terre, de Croylaan 42, Leuven
  JANSSENS Geneviève, Universiteit Leuven,Celestijnenlaan 200 D, Leuven
  JANSSENS Luc, Universiteit Leuven, Celestijnenlaan 200 D, Leuven
  JURGA Stefan, Institut de Physique, Sart-Tilman, Liège
  SALVADOR Pedro, Sciences de la Terre, de Croylaan 42, Leuven
  SANZ Jesus, Sciences de la Terre, de Croylaan 42, Leuven
  SEGEBARTH Christoph, Universiteit Brussel, A. Buyllaan 105, Brussel
  STESMANS André, Universiteit Leuven, Celestijnenlaan 200 D, Leuven
  STONE William, Service Physique-Chim. Min., de Croylaan 42, Leuven
  THIJS Willy, Universiteit Leuven, Celestijnenlaan 200 D, Leuven
  VANDERVORST Serge, Université de Bruxelles, av. A. Buyl 105, Bruxelles
  VERLINDEN Roland, Universiteit Leuven, Celestijnenlaan 200 D, Leuven
  VREYS Herman, Universiteit Leuven, Celestijnenlaan 200 D, Leuven
  WITTERS Jozef, Universiteit Leuven, Celestijnenlaan 200 D, Leuven
CDN STILES James, University of British Columbia, Vancouver
  TOMCHUK Edward, University of Winnipeg, Winnipeg, Manitoba
DK HANSEN Poul, Department of Electrophysics,Technical University,Lyngby
  MOURITSEN Ole, Department of Physical Chemistry, Aarhus
SF TUOHI, Jukka, Wihuri Physical Laboratory, University of Turku, Turku
  YLINEN Eero, Wihuri Physical Laboratory, University of Turku, Turku
F BLANC Jean-Pierre, Université de Clermont-Ferrand, Aubière
  BLOYET Daniel, Institut d'Electronique, Faculté des Sciences, Orsay
  CARON F., Ecole Polytechnique, Paris
  FRIED Françoise, Laboratoire P.M.C., Parc Valrosa, Nice
  GALLICE Jean, Université de Clermont-Ferrand, Aubière
  GALLIER Jean, Faculté des Sciences, Université de Rennes, Rennes
  KORB Jean-Pierre, C.M.O.A. du C.N.R.S., 23 rue du Maroc, Paris
  LAMOTTE Bernard, Centre d'Etudes Nucléaires de Grenoble, Grenoble
  MARUANI Jean, C.M.O.A. du C.N.R.S. 23 rue du Maroc, Paris
  ROUSSEAU Albert, Centre d'Etudes Nucléaires de Grenoble, Grenoble
  SIXOU Pierre, Laboratoire P.M.C., Nice
  SZEFTEL Jacob, Laboratoire de Physique des Solides, Orsay
  THEVENEAU Hélène, Ecole Supérieure Physique et Chimie, Paris
  TRICHET Luc, Faculté des Sciences de Nantes, Nantes
  TROKINER Arlette, Ecole Supérieure Physique et Chimie, Paris
  VIBET Claude, Institut d'Electronique Fondamentale,Orsay
  WEULERSSE Jean-Marc, Commisariat à l'Energie Atomique, Saclay
  JENEVEAU Alain, Laboratoire de Physique Electronique, Orsay

 D ACHLAMA Abraham, Max-Planck-Institut für Medizin. Forsch., Heidelberg
   GOUBEAU Andreas, Physikal. Institut 3, Stuttgart
   HACKBUSCH Wolfgang, Institut für Anorganische Chemie, Heidelberg
   JANNEK Horst, Institut für Physikalische Chemie, Münster
   KNUTTEL Bertold, Bruker Physik AG, Forchheim
   MULLER Detlet, Fachbereich Physik, Universität, Saarbrücken
   GROSSMANN Jürgen, Fachbereich Physik, Universität, Saarbrücken
   SCHICK Elke, Institut für Physikalische Chemie, Münster
   SCHMOLZ Albert, Max-Planck-Institut für Metallforschung, Stuttgart 1
   SCHWARZER Elmar, Institut Theoretische Physik, Stuttgart
   SINNING Gerhard, Universität E III, Dortmund
   VIETH Hans, Max-Planck-Institut, Dep. Molecular Physics, Heidelberg
   OFFERGELD Hans, Institut f. Physikalische Chemie, Aachen
GB BRANSON Peter, University of Nottingham, Nottingham
   COX Stephen, Rutherford Laboratory, Chilton, Didcot, Berks.
   READ Susan, Rutherford Laboratory, Chilton, Didcot, Berks.
   YESINOWSKI James, University Chemical Laboratory, Cambridge
IL GOREN Shaul, Physics Department, Ben Gurion University, Beer Sheva
   IGNER Dan, The Weizmann Institute of Science, Rehovot
   KORN Charles, Physics Department, Ben Gurion University, Beer Sheva
   MEIROVITCH Eva, Weizmann Institute of Science, Rehovot
   POLAK Micha, Department of Chemistry, University, Tel Aviv
   SHAHAM Moshe, Soreq Nuclear Research Centre, Yavne
   VEGA Alexander, Weizmann Institute, Rehovot
   VEGA Shimon, Weizmann Institute, Rehovot
 I ANFOSSA SOMMA Fabrizia, Istituto di Fisica Sperimentale, Napoli
CI CHEZEAU Jean, Faculté des Sciences, Université d'Abidjan, Abidjan
NL FEITSMA Pieter, Solid State Physics Lab., Melkweg 1, Groningen
   HENKENS Leon, Kamerlingh Onnes Lab.,Nieuwsteeg 18, Leiden
   POURQUIE Jean, Technische Hogeschool Delft, Delft
   VAN DONGEN TORMAN, Johannes, Universiteit Nijmegen, Nijmegen
   ZWEERS Aart, Kamerlingh Onnes Lab., Nieuwsteeg 18, Leiden
 N SLOTFELDT-ELLINGSEN Dag, Centr. Inst. for Industrial Research, Oslo
 P JURGA Kazimierz, Institute of Physics, University, Poznan
   SAGNOWKSI Stanislaw, Institute of Nuclear Physics, Krakow
 R GROSESCU Radu, Institute for Atomic Physics, Bucharest
 S BERGLUND Bo, Institute of Chemistry, University of Uppsala, Uppsala
CH ECABERT Marcel, Institut de Physique, Université, Neuchâtel
   MEIER Peter, Swiss Institute for Nuclear Research SIN, Villigen
   PELLISSON Jean, DPMC, Université de Genève, Genève
   PERRIN Bernard, DPMC, Université de Genève, Genève
USA BOOLCHAND Punit, University of Cincinnati, Cincinnati, Ohio
   FOLLSTAEDT, David, University of Illinois, Urbana, Illinois
   KAPLAN Samuel, Massachusetts Institute Technology, Cambridge, Mass.
   LARSEN David, University, St. Louis, Missouri
   MAELAND Arnulf, Worcester Polytechnic Institute, Worcester, Mass.
   MESTDAGH Michèle, Nortwestern University, Evanston, Illinois
YU LUZAR Mariette, Institute "J. Stefan", Jamova 39, Ljubljana
   PRELESNIK Tone, Institute "J. Stefan", Jamova 39, Ljubljana
   ZUMER Slobodan, Physics Department, University of Ljubljana, Ljubljan
   ZUPANCIC Ivan, Physics Department, Institute "J. Stefan", Ljubljana

The Summer School NUCLEAR MAGNETIC RESONANCE IN SOLIDS
on excursion in the Open Air Museum in Bokrijk (Limburg, Belgium), August 29, 1974